Space and Spatiality in Modern German-Jewish History

New German Historical Perspectives

Series Editors: Paul Betts (Executive Editor), Timothy Garton Ash, Jürgen Kocka, Gerhard Ritter, Nicholas Stargardt, and Margit Szöllösi-Janze

Established in 1987, this special St Antony's Series on New German Historical Perspectives showcases pioneering new work by leading German historians on a range of topics concerning the history of Modern Germany and Europe. Publications address pressing problems of political, economic, social, and intellectual history informed by contemporary debates about German and European identity, providing fresh conceptual, international, and transnational interpretations of the recent past.

Volume 1
Historical Concepts between Eastern and Western Europe
Edited by Manfred Hildermeier

Volume 2
Crises in European Integration: Challenges and Responses
Edited by Ludger Kühnhardt

Volume 3
Work in a Modern Society: The German Historical Experience in Comparative Perspective
Edited by Jürgen Kocka

Volume 4
Popular Historiographies in the 19th and 20th Centuries: Cultural Meanings, Social Practices
Edited by Sylvia Paletschek

Volume 5
A Revolution of Perception? Consequences and Echoes of 1968
Edited by Ingrid Gilcher-Holtey

Volume 6
Anti-Liberal Europe: A Neglected Story of Europeanization
Edited by Dieter Gosewinkel

Volume 7
Poverty and Welfare in Modern German History
Edited by Lutz Raphael

Volume 8
Space and Spatiality in Modern German-Jewish History
Edited by Simone Lässig and Miriam Rürup

Space and Spatiality in Modern German-Jewish History

Edited by
Simone Lässig and Miriam Rürup

berghahn
NEW YORK • OXFORD
www.berghahnbooks.com

First published in 2017 by
Berghahn Books
www.berghahnbooks.com

© 2017, 2019 Simone Lässig and Miriam Rürup
First paperback edition published in 2019

Library of Congress Cataloging-in-Publication Data
Names: Lassig, Simone, 1964– editor. | Rurup, Miriam, 1973– editor.
Title: Space and spatiality in modern German-Jewish history / edited by Simone
    Lassig and Miriam Rurup.
Description: New York : Berghahn Books, [2017] | Series: New German historical
    perspectives ; volume 8 | Includes bibliographical references and index. |
    Description based on print version record and CIP data provided by publisher;
    resource not viewed.
Identifiers: LCCN 2017012303 (print) | LCCN 2017012810 (ebook) | ISBN
    9781785335549 (eBook) | ISBN 9781785335532 (hardback : alk. paper)
Subjects: LCSH: Jews—Germany—History—1800–1933—Congresses. | Jews—
    Germany—History—Congresses. | Space perception—Germany—History—
    Congresses. | Space and time—Religious aspects—Judaism—Congresses. |
    Germany—Ethnic relations—Congresses.
Classification: LCC DS134.25 (ebook) | LCC DS134.25 .S687 2017 (print) |
    DDC 943/.004924—dc23
LC record available at https://lccn.loc.gov/2017012303

British Library Cataloguing in Publication Data
A catalogue record for this book is available from the British Library

ISBN 978-1-78533-553-2 hardback
ISBN 978-1-78920-512-1 paperback
ISBN 978-1-78533-554-9 ebook

Contents

List of Illustrations   viii

Preface   ix
   Simone Lässig and Miriam Rürup

Introduction: What Made a Space "Jewish"? Reconsidering a Category
   of Modern German History   1
   Simone Lässig and Miriam Rürup

**I. IMAGINATIONS: Remembrance and Representation
of Spaces and Boundaries**

1. Of Sounds and Stones: The Jewish-Christian Contact Zone
   of a Swiss Village in the Nineteenth Century   23
   Alexandra Binnenkade

2. Imaginations of the Ghetto: Jewish Debates on Ghettos and
   Jewish Society in Late Nineteenth-Century Galicia   40
   Jürgen Heyde

3. Modernization and Memory in German-Jewish History   55
   Nils Roemer

4. From Place to Race and Back Again: The Jewishness of
   Psychoanalysis Revisited   72
   Anthony D. Kauders

5. Jewish Displacement and Simulation in the German Films
   of E. A. Dupont   88
   Ofer Ashkenazi

6. Layered Pasts: The *Judengasse* in Frankfurt and Narrating
   German-Jewish History after the Holocaust   107
   Michael Meng

**II. TRANSFORMATIONS: Emergences, Shifts, and Dissolutions
in Spaces and Boundaries**

7. The Representation and Creation of Spaces through Print Media:
   Some Insights from the History of the Jewish Press   125
   Kerstin von der Krone

8. Out of the Ghetto, Into the Middle Class: Changing Perspectives
   on Jewish Spaces in Nineteenth-Century Germany—The Case
   of Synagogues and Jewish Burial Grounds   140
   Andreas Gotzmann

9. Spatial Variations and Locations: Synagogues at the Intersection
   of Architecture, Town, and Imagination   160
   Sylvia Necker

10. Jewish Philanthropy and the Formation of Modernity:
   Baron de Hirsch and His Vision of Jewish Spaces in
   European Societies   179
   Björn Siegel

11. Reconstructing Jewishness, Deconstructing the Past:
   Reading Berlin's Scheunenviertel over the Course of
   the Twentieth Century   197
   Anne-Christin Saß

**III. PRACTICES: Negotiating, Experiencing, and Appropriating
Spaces and Boundaries**

12. A Hybrid Space of Knowledge and Communication:
   Hebrew Printing in Jessnitz, 1718–1745   215
   Dirk Sadowski

13. Faith in Residence: Jewish Spatial Practice in the Urban Context   231
   Joachim Schlör

14. Photography as Jewish Space   246
   Michael Berkowitz

15. Jews, Foreigners, and the Space of the Postwar Economy:
The Case of Munich's Möhlstrasse   263
Anna Holian

16. Creating a Bavarian Space for Rapprochement: The Jewish
Museum Munich   280
Robin Ostow

17. Real Imaginary Spaces and Places: Virtual, Actual, and Otherwise   298
Ruth Ellen Gruber

Index   317

Illustrations

5.1 Baruch's Hamlet in a newspaper advertisement for *Das alte Gesetz.* Furnished by Deutsche Kinematik   91

9.1 Carl Spitzweg, *In the Synagogue* (1855–60). Used courtesy of Museum Georg Schäfer, Schweinfurt   167

9.2 Max Beckmann, *The Synagogue in Frankfurt am Main* (1919). © 2016 Artists Rights Society (ARS), New York / VG Bild-Kunst, Bonn   168

14.1 Henry Ries, Berliner Kinder beobachten das Landemanöver eines "Rosinenbombers" am Flughafen Tempelhof, July 1948. © bpk / Henry Ries   248

15.1 Shops in the Möhlstrasse as seen from a Munich police car, 30 June 1949. Used courtesy of the Stadtarchiv München   264

16.1 View of the northwest corner of the Jewish Museum Munich. Used courtesy of Jüdisches Museum München   281

16.2 Wall cupboard by Dr. Simon Snopkowski. Used courtesy of Jüdisches Museum München   287

Preface

This volume has a long history—actually too long, but also inspiring. Among other things, it evolved from a scholarly event Simone Lässig organized to a joint editing project with Miriam Rürup. Thus, parallel to the rather unorthodox development of the book project, the remainder of this preface is written in Simone Lässig's voice as she reflects on the foundations of the project, yet the heartfelt thanks to all those who contributed to the project along the way come from both editors.

The roots of this project date back to a visiting professorship I had the pleasure of being invited to hold for the academic year 2009–10 at the European Studies Centre at St Antony's College at Oxford, which was generously sponsored by the Marga and Kurt Möllgaard Foundation in the Stifterverband für die Deutsche Wissenschaft (Association for the Promotion of German Sciences). That was a wonderful time. I profited immensely from the many discussions with colleagues and the tremendous intellectual and library resources available there.

German historians and I personally owe a debt of gratitude to the late Gerhard A. Ritter, who, among so many other achievements, was a very important force in the visiting professor program at Oxford and in fostering dialogue among German and British historians. Likewise, I would like to thank the former program director of the Stifterverband für die Deutsche Wissenschaft, Heinz-Rudi Spiegel, for his great, long-term engagement and interest in supporting the visiting professorship at Oxford and for his activism—he advocated intensively and successfully for the professorship to be continued.

That still today I enjoy reflecting on my productive and inspiring months at Oxford so much is due, not least, to Jane Caplan. She was a wonderful host in a variety of ways, among other things opening doors to the academic culture of England for me. Jane also supported me in organizing the workshop "Jewish Spaces in Modern Societies and Cultures: Germany in Comparative Perspective," which took place at St Antony's College in May 2010. The workshop addressed central questions of a spatially informed historiography for the history of German-speaking Jewry, constructively advancing the field. The discussions were decidedly fruitful, with many of the presenters highlighting problem areas that had hardly even been gauged up to that point from a historical perspective.

Only a week later I was back in Braunschweig at the Georg Eckert Institute, which meant that I had dived once again into an entirely different context—one oriented toward school textbooks and educational media, and soon also the history of knowledge. For historical research on Jewish spaces, there was very little "space" in the schedule, and so it became more and more difficult for me to produce a coherent volume on the topic.

Yet Jane Caplan and her successor at Oxford, Paul Betts, remained interested. And when a conference entitled "Invented Jewish Traditions: Jewish Heritage in Europe between Memory and Staging" was announced at the Institute for the History of German Jews in Hamburg, whose director since 2012, Miriam Rürup, had also participated in the Oxford workshop, it reignited the spark. The conference theme, drafted jointly by Anna Menny and Michael Studemund-Halévy, seemed to be quite complementary to what I had developed at Oxford, and all the scholarly works that had been published in the sphere of our topic suggested that our aim had remained highly relevant—indeed, that its scholarly relevance had even grown in the meantime. Miriam saw things the same way, and so we decided to start a joint venture: we put together approaches oriented toward generating an original and scholarly contribution to the issues of space and boundaries in a historiography that understands Jewish history as a part of, as well as a particular lens on, general history—one that seeks to foster the transcendence of disciplinary rigidity and boundary-setting. To put it another way, we wished to relate scholarly spaces to one another that all too often even now exist merely side by side.

We began to enact this plan in 2014 and have cooperated superbly ever since. Working together with the authors has been a great pleasure, and the contribution of the two editors, too, cannot be praised highly enough; they have supported us in decidedly knowledgeable ways, bringing in several ideas of their own that further developed our project: Dr. Katherine Ebisch-Burton of the Georg-Eckert-Institut–Leibniz-Institut für internationale Schulbuchforschung Braunschweig (GEI, Leibniz Institute for International Textbook Research of Braunschweig) and Dr. Patricia (Casey) Sutcliffe of the German Historical Institute (GHI) in Washington, DC. We would like to express our sincere gratitude to these two colleagues, who edited all the contributions and corresponded with the authors and the publisher with great competence. We would also like to thank the commentators David Feldman of Birkbeck, University of London; Anthony Kauders of Keele University; Joachim Schlör of the University of Southampton; Eliyahu Stern of Yale University; and the panel chairs Abigail Green, Ruth Harris, David Rechter, and Nicholas Stargardt, all of Oxford, as well as the other participants in this event. We owe a special debt of gratitude to Anna Menny and Michael Studemund-Halévy, who also made an important if indirect contribution to the volume as the co-conveners of the conference at the Institute for the History of German Jews in Hamburg (IGdJ).[1] Associated with this, then, tremendous thanks are also due to the four scholarly institutions that supported the above-mentioned events or this volume with their resources: the GEI Braunschweig, the GHI Washington, the IGdJ Hamburg, and the European Studies Centre of St Antony's College, Oxford, where Anne-Laure Guillermain also provided us with outstanding administrative assistance.

We are also grateful to the two anonymous reviewers who gave us all very important suggestions for improving the volume, to Jane Caplan and Paul Betts of Oxford, as well as to all of the coeditors of the series, including Paul Betts at the helm as executive editor, Timothy Garton Ash, Jürgen Kocka, Gerhard Ritter, Nicholas Stargardt, and Margit Szöllösi-Janze, who did not forget the volume despite its long period of "dormancy," but rather actively fostered its completion. Thanks go out, too, to Chris Chappell, the editor in charge of the series "New German Historical Perspectives" at Berghahn Books of Oxford and New York.

Simone Lässig and Miriam Rürup

Note

1. Anna Menny and Michael Studemund-Halévy, eds., "Jüdisches Erbe," *Aschkenas. Zeitschrift für Geschichte und Kultur der Juden* 25, no. 2 (2015): 205–307.

Introduction:
What Made a Space "Jewish"?

Reconsidering a Category of Modern German History

Simone Lässig and Miriam Rürup

Spatial History and Culture

Our awareness of the spatial dimensions of historical processes and interactions among individuals and institutions has grown considerably since the late 1990s. While these developments do not amount to a fundamental paradigm shift—a "spatial turn"—in academic history, perceptions of history are no longer limited to its temporality but encompass its spatiality, too. There is an increased consciousness of both the impact of history and historical actors on spaces and their potential to themselves shape and impact history, in their material existence as well as in their quality as imagined or figurative spaces. Space as a focus of historical research shows unprecedented promise, be it as an epistemological category, as an analytical approach, or as a subject of historical analysis.

Some of the first historical studies to engage with the category of space centered on macrolevel processes—that is, they revolved around the ways societal structures influence and shape spaces and how spaces shape these structures. More recent historical and cultural studies have applied a spatial perspective to micro- and meso-analyses—for example, investigating the formation and stabilization of communities. The results have been substantial, the most fruitful having emerged from studies that conceive of space in a performative sense and analyze how the actions, perceptions, and experiences of individuals and groups in various historical settings have produced social, cultural, or political spaces.[1] Works about practices of spacing or doing spaces have helped to increase historians' awareness of the significance of actions that generate spaces and structures within historical processes.[2]

It may seem surprising that it has taken nearly twenty years for these spatial approaches to be applied to the *history* of minorities. After all, socio-geographic, sociological, and ethnological approaches to the study of minorities, ethnic communities, and diaspora cultures have already shown that foci like "space," "realm," and "place" have the potential to generate particularly rich insights in this field. The construction and depiction of spaces inevitably go along with negotiating and establishing real or imaginary boundaries; to create and interpret social and cultural space always means defining who is included or excluded. Further, the ways that groups occupy, form, and rework space indicate the form and the extent to which certain ethnic or religious groups became part of the majority society; these spatial processes can point, for instance, to the perception of these groups as permanent or temporary. These are central issues researchers confront when they engage with majority/minority relationships, ethnic communities, and the ethnic or faith-based narratives of identity and belonging or exclusion intertwined with such relationships. In the context of diaspora groups such as the Jewish minority, spatial processes grow even more important. Many of its members perceive of themselves as a group exiled from their ancient "homeland" in Palestine, thus generating a close connection to a distant territory. Such an imagined spatial relation to a former homeland forges a vital part of the minority's identity, as does a possible future return.[3]

Despite the crucial role these spatial processes play in identity formation, the boundaries that define spaces, and especially symbolic ones, are not static. Far from being set in stone, boundaries are the subject of—and are subject to—discourses, acts of negotiation, and multilayered processes of cultural translation.[4] In other words, their continuity and change make them amenable, indeed necessarily so, to historicization.

Our perspective thus expands from the history of majority/minority relationships familiar to researchers, including acts of state power such as political or legal decrees, to encompass cultural practices, in their entanglement with one another and with wider developments, as they relate to the formation and dissolution of identities, integration and segregation, identification and distancing. This approach has the potential to explore the dual nature of space: on the one hand, as a given form that shapes the experience and identities of those both inside and outside of it and, on the other, as something fluid and contingent, allowing for appropriations and reconfigurations, as well as giving agency to those within it to shape it anew.

Judaism, Jewishness, and Space in the Face of Modernity

At a general level, spatial issues apply to all historical periods, yet they are especially relevant to the modern era, when a number of transformations occurred simultaneously: toward the primacy of the individual subject, toward greater variation in lifestyles and lifeworlds, and toward nation building. These transformations objectively called nearly all distinct, socially separate groups into question that had often been granted a specific jurisdiction and, in principle, had not clashed with the early modern societal structures. The advent of modernity thus profoundly affected European Jewry as a distinct social group primarily defined by religion and religious law (*Halakhah*) in all spheres of everyday life. The specific character of the Jewish religion, the use of separate non-majority languages (Yiddish/Ladino), and the various forms of discrimination Jews were subject to, over the centuries, had given rise to social spaces and places that fostered particular forms of cohesion and distinctiveness. They formed boundaries and barriers that kept premodern Jewish lifeworlds essentially free of significant external influences, despite the economic and cultural interaction with the "outside" that also took place throughout this period. In this sense, the modern age as it unfolded in Europe was a period of profound transformation for Jewish lifeworlds and the spaces constituting and emerging from them.

The modern age impacted Sephardic and Ashkenazic Jews in different ways because of their distinct relationships to space. The Sephardic Jews of Europe had essentially internalized a repetition of the Diaspora experience; their religiously founded relationship to Zion went hand in hand with viewing Sepharad, the land they had been expelled from, as a place of nostalgia. This imagined space became the focus of their specific longing, while the Sephardic Diaspora generated a unique community with its own spatial points of reference and landscapes, many of which related to real places such as seaports.[5] Ashkenazic lifeworlds, by contrast, remained comparatively autonomous in their structures well into the eighteenth century. Their intellectual foundations rested on the medieval centers of learning represented by the ShUM towns (Speyer, Worms, Mainz) in southern Germany. They had their own primarily faith-based norms and values and clear definitions of who "belonged." Moreover, the specific structures of knowledge that had shaped premodern Ashkenazic spaces entailed a totality and an encompassingly sacred nature that exceeded the general ubiquity of religion characteristic of the early modern period.[6]

The processes of functional differentiation that began to unfold in Western and parts of Central Europe around 1800 resulted in the secular and the sacred becoming differentiated and disentangled from one another. Also, the figure of the modern citizen, rooted in the idea of equality before the law, began to emerge. These developments were essentially incompatible with autonomous social spaces existing beyond society as a whole. The universalist tendencies of the Enlightenment, inherent in the movements toward emancipation and the rise of incipient middle-class systems of thought and being, strongly discouraged all forms of social particularity of religious groups.[7] As a result, modernity jeopardized European Jewry, Judaism, and Jewishness as carried over from the early modern period in many ways. Most importantly, it threatened to destroy the unambiguous signifiers of belonging clearly defined by those on the inside and outside alike and to dissolve the connection between religion and social practices that had heretofore seemed positively solid. From the end of the eighteenth century onward, what defined Jewishness in both an individual and a collective sense, and what a "Jewish place" was and what was perceived as "Jewish space," grew increasingly ambiguous; definitions became more dependent on people's subjective experiences, perceptions, and discourses, and these generated and highlighted certain spaces. Consequently, the transformations Judaism experienced in the modern period were also processes of spatialization.

Against this backdrop, a spatial approach to modern Jewish history—that is, one that recognizes the significance of space in shaping lifeworlds—offers us an opportunity to gain a sense of the changing and increasingly diverse understanding of Jewishness that emerged with modernity and to grasp the wide spectrum of Jewish identities in modern societies. Jews had remained an autonomous minority group on the margins of the premodern state, yet more than a few clearly benefited from the new opportunities that the advent of modernity generated even more rapidly than members of other groups and worked to secure a position for themselves in the center of modern society. This process could be observed on a metaphorical/meta-level and in spatial manifestations such as architecture. Europe's German-speaking territories, above all, are relevant in this regard. They became the birthplace of the *Haskalah,* the Jewish Enlightenment, making them especially fertile ground for developing concepts to meet the challenges of modernity in ways that allowed Jewishness and Judaism to transform and thus survive while facilitating Jews' active participation in forming a modern, middle-class culture. In other words, the German states became a vibrant space for developing and negotiating different concepts of identity, making them an ideal focus for this volume's endeavor to explore Jewish spaces, how they were constituted, how they were perceived, and how those newly created spaces shaped a new understanding of belonging to and identifying with a group.

The German case also allows us to extend our focus to the meta-level—that is, it offers a unique and striking example of how historiography itself has engaged in boundary drawing and thus the construction of historically imagined spaces. Whereas boundaries between Jewish and non-Jewish spaces in modern and postmodern societies have been fluid and contingent, scholars have often obscured this historically open character of Jewish existence by retroactively constructing and projecting boundaries upon it. It is hardly surprising that after the Shoah, boundaries emerged that limited approaches to the interpretation of the history of Jews in Germany; history as an academic discipline, especially in Germany, has long neglected the interaction among Jewish and non-Jewish spaces and thus seems to have avoided analyzing German and Jewish history in the modern and postmodern period as a truly "entangled history."[8] We hope this volume may contribute to overcoming the still common binary division of "Jewish" and "non-Jewish" and raise awareness of interspaces, gray areas, multiple layers of identities and their entanglements, and patterns of boundary construction that were typical of a given period and cultural setting.

Consequently, this volume seeks to adjust this framework by opening a window onto the multifaceted dimensions of Jewish experience and on the spectrum of manifestations of Jewishness, and the Jewish spaces they entailed, in the modern German-speaking territories and beyond. Drawing us to explore spaces constructed or perceived as Jewish either by Jews or by non-Jews are the opposing forces simultaneously at work within them: on the one hand, certain spaces restricted and structured Jewish lives and tended to separate them—actively or passively—from other social groups or even the larger society; but in other ways, these and other spaces facilitated integration, opened up room within which Jews could maneuver, and proved open to changes. Acknowledging the trans-territorial and transnational dimension of Jewish history, we wish especially to contribute to unveiling spatial and temporal structures particular to being Jewish or being defined as such. The concept of Diaspora is one such spatial, and likewise temporal, structure. Indeed, it is often regarded as the key paradigm of Jewish history. Within this view, "Jewish space" has been temporary in nature ever since the Jews were expelled from their historic *place* in ancient Israel, and only Zion, the focus of loss and longing, remained as an explicitly Jewish *space*.[9] We aim here to transcend this perspective and explore the specific properties relating to the "Jewishness" of space and spaces beyond the diasporic context in modern German history.

Definitions of Place, Space, and Boundaries

While "places" bear unambiguous topographical identifiers and tend toward stasis, "spaces" also exist on an imaginary level; they are mutable and largely defined by experience and history.[10] When one considers the history of minorities, this symbolic property of space holds particular appeal for research; "space" can be conceived as a metaphor of social positions and of inclusion, exclusion, belonging, and identity, so that research can focus primarily on the communicative production of spaces—that is, on semantic systems related to space. Indeed, this is what the chapters in this volume do. Yet they go further, too, by drawing on the widely proven assumption that social position is reflected in geographical space and in the materiality of space we encounter in places like houses of worship, museums, and burial grounds—places that both create space and help constitute other spaces, such as a specific subculture within urban lifeworlds. The arrangement of space rarely fails to exert its influence on individual and group behaviors and actions. Material space—the *raumphysikalisches Substrat* in the words of sociologist Markus Schroer[11]—generates and is shaped by numerous social effects. Thus, in this volume we seek, as far as possible, to dissolve the opposition evident in much thinking about space wherein it is conceived either as materiality or alternatively as discourse. The editors and contributors are interested in exploring the social nature of space, how it emerges, the effects space in turn exercises on the social milieu,[12] as well as the relationship between the two.[13] In other words, material space—or place—shapes an imaginary superstratum, which in turn is reproduced in new places when they are built. The contributors to this volume use concepts such as "entangled identities," "hybrid spaces," and "contact zones" in their analyses of Jewish spaces and their interaction with non-Jewish spaces.

Inherent in this perspective is that while neither material nor immaterial spaces are static and immutable containers, there are limits to their dynamism and re-creatability.[14] Material spaces can be institutionalized and set up for the long term, thus representing what is given, established, reliable, not constantly called into question. Schroer asserts that such spaces frequently serve to reduce complexity by bearing pre-inscribed significance and thus relieve us of the obligation to continually attach new significance to them.[15] Spatial structures prescribe specific frameworks for action—indeed, they define power relations. Nonetheless, even seemingly solid, fortified places are susceptible and subject to appropriation, influence, and acts of configuration and reconfiguration from the social sphere. No space, material or otherwise, is an island; spaces acquire their meaning from subjective perceptions and ascriptions, including the symbols and rituals associated with them. One and the same space may be the object of a range of highly divergent perceptions, with different social groups relating in specific ways to it and all regarding it as their own. This phenomenon prompts fascinating questions concerning the significance of spaces to groups' ideas of themselves and to others' perceptions of them, the exclusivity of the structures of spaces that groups relate to in this way, the flexibility of their boundaries, and the historical changes in how people related to these spaces, and, consequently, in the nature of the spaces themselves.

By "boundaries," we mean markers of socially and culturally shaped symbolic spaces, each with specific regimes of recognition, as Bourdieu termed them. This raises questions of the power and resources individuals or groups possess and use to define their place within these spaces and of networks and communities of recognition that promise solidarity, build self-assurance, and enable those who draw upon them to master everyday life. Although boundaries can be as fluid as spaces and often overlap—or constitute transitions between spaces—the term itself tends to smack of static exclusion and "othering." In this book, however, we wish to include rather than exclude open and fuzzy zones of overlapping cultures. It is precisely these liminal, transitional spaces, the spaces at the margins and between distinct entities where change begins through interaction, that we aim to explore. We thus focus both on patterns of exclusion and on more dynamic tensions and practices of contact, interaction, coexistence, and transition without excluding conflicts, fractures, and differences in mutual perceptions or expectations. With this approach, we hope to develop a fresh perspective on German-Jewish history that transcends traditional narratives. The transitional *Sattelzeit* between the mid-eighteenth and the nineteenth century, for instance, generated completely new and sometimes highly unsettling challenges for individuals and social groups concerning social relations, sociocultural contexts, and horizons of experience. Hitherto marginal and largely autonomous groups, such as European Jews, were especially affected by this. Historical studies on this era usually concentrate on the limits of integration but rarely consider this period as a time of first encounters between cultures that had previously had only limited contact. To cite one example, universities were spaces that for centuries had been reserved exclusively for Christians, a practice that went unquestioned. Yet in the *Sattelzeit*, after Prussia and a couple of other German states had permitted males of other religious groups to study much earlier than, for example, England, Jews entered them astonishingly quickly. Thus, academic spaces of the nineteenth century fostered a dynamic interplay between majority and minority cultures. In this context, a broad spectrum of meanings can be applied to the concept of "doing space"—from looking at productive encounters and fear of the unknown to examining practices of social exclusion.

These open concepts of "space" and "boundaries" enable us, then, to approach practices of community building, of distinguishing between the familiar and the uncanny, and of determining what is one's own and what is the "Other's." They also give us access to layers of meaning, structures, and interactions within spaces created by specific social groups.[16] Analogous to the term *Sattelzeit*, with its reference to an in-between period of history on the temporal plane, we could, indeed, identify these liminal spaces and fuzzy boundaries as *Sattelräume*, or "saddle spaces," on the spatial plane.

Although Bourdieu's theory of social and cultural fields as rather stable structures with relatively fixed and closed boundaries, connected to an embodied habitus that cannot be manipulated much by will, gains a great deal of importance when it comes to the social relevance of space, we understand spaces and their boundaries as based on communication and experiences, simultaneously limiting and opening up opportunities. Thus, we see them as less fixed and static than Bourdieu's concept of fields might suggest. Intriguingly, the German language includes terms such as Handlungsspiel*raum* (literally: room or space for action) and Frei*raum* (literally: room for freedom, or room to develop), which distinctly emphasize *actors* and the performativity of spaces.

"Jewish Spaces": Where Historical Research Is Now

After the historical changes that brought an end to socialist Europe, scholarship in the mid-1990s experienced a soaring interest in space as well as a "memory boom" that included the rediscovery of Jewish traditions. Within this overarching context, and specifically in response to phenomena understood as rediscovered traditions, social and cultural studies were among the first contexts in which scholars reflected on Jewish spaces. There was tremendous interest in generating strategies for visualizing and musealizing what had once been Jewish spaces, particularly those that were being revived and appropriated, above all, by non-Jews.[17] This pattern of appropriation was not as new as many initially thought; before 1990, non-Jews in West Germany had increasingly engaged with Jewish history and culture.[18] Yet the rapid emergence and development of interest in Jewish culture in Eastern Europe after 1990 surprised even experts in the field and inspired academics to analyze the new cultures of memory emerging in this period. This research, though centered on "history" and "memory," initially marginalized the "genuinely" historical perspectives on these spaces.[19]

However, given the fact that history takes place in both space *and* time, historical study of these spaces is critically needed. Historical views are essential if we wish to cast light on processes that unfold when long-established practices relating to the use of a space begin to clash with the political or cultural frameworks within which that space had long existed. The case of traditional Ashkenazic religious practices in synagogues, which came increasingly into question from the late eighteenth century by Jews and non-Jews alike, provides a cogent example. After the social and spatial structures that had governed the early modern period dissolved, overlapping manifestations of cultural identification became usual—a strategy that only historical views can reveal. Newly founded Jewish schools, societies, and associations, as well as new regulations and aesthetic standards for the Jewish service, created a unique potential to assure Jews of both their Jewishness and their *Bürgerlichkeit* (middle-class culture and habitus). They provided a forum for acquiring and rehearsing practices that signaled adherence to the middle-class lifestyle taking shape at the same time. Consequently, even religious spaces, or spaces formerly shaped by religion, became experimental settings for a modern way of being Jewish—which by no means always entailed rejection of the tenets of religious law—and, simultaneously, of an emergent middle-class habitus.

Members of the Potsdam-based research group Makom that existed from 2001 to 2007 were among the first scholars to develop a clearly historical focus, although they were much more focused on "Jewish places" than on "Jewish spaces."[20] Whereas spaces initially remained the domain of literature studies, experts in the history of architecture and Jewish culture and religion, anthropologists,[21] and scholars from other humanities and cultural studies disciplines, such as philosophy and history, gradually began to explore them as well.[22] Michael Meng's study *Shattered Spaces: Encountering Jewish Spaces in Postwar Germany and Poland* (2012), for instance, took a broad approach extending beyond the exploration of current tendencies and dynamics involved in the emergence of *new* Jewish spaces; he examined the history of Jewish spaces that had appeared to be lost and the ways they were explored, reexplored, and appropriated after 1945.[23] The historicity of Jewish spaces and the influence of space on the history of Jews and non-Jews alike have also been the focus of scholars like Barbara E. Mann, Vered Shemtov, and Anna Lipphardt, even though they are not historians themselves. Following these early studies, new perspectives such as memory history have likewise attracted the attention of historians.[24]

Scholars of the early modern period were among the first researchers of German-speaking regions to utilize space as an analytical category from social and cultural studies and apply it to their own work.[25] A 2009 conference on Jewish spaces in the early modern period[26] prioritized space for analysis, generating highly interesting insights, one of which was that the cultural spaces constituted in mid-eighteenth-century Europe by educated Jews and non-Jews, while largely separate from one another, emerged along startlingly similar lines. The differences in the structures and sets of rules the Christian and Jewish cultures of learning entailed appear to have been smaller than the differences between the everyday cultures of Sephardic and Ashkenazic Jews in contemporary Europe. Some research has suggested that educated Jews and non-Jews shared a relationship of mutual respect that was often absent between certain Jewish groups, such as Jews from different social classes or of differing educational backgrounds, or Jews from the centers of the Jewish Enlightenment versus Polish Talmudic Jews, at that time.[27]

When we move to the modern period, space becomes an even more promising analytical category for Jewish life because in this era Jewishness and its limits became a lifestyle choice, in modern parlance. A broad spectrum of ways of being Jewish unfolded at this time, resulting in an evident and logical diversity of the spaces perceived or defined as Jewish. With this in mind, we might be forgiven for wondering why academics "doing space" have seemingly not yet addressed Judaism in the modern age and in the period of transformation preceding it.

Compared with the German states, there is a little more research on Western Europe, but this, too, is a relatively recent development, having generated several publications since 2008.[28] More recently, a 2015 collection edited by Alina Gromova, Felix Heinert, and Sebastian Voigt explored the nature of Jewish spaces in various urban contexts across Europe from the early modern period up through more contemporary memories of Jewish spaces in modern literature.[29] While some of the essays in this interdisciplinary volume address Jewish spaces within German states or Germany, the geographical scope of analysis is broad, and the underlying approach focuses on *urban* spaces and, within these, on identity formation through spatial politics and practices.

All of the studies published within the last decade clearly show the growing importance of spatial perspectives in historical analyses, as well as new interest in applying space as an analytical category to the modern period. They build the foundation upon which spatial theory could be developed and contextualized within a broader disciplinary perspective. A principal aim of this volume is to contribute to this. We put forth a distinctly historical approach and, with it, a focus on German-speaking Jewry, contextualizing the analysis, naturally, within a broader comparative and/or transnational perspective wherever it is feasible and useful to do so. Coming from this analytical framework, the various essays explore constructions, experiences, perceptions, and practices around the creation, transformation, and appropriation of social, cultural, and political spaces, places, and (symbolic) boundaries as they relate to Jews and Jewishness. This also includes Jewish experiences and depictions of ruptures or removal of space, "re-spacement," and both voluntary and forced shifts from one space to another.

Imaginations, Transformations, and Practices

In keeping with our sense of the crucial nature of the advent of modernity in uncovering the constituted and constituting properties of spaces in general and for

European Jewry in particular, most of the essays discuss special features of Jewish existence and social interaction with regard to *modern* history. However, we also included contributions on early modern and contemporary history in order to sharpen our view of the major transformations involved in the transition to modernity. These contributions on other eras also highlight underlying inscriptions or hidden spatial scripts that continued to influence behavior.

Along with investigating the historical and sometimes contemporary meanings attributed to physical spaces, several of the contributions transcend the literal to explore the formation of figurative or imagined "Jewish spaces" within practices and how they shaped and were shaped by certain forms of identification and imagination. The contributions encompass a broad range of spatial reference points and manifestations. Some analyze religious and secular spaces that became crucial to emerging Jewish identities or translating them into new contexts; others investigate spaces that were defined and marked by Jews without necessarily being "Jewish spaces"; and still others look at spaces that were perceived, defined, or constructed as "Jewish," although they were actually scarcely different from non-Jewish or inclusive spaces, like the scholarly culture of the nineteenth century or the realms constituted by societies and associations of the educated upper middle class. However, we also take a closer look at places, like synagogues, that were clearly marked as genuinely Jewish but whose forms and meaning could differ significantly depending on time or cultural context.[30] The study of these various spaces and places therefore has the potential to increase our awareness of entangled cultures like *Bürgerlichkeit,* which could not only be culturally translated into Jewish environments but also be co-constituted and created exclusively within them. Furthermore, these spatial studies shed light on internal Jewish practices of social distinction, boundary drawing, and space formation. How spaces came to be perceived as Jewish or otherwise is at the heart of the book; we ask what made a space Jewish by retracing how both Jewish and non-Jewish actors attributed Jewishness to it and what that implied.

The category of space as used in humanities disciplines today is multivalent, and it is difficult—indeed, impossible—to separate the mental, physical, and symbolic layers of space from one another because they are in dialectical interplay, as Lefebvre maintained.[31] Thus, although we arranged this volume along three dimensions of spaces and boundaries—"Imaginations," "Transformations," and "Practices"—we could easily have placed particular essays in a different section. In this, as throughout our endeavor, boundaries have shown themselves to be profoundly permeable and mutable entities.

The essays in part 1, "Imaginations: Remembrance and Representation of Spaces and Boundaries," examine and compare how historical and recent spaces of remembrance and memory (*Gedächtnisorte* and *Erinnerungsräume*) have been imagined in and outside their time. While some of these—like Berlin's Scheunenviertel in Anne-Christin Saß's analysis—have a material point of reference, others, such as the "ghetto," which is at the center of Jürgen Heyde's contribution, were more figurative constructions—a topos. Yet for both material and figurative spaces, historical subjects, in meeting the challenges of their time, attached meaning to them either retrospectively or in reference to a projected future. Due to the close interrelationship between spatial structures (*Raumordnungen*) and the formation of subjects and to the fact that this interrelationship is often produced and mediated via symbols, signs, or cultural codes, this section analyzes both forms of remembrance of spaces as specifically related to Jews as well as representations of such spaces by non-Jews. In this context, the nature of spaces as both mediated and mediating comes to the fore—that is, both the ways that media, such as films and the press, produced spaces and how spaces came to mediate ideas and identities. This duality demonstrates the potential of media to be both the source and object of spatial analysis.

Alexandra Binnenkade, whose chapter is concerned with the Swiss village of Lengnau in the nineteenth century (chapter 1), takes a detailed look at the everyday practices that united or distinguished Christians and Jews within the social structure of the village space and the ideas of this space manifested in its architecture. Reiterating the concept of "contact zones" as elements of the spatial experience of minorities, she emphasizes physical components in the creation of space, such as streets and residential buildings, pointing out their influence on the processes of defining identities and states of belonging.

The ghetto, or rather its image in an era when it had (temporarily) ceased to exist, is at the heart of Jürgen Heyde's contribution (chapter 2). As the walls of real ghettos came down, the imagined space gained currency as a metaphorical point of reference. Heyde's analysis revolves around a journalistic debate in nineteenth-century Galicia whose participants evinced a fascination with the idea of the ghetto; the debate also drew the attention of many Jews in German states as well.

Nils Roemer (chapter 3) explores the emergence of a Jewish culture of remembrance in German cities, particularly since the Kaiserreich. On the basis of a wide range of characterizations, from travelogues and literature to museums and archives, he analyzes how remembrance is derived from physical remains and how these remains are presented in cities.

Anthony D. Kauders (chapter 4) explores the widespread idea that psychoanalytic theory and practice were particularly "Jewish" fields. Kauders establishes that Jewish and non-Jewish historians of the field, in their attempts to retrospectively trace the real or imagined origins of the practice in Jewish ethnospaces, were the ones who constructed psychoanalysis as a "Jewish" discipline.

The Jewish origins of key figures in a creative process are likewise central to the analysis presented by Ofer Ashkenazi in his essay on Jewish displacement in Weimar film (chapter 5). Ashkenazi looks at two films by Jewish director Ewald A. Dupont, *Peter Voß, der Millionendieb* and *Das alte Gesetz*. Both films engaged with the situation of Jews in the later years of the Weimar Republic and created a language of imagery from which to implicitly address Jews' position as a minority at that time.

Like Nils Roemer, Michael Meng (chapter 6) retraces the discourse around the literal unearthing of remains, including a *mikveh,* of Jewish life in Frankfurt's *Judengasse* during postwar construction work. His contribution narrates the development of a broader debate that emerged from this discovery revolving around the politics of memory, ways of approaching the Nazi past in postwar Germany, and how to appropriately remember and commemorate the persecution suffered by Frankfurt's Jews.

Part 2, "Transformations," is dedicated to the exploration of movements of emergences, shifts, and dissolutions in spaces and boundaries. The chapters in this section uncover ways in which Jews and non-Jews created spaces that came to be labeled Jewish—both consciously and unintentionally—at varying points in time. The authors investigate how various actors occupied and appropriated existing spaces that had seemed stable before, tracing how these actors arranged and rearranged these spaces, recoded them, or indeed opened them up to completely new sets of ascriptions and associations.

One component appears to be the enlargement of spaces by expanding purviews. Looking beyond national borders, Kerstin von der Krone (chapter 7) elucidates the emergence of a Jewish press from a transnational perspective. The Jewish press paid attention to issues of interest to a Jewish audience and provided the frame for a modern Jewish public space.

Andreas Gotzmann's contribution (chapter 8) revolves principally around two exclusively Jewish places and their concomitant spaces—the synagogue and the Jewish burial ground—relating the transformations the Jewish minority experienced to changes in Jewish space at the threshold of modernity. He discusses both shifts in the societal and legal frameworks of these spaces and their use, along with increasing diversity within Judaism, all of which provided Jews with a range of options for practicing their faith.

The transformation of spaces and their purposes in the transition to modernity is also at the core of Sylvia Necker's analysis of the changes in the siting and design of synagogues in German cities (chapter 9). Necker argues that Jewish communities, being involved in processes of urbanization, relocated synagogues from the peripheries or run-down parts of town to the old or newly emerging centers, making them increasingly prestigious and visible and thus de-marginalizing them as a symbol of Jewishness. In this, they functioned as a place and a space at the same time.

Björn Siegel's essay (chapter 10) investigates the creation of an increasingly transnational ideal Jewish space in the shape of the Jewish philanthropic movement that came into being in the nineteenth century. Utilizing the example of the humanitarian endeavors organized by Baron de Hirsch, Siegel explains how a Europe-wide Jewish network of assistance arose in response to episodes of Jewish persecution.

Anne-Christin Saß (chapter 11) casts light on the discursive production of associations between material places and imagined spaces by analyzing the ways that Berlin's Scheunenviertel has been perceived and utilized over the long view. She argues that it has been seen as representative of Jewish Weimar Berlin and of the resurrection of Jewish life postwar and particularly post-1989. Concerning the most recent era, Saß points to the entanglement of topographical depictions with concrete political aims and wish fulfillment, which is inherent in specific nostalgia-driven associations with the space.

Part 3, "Practices: Negotiating, Experiencing, and Appropriating Spaces and Boundaries," focuses on the things that those who live and act within spaces do. Thus, it zeroes in on the question of what practices give rise to and change spaces and how shifting spaces, in turn, shape practices. Not all of these practices are necessarily dynamic; many historical spaces are stable, influenced by specific power structures, knowledge formations, or rituals passed down through a number of generations. Such spaces can shape social processes over long periods of time. People, individually or collectively, hold particular perceptions of such spaces, acquire experience in and with their structures, and take part in processes of negotiation, interpretation, and translation of them. Although—or even because—these processes frequently give rise to conflict, new spaces or new interpretations of them can also emerge from this. In making use of these spaces, people may perhaps attempt to shift their boundaries, yet—as we see when we consider spaces such as state education systems, workhouses, the military, and places of worship—often find themselves, their agency, and dynamism subject to significant political, religious, or social limitations.[32] Conversely, physical and symbolic boundaries—especially those that became typical of the modern age—sometimes dissolved in the course of historical developments. What happened in these instances? Did new spaces of interaction emerge or fail to do so?

Dirk Sadowski (chapter 12) provides a case study of a specific profession, investigating the degree to which a Hebrew print shop in the town of Jessnitz in the eighteenth century constituted a Jewish space. He demonstrates the reciprocity at the heart of the emergence of space in tracing how the professional practice itself gave rise to an arena of what was perceived as Jewish.

Material components of the production and formation of space form the center of chapter 13, by Joachim Schlör. The threshold, in its capacity as simultaneously a material and metaphorical point of separation between the inside and the outside of a Jewish lifeworld, is the focus of his examination of the significance for Jews of moments of arrival and settling in, of departure and migration.

Michael Berkowitz's essay (chapter 14) centers around another professional and key cultural medium, photography, exploring both the practice and its practitioners. The chapter recounts the formation of a network of Jews within modern photography that continued to exist in the National Socialist period, when the Nazis exploited the group for their own photographic purposes.

Taking us to the immediate postwar period, chapter 15, by Anna Holian, examines practices of the Jewish lifeworld that then began to arise in Munich's Möhlstrasse. The multilayered interactions and negotiations within the postwar black market created a Jewish space, yet also gave rise to perceptions from the outside of the space as "foreign" and hence a threat.

A museum and the communicative practices it engenders are at the heart of chapter 16, by Robin Ostow. Focusing on Munich's Jewish Museum, opened in

2007, she analyzes the production of a space that not only reflects upon Jewish history and culture but simultaneously seeks to create a space of dialogue between Jews and non-Jews about Jewish life in the present day. Ostow refers to the debates that erupted both concerning the museum's location within the Munich cityscape and concerning the practices that have begun to unfold in this space.

In the concluding essay (chapter 17), which reflects the discourse about another Jewish museum that opened only recently in Warsaw (POLIN), Ruth Ellen Gruber reconsiders her previously developed and highly influential concept of "virtual Jewishness"—essentially an artificial filling of the gap left by the Jews murdered under Nazi persecution at formerly Jewish places—and its transformations. As the idea has generated considerable debate, her contribution to this volume elaborates on "misinterpretations," providing a positive assessment of the phenomenon: it opens up new possibilities of communication between Jews and non-Jews in spaces that have been rediscovered as Jewish or re-formed and reappropriated as such. In a sense, then, Gruber's essay also reflects directly on one of the central purposes of this volume—to call into question the still widespread dichotomy between Jews and non-Jews and their respective spaces of living, actions, and experiences and to expose the complex and intertwined nature of "the majority" and "minorities" as they manifest themselves in acts of boundary drawing.

Outlook

Applying a spatial perspective to modern German-Jewish history, this book explores concepts and theoretical frameworks that might help scholars in and beyond the field of Jewish history to conceive of and reconsider the complexity, the interconnectedness, and the historical variability of spaces, of identities and their markers, of symbolic boundaries, and of social practices and cultural patterns of belonging. We identify spaces where identities were more fluid—spaces whose ascribed characteristics and definitions were more varied and subjected to processes of negotiation—and where diverse interrelationships emerged. The essays discuss mutual perceptions and expectations between various groups and in cultural practices of integration and segregation and of the construction, translation, and negotiation of identities in modern societies and cultures. Some of the authors focus on spaces that breached the solidity of real or symbolic boundaries or translated them into other cultural contexts; some elaborate on spaces whose limits proved immovable and thus impossible to overcome; and some look at spaces that Jews and non-Jews, acting from a range of motives, re-created and transformed. Performative aspects of the creation of new spaces and the appropriation, affirmation, and legitimation of existing ones are key to the endeavor of the whole volume,[33] which pays tribute to the fact that cultural and social spaces can contain multiple subspaces representing various sets of asymmetrical power relations or social hierarchies.

Although we have chosen case studies we found particularly significant for our analytical approach, this volume cannot be comprehensive. The list of questions relevant to but also beyond the German-Jewish context, which itself has still scarcely been researched, is as long as it is stimulating to further research. For instance, we still know relatively little about what made a space exclusive and "closed," inclusive and "open," or rather transparent and "mixed"—both in its essence as well as in how it was perceived. It is not always clear what spaces were perceived as Jewish, at what times, and by whom, nor why these perceptions changed when they did and to what extent. More work is needed to understand to what degree space-forming attitudes, perceptions, and practices were conditioned by shared religious or ethnic (Diaspora) roots, as well as how inclusive spaces of interethnic interaction and mutual recognition were constructed and supported or drained and destroyed. Moreover, it would be enlightening to investigate the ways and the extent to which Jews and non-Jews engaged in boundary drawing or attempted to transcend previously held positions and to explore how Jews have managed (or perhaps failed) to preserve what was their own in a space defined from without, in a form respected by those both within and without.

Across all facets of the issue, this book demonstrates that historical spatial perspectives uniquely enhance our awareness of the complexity and diversity of constructions of difference and strategies for negotiating it. It illustrates that cultural difference is not a solid and immutable essence but a product and subject of discourses, negotiations, and translations related to and resulting from specific temporal and spatial contexts. Thus, the dissolution of boundaries such as those between legal jurisdictions, so prevalent in the early modern period, ushered in and made space for new boundaries and new spaces. Such boundaries, and the group-related identities they engender, are invariably shaped by a range of strategies and potential courses of action that notably call for further historical research. Perspectives dominated by dichotomies, such as studies of Jews and other minorities in Germany that have focused on the opposition between assimilation and resistance thereto, have long blocked scholarship from seeing that boundaries are not fixed and inflexible but are mutable, produced as they are by political, societal, and cultural factors in specific historical contexts. Similarly, boundaries can be exclusively imagined, manifestations of processes that distinguish between who is "in" and who is "out," even where everyday practices resist or soften these dichotomies. And boundaries as well as the spaces they mark are, finally, products of conflict-laden interaction between various groups with contrasting interests who, in drawing them, give expression to hierarchical and power relations. In consideration of all this, those engaged in historical scholarship may feel called to make boundaries, the shifts they undergo, and the spaces and spatial changes they prompt the subjects of their research, especially when their work relates to minorities such as the Jews. This perspective might enable scholars to scrutinize the nature of the majority-minority boundary itself more closely and, by acknowledging its strictly historical essence—that is, that it has never been fixed and immutable—develop an alternative to the widely assumed characteristics frequently attributed to it.

Notes

1. Giddens, *The Constitution of Society*; Löw, *Raumsoziologie*. See also Jerram, "Space: A Useless Category?"; and Schwerhoff, "Spaces, Places, and the Historians."
2. Schlögel, *Im Raume*; Schlögel, "Räume und Geschichte"; Dipper and Raphael, "Raum." There is a good overview of the spatial in historical research, albeit focused on France and German-speaking regions, in Rau, *Räume*; and multidisciplinary reflections on developments in this arena of research in Döring and Thielmann, *Spatial Turn*.
3. Silberman, Till, and Ward, *Walls, Borders, Boundaries*; Brah, *Cartographies of Diaspora*; Hastings and Wilson, *Borders*. See also Agnew, "Borders on the Mind."
4. Bachmann-Medick, "Translation"; a discussion of the concept's value for historical research is in Lässig, "Übersetzungen in der Geschichte."
5. Examples are Dubin, *The Port Jews of Habsburg Trieste*; Schapkow, *Role Model and Counter-Model*; Menny, *Spanien und Sepharad*.
6. On the challenges the various Christian denominations faced at the advent of modernity, see Schlögl, *Alter Glaube*.
7. Eisenstadt, "Minorities."
8. This is related to two new perspectives on the subject. First, there is the approach championed by Dan Diner of using Jewish history as a promising route, within theories of knowledge, to accessing issues of history in general. Secondly, there is a new awareness that a number of phenomena interpreted by researchers of Jewish history in terms of Jewish "adaptation" or "conformity" were not fixed entities that Jews could "conform" to but were rather emergent in character. Thus, Jews helped to shape them; the new middle-class habitus of the period is one example. Weiss and Gross, *Jüdische Geschichte*.
9. See, e.g., Wettstein, *Diasporas and Exiles*; and Tölölyan, "Rethinking Diaspora(s)." A discussion of the category of diaspora in historical analysis can be found in Rürup, "Von der religiösen Sehnsucht zur kulturellen Differenz."
10. Lefebvre, *The Production of Space*; Tuan, *Space and Place*. On the nature of "space" as a construct and the range of perspectives research has brought to bear on it, see, *inter alia*, Günzel, *Raum*, 121–320; Rau, *Räume*.
11. Schroer, *Räume, Orte, Grenzen*, 176.
12. Ibid., 145.
13. Döring and Thielmann, *Spatial Turn*, 18–19, 26, 81.
14. Schroer, "Bringing Space Back In," in Döring and Thielmann, *Spatial Turn*, 137, argues against spatial determinism just as much as against spatial volunteerism.
15. Ibid., 142.
16. Bourdieu, "Physischer, sozialer und angeeigneter physischer Raum."
17. Pinto, "A New Jewish Identity"; Gruber, *Virtually Jewish*; Lehrer, *Jewish Poland Revisited*; Gantner and Kovács, "The Constructed Jew"; Waligórska, "Jewish Heritage Production and Historical Jewish Spaces"; Coles and Timothy, *Tourism, Diasporas and Space*; most recently: Lehrer and Meng, *Jewish Space in Contemporary Poland*.
18. Gilman and Remmler, *Reemerging Jewish Culture in Germany*.

19. Fonrobert and Shemtov, "Introduction: Jewish Conceptions and Practices of Space." The special issue of *Jewish Social Studies* 1, no. 3 (2005) introduced collected essays focusing on biblical times and the present. Consequently, it does not have a historical focus.
20. A key interdisciplinary publication evolving from the work of this research group is Brauch, Lipphardt, and Nocke, *Jewish Topographies*; see also the essay collection Kümper et al., *Makom*.
21. Gantner and Oppenheim, eds., "Jewish Space Reloaded"; Mann, *Space and Place in Jewish Studies*; Lipis, *Symbolic Houses in Judaism*; Meitner, "The Mezuzah; Shemtov, "Between Perspectives of Space," 142; Fonrobert and Shemtov, "Introduction: Jewish Conceptions and Practices of Space"; Fonrobert, "The New Spatial Turn in Jewish Studies"; Baumel and Cohen, *Gender, Place and Memory*; Bodemann, *Gedächtnistheater*.
22. Cohen, *Visualizing and Exhibiting Jewish Space and History*; Ernst and Lamprecht, *Jewish Spaces*.
23. Meng, *Shattered Spaces*.
24. Baumel and Cohen, *Gender, Place and Memory*; with a clear focus on place and displacement: Cesarini, Kushner, and Shain, *Place and Displacement*; Kauders, "Emotional Geography."
25. Kießling et al., "Räume und Wege."
26. A report by Rotraud Ries can be found at http://www.forum-juedische-geschichte.de/ForumBericht09.pdf (accessed 23 March 2015).
27. Ries cites the example of the Sephardic traveler Hayyim Yosef David Azulai (1724–1806), who perceived Ashkenazic regions as hostile, uncultivated, and unwelcoming; ibid., 8.
28. See, e.g., on France, Malinovich, *French and Jewish*; Haus, *Challenges of Equality*; on England, Kushner, *Anglo-Jewry since 1066*; and on the Netherlands, Frishman et al., *Borders and Boundaries*. For a comparative perspective on London, Paris, and Berlin, see Laguerre, *Global Neighborhoods*.
29. Gromova, Heinert, and Voigt, *Jewish and Non-Jewish Spaces in the Urban Context*.
30. Kümper et al., *Makom*.
31. Schroer, "Bringing Space Back In," in Döring and Thielmann, *Spatial Turn*, 138, has a detailed discussion.
32. Döring and Thielmann, *Spatial Turn*, 26, point out that while spaces come into being in a social context, we cannot specifically generate every space socially and at times find ourselves socially placed/located within specific spaces.
33. Cf. also Hödl, *Wiener Juden*.

Bibliography

Agnew, John. "Borders on the Mind: Re-framing Border Thinking." *Ethics & Global Politics* 1, no. 4 (2008): 175–91.
Bachmann-Medick, Doris. "Translation—A Concept and Model for the Study of Culture." In *Travelling Concepts for the Study of Culture*, edited by Birgit Neumann and Ansgar Nünning, 23–43. New York: de Gruyter, 2012.
Baumel, Judith T., and Tova Cohen. *Gender, Place and Memory in the Modern Jewish Experience*. London: Vallentine Mitchell, 2003.
Bodemann, Y. Michal. *Gedächtnistheater. Die jüdische Gemeinschaft und ihre deutsche Erfindung*. Hamburg: Rotbuch Verlag, 1996.

Bourdieu, Pierre. "Physischer, sozialer und angeeigneter physischer Raum." In *Stadt-Räume: Die Zukunft des Städtischen,* edited by Martin Wentz, 25–34. Frankfurt am Main: Campus, 1991.

Brah, Avtar. *Cartographies of Diaspora: Contesting Identities.* 3rd ed. London/New York: Routledge, 2002.

Brauch, Julia, Anna Lipphardt, and Alexandra Nocke, eds. *Jewish Topographies: Visions of Space, Traditions of Place.* Aldershot: Ashgate, 2008.

Cesarini, David, Tony Kushner, and Milton Shain, eds. *Place and Displacement in Jewish History and Memory: Zakor v'Makor.* London: Vallentine Mitchell, 2009.

Cohen, Richard I., ed. *Visualizing and Exhibiting Jewish Space and History.* Oxford: Oxford University Press, 2012.

Coles, Tim, and Dallen J. Timothy, eds. *Tourism, Diasporas and Space.* New York: Routledge, 2004.

Dipper, Christoph, and Lutz Raphael. "Raum in der europäischen Geschichte." *Journal of Modern European History* 9 (2011): 27–41.

Donnan, Hastings, and Thomas M. Wilson. *Borders: Frontiers of Identity, Nation and State.* New York: Bloomsbury Academic, 1999.

Döring, Jörg, and Tristan Thielmann, eds. *Spatial Turn. Das Raumparadigma in den Kultur- und Sozialwissenschaften.* Bielefeld: transcript, 2008.

Dubin, Lois C. *The Port Jews of Habsburg Trieste: Absolutist Politics and Enlightenment Culture.* Stanford, CA: Stanford University Press, 1999.

Eisenstadt, S. N. "Minorities, the Formation and Transformation of Nation-States, and Inter-civilizational Relations—Jewish and Christian Minorities in Germany." *Tel Aviver Jahrbuch für deutsche Geschichte* 37 (2009): 23–32.

Ernst, Petra, and Gerald Lamprecht, eds. *Jewish Spaces. Die Kategorie Raum im Kontext kultureller Identitäten.* Schriften des Centrums für Jüdische Studien 17. Innsbruck: Studienverlag, 2010.

Fonrobert, Charlotte Elisheva. "The New Spatial Turn in Jewish Studies." *AJS Review* 33, no. 1 (2009): 155–64.

Fonrobert, Charlotte Elisheva, and Vered Shemtov. "Introduction: Jewish Conceptions and Practices of Space." *Jewish Social Studies* 1, no. 3 (2005): 1–8.

Frishman, Judith, David J. Wertheim, Ido de Haan, and Joel Cahen, eds. *Borders and Boundaries in and around Dutch Jewish History.* Amsterdam: Uitgeverij Aksant, 2011.

Gantner, Eszter, and Mátyás Kovács. "The Constructed Jew: A Pragmatic Approach for Defining a Collective Central European Image of Jews." In Šiaučiūnaitė-Verbickienė and Lempertienė, *Jewish Space,* 211–25.

Gantner, Eszter B., and J. (Koby) Oppenheim, eds. "Jewish Space Reloaded." Special issue, *Anthropological Journal of European Cultures* 23, no. 2 (2014).

Giddens, Anthony. *The Constitution of Society.* Los Angeles: University of California Press, 1984.

Gilman, Sander, and Karen Remmler, eds. *Reemerging Jewish Culture in Germany: Life and Literature since 1989.* New York: New York University Press, 1994.

Gromova, Alina, Felix Heinert, and Sebastian Voigt, eds. *Jewish and Non-Jewish Spaces in the Urban Context.* Berlin: Neofelis Verlag, 2015.

Gruber, Ruth E. *Virtually Jewish: Reinventing Jewish Culture in Europe.* Berkeley: University of California Press, 2002.

Günzel, Stephan, ed. *Raum. Ein interdisziplinäres Handbuch.* Stuttgart: J. B. Metzler, 2010.

Haus, Jeffrey. *Challenges of Equality: Judaism, State, and Education in Nineteenth-Century France*. Detroit: Wayne State University Press, 2009.

Hödl, Klaus. *Wiener Juden, jüdische Wiener. Identität, Gedächtnis und Performanz im 19. Jahrhundert*. Innsbruck: Studien Verlag, 2006.

Jerram, Leif. "Space: A Useless Category for Historical Analysis?" In Kümin and Usborne, "Forum," 400–19.

Kauders, Anthony D. "The Emotional Geography of a Lost Space: Germany as an Object of Jewish Attachment after 1945." In *Heimat: At the Intersection of Memory and Space*, edited by Friederike Eigler and Jens Kugele, 193–207. Berlin: de Gruyter, 2012.

Kießling, Rolf, Peter Rauscher, Stefan Rohrbacher, and Barbara Staudinger, eds. *Räume und Wege: Jüdische Geschichte im Alten Reich 1300–1800*. Berlin: Akademie Verlag, 2007.

Kümin, Beat, and Cornelie Usborne, eds. "Forum: At Home and in the Workplace: Domestic and Occupational Space in Western Europe from the Middle Ages." Special issue, *History and Theory* 52, no. 3 (2013).

Kümper, Michal, Barbara Rösch, Ulrike Schneider, and Helen Thein, eds. *Makom. Orte und Räume im Judentum. Real—Abstrakt—Imaginär*. Hildesheim: Georg Olms Verlag, 2007.

Kushner, Tony. *Anglo-Jewry since 1066: Place, Locality and Memory*. Manchester: Manchester University Press, 2009.

Laguerre, Michael S. *Global Neighborhoods: Jewish Quarters in Paris, London, and Berlin*. Albany: State University of New York Press, 2008.

Lässig, Simone. "Übersetzungen in der Geschichte—Geschichte als Übersetzung? Überlegungen zu einem analytischem Konzept und Forschungsgegenstand für die Geschichtswissenschaft." *Geschichte und Gesellschaft* 38, no. 2 (2012): 189–216.

Lefebvre, Henri. *The Production of Space*. Oxford: Wiley, 1991.

Lehrer, Erica T. *Jewish Poland Revisited: Heritage Tourism in Unquiet Places*. Bloomington: Indiana University Press, 2013.

Lehrer, Erica, and Michael Meng, eds. *Jewish Space in Contemporary Poland*. Bloomington: Indiana University Press, 2015.

Lipis, Mimi Levy. *Symbolic Houses in Judaism: How Objects and Metaphors Construct Hybrid Places of Belonging*. Aldershot: Ashgate, 2011.

Löw, Martina. *Raumsoziologie*. Frankfurt am Main: Suhrkamp, 2001.

Malinovich, Nadia. *French and Jewish: Culture and the Politics of Identity in Early Twentieth-Century France*. Oxford: Littman Library of Jewish Civilization, 2008.

Mann, Barbara E. *Space and Place in Jewish Studies*. New Brunswick, NJ: Rutgers University Press, 2012.

Meitner, Erika. "The Mezuzah: American Judaism and Constructions of Domestic Sacred Space." In *American Sanctuary: Understanding Sacred Spaces*, edited by Louis P. Nelson, 182–202. Bloomington: Indiana University Press, 2006.

Meng, Michael. *Shattered Spaces: Encountering Jewish Ruins in Postwar Germany and Poland*. Cambridge, MA: Harvard University Press, 2011.

Menny, Anna L. *Spanien und Sepharad*. Über den offiziellen Umgang mit dem Judentum im Franquismus und in der Demokratie. Göttingen: Vandenhoeck & Ruprecht, 2013.

Pinto, Diana. "A New Jewish Identity for Post-1989 Europe." *JPR Policy Paper*, no. 1 (1996). Accessed 20 January 2017, http://www.jpr.org.uk/documents/A%20new%20Jewish%20identity%20for%20post-1989%20Europe.pdf.

Rau, Susanne. *Räume. Konzepte, Wahrnehmungen, Nutzungen*. Historische Einführungen 14. Frankfurt am Main: Campus, 2013.

Rürup, Miriam. "Von der religiösen Sehnsucht zur kulturellen Differenz. Diasporakulturen im historischen Vergleich." In *Praktiken der Differenz. Diasporakulturen in der Zeitgeschichte*, edited by Miriam Rürup, 9–39. Göttingen: Wallstein, 2009.

Schapkow, Carsten. *Role Model and Counter-Model: Iberian Jewry in German-Jewish Remembrance Culture 1779–1939*. Cologne: Böhlau, 2011.

Schlögel, Karl. *Im Raume lesen wir die Zeit. Über Zivilisationsgeschichte und Geopolitik*. Munich: Carl Hanser, 2003.

———. "Räume und Geschichte." In *Topologie. Zur Raumbeschreibung in den Kultur- und Medienwissenschaften*, edited by Stephan Günzel, 33–52. Bielefeld: transcript, 2007.

Schlögl, Rudolf. *Alter Glaube und neue Welt. Europäisches Christentum im Umbruch 1750–1850*. Frankfurt am Main: S. Fischer, 2013.

Schroer, Markus. *Räume, Orte, Grenzen. Auf dem Weg zu einer Soziologie des Raums*. Frankfurt am Main: Suhrkamp, 2006.

———. "Bringing Space Back In. Zur Relevanz des Raums als soziologischer Kategorie." In *Spatial Turn*, edited by Döring and Thielmann, 125–48.

Schwerhoff, Gerd. "Spaces, Places, and the Historians: A Comment from a German Perspective." In Kümin and Usborne, "Forum," 420–32.

Shemtov, Vered. "Between Perspectives of Space: A Reading in Yehuda Amichai's 'Jewish Travel' and 'Israel Travel.'" *Jewish Social Studies* 11, no. 3 (2005): 141–61.

Šiaučiūnaitė-Verbickienė, Jurgita, and Larisa Lempertienė, eds. *Jewish Space in Central and Eastern Europe: Day-to-Day History*. Cambridge: Cambridge Scholars Publishing, 2007.

Silberman, Marc, Karen E. Till, and Janet Ward, eds. *Walls, Borders, Boundaries: Spatial and Cultural Practices in Europe*. New York: Berghahn Books, 2012.

Tölölyan, Khachig. "Rethinking Diaspora(s): Stateless Power in the Transnational Moment." *Diaspora* 5, no. 1 (1996): 3–36.

Tuan, Yi-Fu. *Space and Place: The Perspective of Experience*. Minneapolis: University of Minnesota Press, 2001.

Waligórska, Magdalena. "Jewish Heritage Production and Historical Jewish Spaces." In Šiaučiūnaitė-Verbickienė and Lempertienė, *Jewish Space*, 225–50.

Weiss, Yfaat, and Raphael Gross. *Jüdische Geschichte als allgemeine Geschichte*. Göttingen: Vandenhoeck & Ruprecht, 2006.

Wettstein, Howard, ed. *Diasporas and Exiles: Varieties of Jewish Identity*. Berkeley: University of California Press, 2002.

PART I

Imaginations

Remembrance and Representation
of Spaces and Boundaries

1

Of Sounds and Stones

The Jewish-Christian Contact Zone of a Swiss Village in the Nineteenth Century

Alexandra Binnenkade

Space. The Final Frontier.[1]

The history of Jewish-Christian coexistence is a history of spatial orders.[2] In the eighteenth and nineteenth centuries the mainly rural populations of Lengnau and Endingen, two villages in the northeast of Switzerland, experienced belonging and exclusion, power and powerlessness, center and periphery, success and failure intimately by means of space. Owning land, living on it, and being named after it, in particular, were pivotal points of reference in an agricultural society's coordinate system of values. The right to settle—this existential legal, economic, and at the same time social value—was denied to Switzerland's Jewish population until 1879. Communal and subsequently cantonal (later national) citizenship rights on both the collective and individual levels were reserved for landowners. It is no surprise, therefore, that the debate on Jewish civil rights ultimately turned out to be a question of the land in which it literally played out.[3] As elsewhere around the world, in Lengnau and Endingen the distribution of land signified the distribution of power. Conflicts evolved around spatial relations, while social relations simultaneously molded the physical sphere.

Material and Immaterial Aspects of the Contact Zone

Thus far, scholars have primarily used written sources—normative and administrative texts—to describe everyday life in Lengnau or Endingen. The majority of these refer to separation and conflict rather than contact. Over the years, the narrative scholars have derived from analyzing these has solidified into an account of two distinct societies within one spatial frame. In so doing, scholars have implicitly reinforced the impossibility of successful Jewish-Christian coexistence, omitting the unspectacular but peaceful side of contact, which in fact was the rule, not the exception. This hegemonic narrative needs to be changed. To this end, a different category of source needs to be taken into account.

Reading space as such a source opens a window onto hitherto undiscovered aspects of everyday life in the village and changes the meaning of contact. Space is intrinsically rooted in social practice; it is being done: performed, constructed and reworked; an interactive and entangled core element of the contact zone. Taking its material character as seriously as its immaterial qualities, I seek to test a previously uncontested narrative (written and oral) against the presence and enactment of physical elements such as streets, woods, and, most of all, houses.[4] In this chapter, I present two approaches to the analysis of space: the first one begins with the materiality of a specific place, a very special kind of house, and links it to statistical data and narrative identity politics; the second is captured with the notion of layers and discusses immaterial factors, like sounds or times, that create Jewish and Christian spaces.

The capstone of the prevailing narrative is Lengnau's and Endingen's historical denomination as *Judendörfer*. The expression suggests a Jewish majority or cultural dominance. In fact, however, the Jewish population never accounted for more than one-third of Lengnau or 50 percent of Endingen over the two hundred years of Jewish presence. Today *Judendorf* is a name the two villages chose to apply to themselves to demonstrate (and promote) their cultural uniqueness in the area. Specific buildings such as the synagogues, the cemetery, the Jewish bakery, and the *mikveh,* but particularly houses with dual doors, have become the architectural signifiers of these *Judendörfer* and a shared Jewish-Christian lifeworld. These houses will be the topic of a close analysis later in the chapter.

Researching Space

After a Marxist-oriented initial surge of attention in the 1960s and 1970s, which ebbed together with the influence of the Annales, a second wave of interest in "space" as a research focus arose in the late 1990s within the context of cultural studies. The oft-cited topographical, topological, or spatial turn has been widely discussed.[5]

Within this scholarly context, I conceptualize space as a dialogical category. I link texts or images about social interaction and symbolic representation to the analysis of material, haptic objects in such a way that reading them together can generate productive analytical back-and-forth movement. Such an analytical movement allows me to draw conclusions about the micro-practices of contact and reveals a much more complicated fabric of actors and interactions than earlier historiography had suggested. The result is no longer a story about "Christians" and "Jews" but also one about Catholics and Protestants, citizens and villagers without communal rights, about representatives of the canton and local dignitaries, and, most of all, about the individual women and men in the village.[6] Therefore, I use a vocabulary that distinguishes different levels of agency within a spatial perspective.

Spaces, as Michel de Certeau has said, come into being. Space creates connections, a social fabric of moving elements. It is imbued with the entirety of all the unfolding movement therein. De Certeau describes space as "an act of a present (or of a time)" that "is modified by the transformations caused by successive contexts."[7] In his understanding, spaces are the living or at least changeable results of relationships; spaces are being done in time. De Certeau's spotlight on time and the constitutive role of contexts makes spaces particularly interesting for historians. Spaces are articulated, organized, lived, felt. They depend on activity and variation. At the same time, they are also described, interpreted, represented. They are, therefore, excellent media for capturing and describing social change, as I will do in the second section of this chapter.

Places are the building blocks of space. A place, like a tree or a house, is the manifestation of an order "in accord with which the elements are distributed in relations of coexistence."[8] De Certeau explains the characteristics of places based on the possibility of referring to something as "behind," "above," "to the left of" something else. Places are fixed points in a specific constellation. They structure spaces.

Stones

One such constellation, for example, consists of the buildings of Lengnau and Endingen. People actively determined their distribution throughout the village. Once built, the houses defined the village space. Geographical maps from different points in time show that most of the houses built in the eighteenth and nineteenth centuries belonged to Jewish owners. These new houses were doubly separated from the older "Christian" houses, by both a road and a river. In this process Lengnau underwent a de facto spatial segregation into a mostly Christian and a mostly Jewish part. But it would be too easy to conclude simply from this architectural arrangement that everyday life in Lengnau was indeed heavily segregated.

Exploring the physical qualities of the site today, one can observe that the new side of the village is located on the sunny—that is attractive and valuable—flank of a little hill. In this area Lengnau's farmers, who were all Christian at the time, produced wine and cultivated crops, working their land on a daily basis. Thus, from the beginning Christian farmers spent time on both sides of the river. Written sources and a short walk through the village further show that the public house where the municipal council, the village's highest political body, held its weekly meetings was located in the new part of the village. The public house was not the only important public structure. The post office, which acted as a central communication hub for the local community, was also located directly next to the synagogue. Also, most merchants sold their goods in this part of the village. The communal public scales, which were used to measure crop yields, and some wooden benches, mounted under shady trees at the turn of the twentieth century, were also located here. These benches embodied the material invitation for villagers to sit and chat with others or simply to watch the daily business. The places in the village that seemed "Jewish" at first glance were, in fact, jointly created by Christians and Jews, they were rural and a little urban at the same time.[9] This contact zone offered pivotal services and opportunities for interaction. Growth on this newly developed side of the River Surb did not primarily signify marginalization. Progressively, the synagogue square became the new center of Lengnau.

Written documents do not convey such information about the complex relationship between Jewish and Christian villagers. It is the *places* that reveal these secrets—and, along with them, the meaning of contact. Following the definition given by the *Oxford English Dictionary*, bodies in contact touch each other's external surfaces. Transferred to social relations, contact as a cultural form means neither blending nor a distinct and sharp separation. Contact implies (only) partial involvement with the Other, an involvement that emerges from the material and immaterial practices of daily interaction.

What Makes a Space Jewish? Houses with Dual Doors

Today Endingen and Lengnau represent their character as former *Judendörfer* not by pointing to their synagogues but by highlighting specific private houses. These buildings, which were essentially semidetached but each of which might house multiple families, are easy to spot: they have two identical entrance doors built into one frame, so-called dual doors (*Doppeleingänge*). Most of these houses were taken down over time. But they are one of the first distinctive architectural landmarks mentioned in descriptions of the history of Lengnau and/or Endingen. Dual doors have become the most important symbol for the local Jewish-Christian culture of the past. One of the doors is usually identified as the entrance for Christians, while the other is shown as the one the Jewish inhabitants used. The reason given for this phenomenon in talks, in books, or, lately, on walking tours of the locality is a historic prohibition on "Jews and Christians living together under one roof," which dates back to a decree codified in eighteenth-century law.[10]

For local villagers, dual doors signify segregation. Apparently paradoxical, these houses are labeled as Jewish (*Judenhäuser*) and thereby assigned to only one of the two groups that lived in Lengnau and Endingen. The current narrative does not specify whether the local Catholic citizenry demanded that this cantonal prohibition be enforced or whether it came exclusively from the cantonal administration and was thus imposed from the outside. Each case makes a different statement about the extent to which Jewish men, women, and children were accepted by their Christian neighbors. The prohibition implies that those who should not live "under one roof" should also have little contact in everyday life, which sheds an unflattering light on the Christian majority, local or regional. However, the material result of the order suggests that the relations were better than expected, because very obviously the dual doors bypassed an order and allow it to be interpreted as potentially unpopular. If we were to hold the image of this rural Jewish-Christian neighborhood next to descriptions of contemporary German towns where entire Jewish neighborhoods were closed off—sometimes even by a wall and gates, closed in the evenings as well as on Sundays and holidays like Christmas or Easter—the contact zone of Lengnau and Endingen was quite permeable by comparison.

There is no stronger symbol of the ambivalence of the Lengnau contact zone than these dual doors. However, scholars and villagers, so far, have exclusively associated dual doors with segregation and Jewishness. These explanations come into question once the materiality of the dual doors is more closely inspected. The following section explores the connection between narrative and materiality in the construction of a Jewish space. Space and place are, indeed, entangled, and there are historiographical gains to be made from such a spatial approach to everyday life.

Entering a "Jewish House" Raises Questions

Imagining everyday life behind these doors means looking at how the houses were built and what their architectural arrangements initiated, fostered, or prevented. In the house I entered, the doors led into a shared space. Seen from the inside, the doors look like a pretense. But what for? How far did this separation go in practice, how far could it go, architecturally and culturally, and why did the people in Lengnau and Endingen build those doors so close together instead of, say, using the two opposite narrow sides of the house (such houses exist in other villages) if they wanted to make sure people were really segregated? Why live in one house all together?

There are no written sources for what housing looked like in the two villages in the eighteenth and nineteenth centuries except for inventories: lists of the possessions owned by the deceased. The only testimony of everyday practices inside this particular contact zone are the houses themselves. What these stones could tell us needs to be examined against the backdrop of the written and oral tradition, statistical surveys, the expertise of current heritage conservationists, and anthropological scholarship.

Normative Sources

The decree forbidding Jews and non-Jews from living "under one roof" is recorded in one of the oldest legal agreements about the right to settle that was granted to Jewish people in the Surbtal. It dates back to 1744 and was repeated in 1774. Article one stipulates that the 108 Jewish households then under the direct protection of the so-called *Alte Orte*—regions that had banded together and would later become the Swiss nation-state in a different legal and geographical form—were restricted to settling in Lengnau and Endingen. The importance of spatial regulation is demonstrated already in the second paragraph, which states that houses in which Jewish people lived were not permitted to be "increased in their number, that is, new [ones] bought or built, and [the existing ones may be] increased neither in height nor in extent, yet [they may be] changed in their interior."[11] Jews were permitted neither to buy or sell nor to own a house. The intention of the authors is obvious: to limit the Jewish population.[12] But there is more to this regulation than meets the eye: Only those who owned a house had access to common land (*Allmendnutzen*), an exclusive and existential resource that was particularly guarded. The local authorities tried to prevent what they considered overuse of the common land, applying some kind of a protectionist first-come-first-served attitude against the Jewish immigrants. Therefore, the only housing option for Jewish settlers in the Surbtal was rental, which was widely practiced.[13] The legal order concludes with the now famous provision that "no Christian and Jew should live together under one roof."[14]

In current historiographical literature about Lengnau and Endingen, this legal text is understood as the normative blueprint for local architecture. The local historian Franz Laube-Kramer traced it back to a regulation of 1658[15] and suggested that in the eighteenth century the original ruling that Jews and Christians should "not live together" got interpreted as "not under one roof." In his reading, doors stand for "roofs" as the synecdochal symbol of the household: two doors stand for two houses, regardless of the actual number of roofs. He interprets the construction of dual doors as a local attempt to evade imposed segregation. For Laube-Kramer, the doors are a sign of friendly relationships and collaboration in the village against ignorant, probably even anti-Jewish, cantonal authorities outside of it. He explains the social side of their construction as follows: Familiar with the ban, a Jewish developer would commission a Christian villager to build a house with two apartments. Later on, he would formally rent one apartment from the Christian who built it, which, in fact, meant renting from himself. Of course, he would not have to pay rent for it. Meanwhile, the Christian party to the arrangement would live in the other apartment. Whether or not rent was paid remains unclear.

Who would have been eligible for such a deal on the Christian side? Another look at the houses shows that they were usually built without barns and had only small gardens, which, in a rural society, marked a conscious renunciation of an agricultural way of life. Whoever lived in these houses did not farm. This also indicates that these buildings did not house prosperous farmers and well-off tenants but rather those who earned their living as craftsmen, day laborers, or in other occupations with low incomes. Thus, following Laube-Kramer's explanation, the living "under one roof" of Jewish and Christian neighbors at that time must have been more than just a matter of religious interactions; it was equally structured by class.

Statistics: Sharing a Roof in the Surbtal between 1700 and 1900

One of the oldest descriptions of the village originates from the year 1794. Hans-Rudolf Maurer, a Swiss priest and deacon who wrote much-cited travel books, like the one I quote from below, also reported on Lengnau and Endingen. Although he had a sharp eye for everything in the seemingly exotic (Jewish) village, he never mentioned houses with dual entrances. But he did describe the blue woodwork that distinguished Jewish houses from Christian ones, indicating that he did pay attention to the specifics of Christian/Jewish cultural everyday life.[16] If he had noticed houses with dual doors, he would, most probably, have mentioned them. Thus, it must be concluded that houses with double doors must have been established only after 1800 or, at least, that they were not characteristic of the interreligious contact zone. But why would residents suddenly have followed a legal decree from the mid-seventeenth century in the early nineteenth century? Most of the buildings Maurer mentioned are not preserved today. What we do have, however, are statistical surveys that allow some conclusions about housing arrangements in Lengnau and Endingen.

The earliest survey of the housing situation in the Surbtal dates from 1778.[17] It was meticulously made by pastor Fridolin Stamm, who lived in the area. At the time, about 70 percent of the Christian households were in single-family homes. The rest of the villagers lived in shared dwellings. The situation was slightly different for Jewish families: 77 percent occupied a single-family home, while multiple-family Jewish homes typically had more families in them than their Christians counterparts. The 1779 survey does not mention any houses shared by Christian and Jewish tenants. This means that at the end of the eighteenth century, Jews and Christians did not actually live "under one roof." These arrangements did not change much over the next half century. The census of 1850 lists each house and its inhabitants, their names, occupation, status, including even former residents who had left the family home.[18]

In 1850, only 7 of the total of 184 houses in the locality were recorded as having both Jews and Christians sharing a dwelling; this amounted to 4 percent of the population. But about a third of the Christian and almost three-quarters of all Jewish families lived in multi-family abodes. Yet the sharing did not, in most cases, cross the religious line. The fact that the majority of the shared buildings were occupied by Jewish residents explains why houses with dual doors were considered Jewish (*Judenhaus*). In essence, they communicated more about cramped housing conditions than about segregation. The dual doors give us clues about the economic stratification of the village: a clear majority of Lengnau residents who lived in multiple-family homes were Jewish. This, of course, is related to the professional structure of a society based on crafts and farming, heritage traditions, and the number of houses available on the market. Yet it also indicates the economic means of the majority of the Jewish community did not differ much from those of the Christian villagers. They all had to live on small incomes.

Insights from Heritage Conservation Studies

These findings echo analyses from modern farmhouse research, which has identified similar houses in other areas of the canton, even though this exact kind of architecture could certainly not be found in every village. As a general trend, though, house partitions were a common local practice in rural societies in this area at the time. In fact, from the seventeenth century onward, splitting houses was more the rule than the exception as the population grew and the laws governing access to common land changed. Henceforth, land was no longer tied to one family but rather to one house. Under these conditions, some houses could be split among two to nine families, which meant that a total of forty-five people might call one and the same house their home. In general, these were not tenants but co-owners. In other words, house partitions solved potential family strife borne of economic hardship. It was an intimate form of coexistence with only minimal social and cultural differences between the inhabitants. No Catholics, for instance, would have shared a house with Protestants.

Looking at floor plans and exploring existing houses today makes the closeness in which people lived in these houses tangible. Narrow rooms, thin clay walls, and small windows structured the movements and interactions of a large number of people. Whoever needed light or air left the doors open, so everybody could hear what was going on in the other rooms, in the other family. Cooking yielded similar proximity. Until the middle of the eighteenth century, the majority of such simple houses were furnished with an open-fire kitchen without a chimney. The smoke escaped directly through the roof. Kitchens could be split, be it on the same floor or through the establishment of another hearth on the second floor, but this worked only if there were enough openings between the floors for the smoke to exit. Dual doors, then, indicated cramped and intimate living conditions rather than a clear separation of culturally distinct people.

In apparent contradiction to this finding, the pastor Fridolin Stamm, who listed the houses in Endingen in 1779, noted several that were indeed inhabited by Jewish and Christian villagers side by side. Registers and floor plans show that almost every house in Endingen was divided at some point. But of the 150 houses under scrutiny, only a few were built as homes for multiple parties. The vast majority were only gradually partitioned into up to eight units by their Christian owners, and it was only with the passage of time that some of those apartments were rented out to Jewish villagers. Multi-family homes originally built for this purpose almost exclusively belonged to Jews.

By examining the material reality of this specific place, I have shed new light on the meaning of dual doors within the Jewish-Christian contact zone. The combination of physical appearance and floor plans with statistical data cast doubt on a hitherto prevailing explanation of these entrances: dual doors did not arise in response to a legal decree or as a way to outwit one. They do not stand for segregation but are signifiers whose meaning has changed over time. In the seventeenth and eighteenth centuries, they helped the local Jewish community solve the legal dilemma generated by the fact that their number was not to increase in entities counted by roofs, which meant that they had to close ranks and share the houses that were available. The partition of houses was a canton-wide practice among Christian families not wealthy enough to settle hereditary succession by building new homes for their sons. Thus, houses with dual doors displayed economically straitened circumstances and lack of space—in the entire village and for everybody. Both the majority of the Jewish and a large percentage of the Christian population lived their everyday lives under these conditions.

The doors retained their signifying qualities but shifted their meaning in the nineteenth century, when the majority of the Jewish population emigrated from the Surbtal. With this migration, the doors increasingly became an element of identification, more precisely, a symbol of Jewish presence. The Christian community associated the dual doors with Jewishness (*attributive* identification), turning them into a signifier of the Other. The doors thus became a material and spatial means of distinction.

Toward the end of the twentieth century, the doors became part of the self-perception of the people of Lengnau and Endingen as inhabitants of a former *Judendorf*. The fact that these dual-door houses exist sets the two villages apart from others in the area and points to a different history from those of the neighboring villages. Now the doors enabled *affirmative* identification with a Jewish-Christian past. The material relic that once organized and created the living space of the village changed from being a signifier of the Other into an attribute of the Self. At the same time, it reveals this own past always as a partly shared, if not indeed Other, one. It may even have not been until the twentieth century that the doors were narratively linked to the prohibition for Jews and Christians to live "under one roof." Their existence thus perhaps served not so much to explain the actual use of the houses but rather played an important role in local memory practices. For this reason, houses with dual doors remain a relevant physical presence that points to the close association of Jewish and Christian villagers living simultaneously closely together and separately. They mirror ambivalence, a central experience of the contact zone. Even though *Doppeltürenhäuser* (houses with dual doors) explicitly name a space to which exclusive Jewishness is ascribed, their narrative texture has revealed far more heterogeneous realities.

Space is something that is always "being done"; it is not a fixed, observable entity made of solid elements. Both cultural contexts and what we refer to as a natural environment are co-produced by the spatial practices performed by those who occupy the space, and vice versa. The fields of reference thus uncovered are too manifold to enable us to determine with any finality what is a "Jewish" or a "Christian" space. The houses with dual doors in Lengnau and Endingen are striking illustrations of exactly this kind of space "becoming by doing across time." They are not the only such examples. The analysis of space can go even further and explore the extent to which also immaterial elements of the contact zone create spaces and even change the materiality of the places.

Layers of the Contact Zone

Beyond its houses, barns, religious buildings, gardens, paths, and places, a village has another level of space. It is made of varying layers of sounds, periods of time, rituals, smells, and movements. Little research has been done on these subtle elements of the village space to find out more about the qualities of the contact zone. In Lengnau and Endingen, space emerged from the routine of daily tasks, from what its Jewish and Catholic inhabitants did in between and within their houses. Their individual activities interweaved and thereby created yet another quality of space.

Layers form in different combinations and in varying densities. They mix in ways perhaps similar to the polyvalent potential of layers of color. Each particular mixture is situative: it depends on the cumulative micropractices of the villagers and is tied to the individual perception of those who move within these layers. Although there were many fundamental components of the Surbtal contact zone, they were not all equally important to all the people living in the village. Some elements of everyday life were largely taken for granted and therefore rarely actively noticed, with the result that they have been ignored by historians.

Soundscapes

> Autumn has come and mist is rising over the Surbtal valley. ... Lippel, the faithful *shames*, is passing light-footed through the village streets straight after the synagogue clock has struck six. Soon, from his call, all the members of the community wake up, Samperle, Selig, Lieber, Aisig, Benion, Rebmauschele, Judel, Izigle Jischele, Gutel Michel and all the others, the *balbatim* of the old *kile* of Endingen. Drowsily, like mysterious figures, they slip through the streets to their temple, where Almes Mausche ... has started filling the space of the half-empty synagogue with the sounds of the *selichot* prayers and songs. Gradually, the synagogue fills with the faithful, who, by the end of the morning service, are all joining powerfully in with the prayers, until, at the end, Rebeilingens Schlome blows his horn and the old, dutiful *shofar* spreads its stirring *Tekios* and *Tekio Gedaulos* from the synagogue all over the valley.[19]

Like a veil of sounds, the Jewish space lay over the chilly Surbtal valley. In this description, from Emil Dreyfus's romanticized childhood memories published in the 1920s, it spread throughout the Christian as well as the Jewish homes and encompassed everything that was touched by the sound of the shofar, including those who knew neither the horn, its melody, nor its meaning. For Dreyfus, who published his memoir at the age of seventy-seven, the (acoustic) Jewish space seemed almost without boundaries at that time. Dreyfus and his community knew whose religion and culture were being represented. The *tekia* created a specifically Jewish space, corresponding to the extent of the sound.

On that same morning, the Catholic church of Lengnau, St. Martin's, likewise rang its bells to stake its claim. In a way similar to the sound of the shofar horn, they created a Catholic space throughout the same area. The existence and use of the church bells expressed the faith, tradition, and, last but not least, wealth of the Christian community. At the time of Dreyfus's account, five different bells were rung in St. Martin's, each with a distinctive tone and its own name and significance. Each sound was an individual force, structuring every villager's day, Jewish or Catholic, communicating unique messages all villagers could understand regardless of their religious tradition. "Insiders" were, therefore, not just Catholics but also members of Lengnau society.

The bells bore what we might consider to be magical powers. The messages inscribed into the bells spread over the village through the sound, covering the space like a protective coat. The connection between religion and magic was indisputable to many people, both Jewish and Christian, at that time.[20] The bells guarded their faithful physically and socially, from poor weather, illness, and bad luck, and fostered the spiritual cohesion of the community. They were protection and mission in one. Remarks in written documents indicate that the Jewish population was considered to be part of this inner village circle. The only people these bells were meant to exclude were non-villagers.

Sounds generated a spatial contact zone, and there was one place where they mingled in a very specific way: the synagogue bells in Endingen. These bells are a rare feature in the cultural landscape of southern Germany and Switzerland.[21] Only one other synagogue with two bells exists in these regions—in Buchau in Germany. Like church bells, they carried inscriptions intended to be imbued with power, and they were rung daily to summon the community to prayer.[22] A synagogue with bells or, as in Lengnau, with a prominent clock adopts elements and features of church towers, self-consciously representing societal presence and success. The Endingen synagogue was built in 1852, with the bells giving signals to both the Jewish and the Christian inhabitants as part of the communicative soundscape of the village. They engendered or symbolized a contact zone that was immaterial and tangible at the same time.

Timescapes

Another layer consisted of time. Lengnau had Jewish times and Christian ones. This is a reference not to the two parallel calendars used in Lengnau or the partial translation of Jewish holidays into Christian ones (like referring to "Weyhnachten" for Chanukah in the minutes of Jewish meetings or correspondence), but to temporal practices that changed the village space sustainably. During Sukkot, the Feast of Tabernacles, when Jewish families built their ceremonial booths, Lengnau and Endingen counted more "houses" than usual. A similar temporal space was created by the *eruv*, a boundary within which Jewish residents are able to carry objects outside their homes on the Sabbath without violating religious law.[23] It was created more by symbolic signs, like wires and wooden poles, and Jewish community practices than by real boundaries, so that outsiders might easily disregard the physical markers of this space,[24] even as they clearly noticed them.

In other locations and to this day, *eruvim* have been sources of conflict.[25] But for Lengnau or Endingen no complaints were recorded. From a social point of view, this suggests that this contact zone had an exceptionally peaceful quality—a social fact that needs to be noted very positively. From a historiographical point of view, however, this absence of conflict generated very few historical sources; Emil Dreyfus's memoir provides the only report about an *eruv* in Lengnau.

In spite of the written sources' silence on the *eruv*, it cannot have gone unnoticed in practice, given how it altered the village space. During its existence, some men and women would move busily, while others walked in a more solemn way; some would be wearing everyday clothing, and others would have donned more festive attire. Some people would have carried and transported food or goods, while others would have made it explicitly clear they were doing no such thing; Dreyfus describes Jewish neighbors who pointedly tied their handkerchiefs around their waists, turning them into belts, to demonstrate they were not even carrying an item as mundane as this.[26] Every week the *eruv* orchestrated movement in the village so that, in this sense, time made space.

How did the Christian villagers respond to the *eruv*? For them Sabbath was a regular working day. Their concept of time and related spatial practices was different from that of their Jewish neighbors. Did they avoid or ignore the *eruv* on the way to their gardens, fields, or shops? Did they act from a position of understanding or ignorance toward its significance? We cannot tell. But at the very least, the obvious change in the nature of the space on Sabbath impacted the village as a *Judendorf*.

Time, obviously, generated contact zones yet also parallel social universes, which in their distinctness helped (re-)establish boundaries demarcating what was one's "own" from what was "Other." Sabbath was a day of contact and inclusion. On this day, Christian neighbors and their children helped Jewish families to abide by their religious laws and traditions. Even if only every second Jewish family had somebody come to their home to kindle the fire or light the lamps, then in 1805 at least 196, in 1850 about 260, and in 1870 about 188 Christian Sabbath maids (*Schabbatmägde*) would have entered their neighbors' houses weekly. They overheard conversations, saw what was being served at table, and participated with all their senses in a Jewish household. This unspectacular contact zone, bound to the physical place of the house and to a specific and repeating time, prompted reciprocal perceptions of otherness, but remains, again, mostly unexplored in the historiography.

The village was altered not just on Saturdays, by the *eruv* or the Sabbath service. Sundays, too, saw the village looking different and, above all, more populous than on weekdays. The (mostly male) Jewish traders who worked elsewhere during the week returned for Sabbath, and their presence attracted farmers from other villages.[27] Sunday was a business day in Lengnau and Endingen. In rural regions in the late eighteenth and early nineteenth centuries, money transactions, like any business agreement, were still handled face-to-face.[28] Under these circumstances, some Jewish houses were public to a certain degree, as Christian customers entered them to discuss business. Frequently, traders and their customers negotiated and/or sealed their agreements in public houses, turning these places into what we might like to imagine to be buzzing contact zones.

Sundays also altered the face of the village for another reason. Between 1717 and 1808, Lengnau was a so-called parity parish (*Paritätsgemeinde*), with a shared church. Over a period of almost one hundred years, on Sundays, while the Jewish men and women dealt with their paperwork and Catholic families walked back from the early mass in their Sunday best, some dozen local Protestants, accompanied by between 50 and 150 fellow believers from the Zürich area, proceeded through the village, making sure they were seen and heard.[29] In St. Martin's, one of the oldest Catholic churches in the region, they attended their regular Protestant service. For a couple of hours the Protestant churchgoers physically and symbolically took possession of the church, thus altering the space of the Catholic community and of the village as such.

Movements

Movements within and around the village generated layers as well. They originated from religious processions and traditions, or they were based on the space-related activities of the Jewish traders in their *medinot*, a term that describes the trading territory of one person. *Medinot* laid out an alternative cultural geography atop the one represented in physical maps of the region.[30] This economic map transgressed and to a certain degree ignored national boundaries and created individualized Jewish regions of commerce. The traders, walking and circulating between farms and villages, by virtue of their movement and their goods, connected the villagers to greater social streams such as politics or important regional knowledge, but also consumption patterns. Their return to the village on Fridays was an event much observed and seldom missed by the villagers. Along with their cattle or backpacks, they brought all this news back home, where it was exchanged, evaluated, connected, and turned into social capital.

Outsiders traveled to the villages as well. Brides, family members, and Jewish visitors visited or moved in, thereby subtly changing the villages' character. Some wore more urban, more "French," or more "Austrian" clothes; others decorated their houses with unfamiliar furniture. Again others spoke Yiddish with a different accent or vocabulary. The Jewish authorities of the local community—the precentor, teacher, and rabbi—had at least been taught abroad, outside Switzerland. Having moved between cities and villages, between nation-states and languages, between liberal and conservative attitudes, their cultural movement influenced the particular village culture of Lengnau—in its Christian as well as in its Jewish characteristics (songs, food recipes, manners, language, to name a few more).

These constitutive movements created spaces of unifying belonging. Outsiders perceived Endingen and Lengnau as cultural entities, as *Judendörfer*. And some of the layers just sketched created a common rather than a segregated space. At the same time, to the insiders these practices separated what, in the historiography, was then and later termed "the Jewish community" or, for this purpose, "the Jewish space" of Lengnau or Endingen from "the Christian" one. However, the closer one zooms in on the villages, the more quickly these latter categories crumble in the face of the ways in which space was actually generated, used, and represented by those belonging to each of these supposedly homogeneous groups.

In addition to the social distinctions applied to the Jewish communities within Lengnau and Endingen, the two villages identified very distinctly as Catholic and Protestant, respectively. These observations question the idea of them being *Judendörfer*. In other words, the epithet *Judendorf* signifies more than anything else that a special character, a distinct history was attributed to the villages. The villages were not made of two separate kinds of singular cultures ("a Jewish culture" and "a Christian culture"); instead, in and through shared space, the men, women, and children of Lengnau and Endingen practiced a distinctly and unique Lengnauean or Endingean everyday lifeworld.

In all features of village life just examined, spaces expanded beyond religious boundaries and changed existing concepts of what Endingen or Lengnau looked like. These spaces were not only created in relation to time or movements. Their overlapping and reciprocal character shows the contact zone as entangled, with blurry edges. This was true of the material hardware the houses were made of, and equally of the layers, those mobile elements of space, visible and invisible at the same time, perceptible equally by individuals and by the village society as a collective body. Both material and immaterial approaches to spatial analysis thus potentially integrate simultaneity and contradictions into new narratives.

Alexandra Binnenkade is an independent scholar at the University of Basel in Switzerland. Her current research interests include practices of knowledge circulation/history of knowledge regarding narratives of violence and race in education, the U.S. Civil Rights Movement, visual studies and cultures in contact, and working with emerging methodologies in cultural studies. Her publications include *KontaktZonen. Jüdisch-christlicher Alltag in Lengnau* (2009), *Doing Memory: Teaching as a Discursive Node* (2016), and *Voicing Dissonance: Teaching the Violence of the Civil Rights Movement in the U.S.* (2016).

Notes

1. This is the first line of the opening voiceover in *Star Trek*.
2. This chapter is based on Binnenkade, *KontaktZonen*.
3. I am referring to the so-called Mannlisturm; see Binnenkade, *KontaktZonen*, 177–238. Mattioli, "Der 'Mannli-Sturm'"; Schaffner, *Demokratische Bewegung*.
4. For the meaning of the woods for the local community, see Binnenkade, *KontaktZonen*, 197–210.
5. Brauch, Lipphardt, and Nocke, *Jewish Topographies*.
6. Binnenkade, *KontaktZonen*.
7. De Certeau, *Practice of Everyday Life*, 117.
8. "… *Selon lequel les éléments sont distribués dans les rapports de coexistence*"; ibid., 173.
9. Müller, *Franz Jakob Müller*.
10. See below.
11. Krütli et al., *Amtliche Sammlung*.
12. Roming, "Topographie eines 'Judendorfs,'" 303; Weldler-Steinberg, *Geschichte der Juden in der Schweiz*, 33–35; Steinhauser, *Zugrecht*, 37–39.
13. Maurer, *Kleine Reisen*, 168.
14. "[K]ein Christ und Jud bei einander unter einem Dach wohnen" (translated by the author); Krütli et al., *Amtliche Sammlung*, 871.
15. Laube-Kramer, "Judenhäuser," 13–40.
16. Maurer, *Kleine Reisen*, 168–69.
17. Fridolin Stamm, "Oeconomische Tabellen," Staatsarchiv Zürich StAZH BIX 6.
18. Volkszählung 1850 Staatsarchiv Aargau, StAAG MF1.A11.
19. Dreyfus, "Erinnerungen," 22 (translated by the author).
20. Breuer, "Jüdische Religion," 76.
21. Hahn, *Synagogen in Baden-Württemberg*; and Hahn, *Erinnerungen*.
22. See the entry on the Buchau Synagogue on the website of Alemannia Judaica, quoting the *Bayerische Israelitische Gemeindezeitung* from June, 28 1929, http://www.alemannia-judaica.de/synagoge_buchau.htm (accessed January, 15 2017).
23. Cousineau, "Rabbinic Urbanism in London."
24. See, e.g., Dreyfus, "Erinnerungen," no. 19, 5. J.: "This *eruv* was shown at the various ends of the village by a wire (like one of today's telegraph wires) stretched from roof to roof" (translated by the author).
25. Herz, "'Eruv' Urbanism"; Watson, "Symbolic Spaces"; Fonrobert and Shemtov, "Introduction," 5; Rapoport, "Creating Place, Creating Community"; Cousineau, "Rabbinic Urbanism." See also chapter 13 (by Joachim Schlör) in this volume.
26. Dreyfus, "Erinnerungen," no. 19/5. Earlier sources describe how particularly women, and among them mostly those who had married "into the village," wore their more urban-style dresses on these days, clearly showing their fashionable side.
27. Lengnauer Pfarr-Chronik des Deutschordenspriesters Josef Anton Bröchin 1766, KAL, Nr. 53, fol. 186.
28. Ibid.
29. Meyer, "Geschichte der Pfarrer von Lengnau," 17.
30. The singular form of *medinot* is *medinah*.

Bibliography

Binnenkade, Alexandra. *KontaktZonen. Jüdisch-christlicher Alltag in Lengnau.* Cologne: böhlau, 2009.

Brauch, Julia, Anna Lipphardt, and Alexandra Nocke, eds. *Jewish Topographies: Visions of Space, Traditions of Place.* Aldershot: Ashgate, 2008.

Breuer, Mordechai. "Jüdische Religion und Kultur in den ländlichen Gemeinden 1600–1800." In *Jüdisches Leben auf dem Lande,* edited by Monika Richarz and Reinhard Rürup, 69–78. Tübingen: Mohr Siebeck, 1997.

Cousineau, Jennifer. "Rabbinic Urbanism in London: Rituals and the Material Culture of the Sabbath." *Jewish Social Studies* 11, no. 3 (2005): 36–57.

de Certeau, Michel. *The Practice of Everyday Life.* Berkeley: University of California Press, 1984.

Dreyfus, Emil. "Erinnerungen. Aus den Memoiren eines aus Endingen stammenden israelitischen Schweizers." *Israelitisches Wochenblatt der Schweiz* 14, no. 44/22 (1924).

Fonrobert, Charlotte Elisheva. "The Political Symbolism of the Eruv." *Jewish Social Studies* 11, no. 3 (2005): 9–35.

Fonrobert, Charlotte E., and Vered Shemtov. "Introduction: Jewish Conceptions and Practices of Space." *Jewish Social Studies* 11, no. 3 (2005): 1–8.

Hahn, Joachim. *Erinnerungen und Zeugnisse jüdischer Geschichte in Baden-Württemberg.* Stuttgart: Theiss, 1988.

———. *Synagogen in Baden-Württemberg: "Hier ist nichts Anderes als Gottes Haus. ..."* Stuttgart: Theiss, 2007.

Herz, Manuel. "'Eruv' Urbanism. Towards an Alternative 'Jewish Architecture' in Germany." In *Jewish Topographies,* edited by Brauch et al., 43–62.

Krütli, Joseph K., et al., eds. *Amtliche Sammlung der älteren Eidgenössischen Abschiede.* Lucerne: [no publisher], 1871.

Laube-Kramer, Franz. "Judenhäuser–Christenhäuser in Lengnau und die Geschichte der Doppeltüren." *Beiträge zur Geschichte des Bezirks Zurzach* 3 (2004): 139–40.

Mattioli, Aram. "Der 'Mannli-Sturm' oder der Aargauer Emanzipationskonflikt 1861–1863." In *Antisemitismus in der Schweiz,* edited by Aram Mattioli, 135–70. Zürich: Orell Füssli, 1998.

Maurer, Hans-Rudolf. *Kleine Reisen im Schweizerland. Beyträge zur Topographie und Geschichte desselben.* Zürich: Orell Füssli, 1794.

Meyer, Josef. "Geschichte der Pfarrer von Lengnau. 1114–1932: 1000-jährige Kulturgeschichte der Pfarrei Lengnau." *Erb und Eigen,* Beilage der Botschaft, Klingnau 6/9 (1942)—7/12 (1943).

Müller, Andreas. *Franz Jakob Müller. Eine kleine Wirtschaftsgeschichte von Lengnau 1848 bis 1890.* Gontenschwil: [no publisher], 2004.

Rapoport, Michele. "Creating Place, Creating Community: The Intangible Boundaries of the Jewish 'Eruv.'" *Environment and Planning D: Society and Space* 29 (2011): 891–904. Accessed January 15, 2017. doi:10.1068/d17509.

Roming, Gisela. "Topographie eines 'Judendorfs': 'christliche Bauernhäuser' und 'jüdische Bürgerhäuser.'" In *Geschichte einer Hochrhein-Gemeinde,* edited by Franz Götz, 291–380. Gailingen: Gemeinde Gailingen, 2004.

Schaffner, Martin. *Die Demokratische Bewegung der 1860er Jahre.* Basel: Helbing und Lichtenhahn, 1982.

Steinhauser, Alois. *Das Zugrecht nach den bündnerischen Statuarrechten.* Chur: Manatschal Ebner & Cie., 1896.

Watson, Sophie. "Symbolic Spaces of Difference: Contesting the Eruv in Barnet, London, and Tenafly, New Jersey." In *City Publics,* 20–40. New York: Routledge, 2006.

Weldler-Steinberg, Augusta. *Geschichte der Juden in der Schweiz.* Zürich: SIG, 1966.

2

Imaginations of the Ghetto

Jewish Debates on Ghettos and Jewish Society
in Late Nineteenth-Century Galicia

Jürgen Heyde

Since the advent of the early modern period, the "ghetto" has been part of the Jewish experience in Europe. Initially the term was used to denote Jewish quarters established and controlled by non-Jewish authorities in Italian towns. During the nineteenth century, when municipalities everywhere in Europe tore down the walls that had surrounded their cities, the confines of the Jewish quarters likewise vanished. The term "ghetto," however, did not pass into oblivion but became a metaphor of the relationship between the Jewish population and the surrounding non-Jewish society in these cities.

This chapter examines the role of the "ghetto" as a concept in nineteenth-century debates, mostly in the Jewish press in Galicia, where it was discussed with great intensity. At their heart, these debates were not about ghettos per se, but about Jewish society in a situation in which emancipation had not yet been fully achieved or in which certain parts of the Jewish population could not or would not embrace post-emancipation modernity. In this context, emancipation and modernity appeared to be synonymous, while the ghetto symbolized a space characterized by the lack of both. In this way, "ghetto" became inscribed into a temporal framework, but—alluding to its original context—was envisioned also as a space, linked with notions of confinement and darkness. What makes the debates between "Assimilationists" (at that time still used as a self-designation) and Zionists in late nineteenth-century Galicia interesting is their significance for our understanding of the spatial, temporal, and emotional connotations the word "ghetto" carries to this day.

Definitions and Connotations

Adolf Gaisbauer, in his groundbreaking 1988 study of Zionism in the Habsburg Empire, names the aspired-to set of values to which "ghetto" acted as a counterpoint: "The lawlessness and confinement of the ghetto was replaced by 'emancipation,' the full legal equality of Jews and non-Jews and (at least as a hope, as an expected second stage) the social integration of Jews as people of equal worth and esteem into non-Jewish society."[1] Even apart from its Jewish context, "ghetto" is often used today as a general spatial trope demarcating social or cultural boundaries of all kinds. This usage leads historians and sociologists to deplore the terminological "blurring," which renders it useful only as a descriptive rather than analytical term.[2]

In their quest for a robust academic definition of the term "ghetto," historians and sociologists alike have generally neglected the problem of imagination. Most of the studies aiming to characterize the ghetto focus on physical places referred to as "ghettos" by their contemporaries. This means that the phenomenon has received extensive and close attention in the historiography of early modern Italy[3] and of the regions under German occupation during World War II.[4] The reality of life in Nazi ghettos was the focus of a recent study that sought a legal definition of "ghetto."[5] By contrast, a number of historical studies use the term "ghetto" in a strictly figurative way, as they primarily address issues of nineteenth-century Jewish history.[6]

Louis Wirth, in his pioneering sociological study *The Ghetto* (1928), envisioned the Frankfurt *Judengasse* as an archetypical space of exclusion and marginalization.[7] In transferring his historical observations to immigrant quarters in the contemporary United States, he paved the way for modern sociology to apply the term "ghetto" to a wide range of social, economic, and cultural problems. To keep "ghetto" functional as an academic (albeit descriptive) term, sociologists often add distinctive markers, generating phrases such as "hyperghetto" and "iconic ghetto."[8]

Studies on the ghetto as an imagined space have mostly been conducted from a literary point of view. The "ghetto literature" of the nineteenth century has been analyzed as a literary genre[9] and as a means of communicating traditional Jewish values to a "modern" readership.[10] Kristiane Gerhardt[11] and Joshua Shanes[12] looked at usages of "ghetto" in journalistic writing. Recent studies on the history of the term "ghetto" have also sought to bring these various strands of research into a common focus.[13] This work has shown that the uses of the term found in the nineteenth and early twentieth centuries fulfilled a dual function as a designation (of specific places) and a metaphor (charged with meaning).

The Early Modern Ghetto

The origins of the term "ghetto" can be traced back to 1516, when the Senate of Venice selected an old abandoned foundry—hence the name "ghetto"[14]—as the site upon which to erect a residential area for Jews. The Jews, who had been residing in the city since 1503, needed to leave their quarter in Mestre on the Venetian *Terraferma* because of the ongoing war. The new residential area was surrounded by walls and connected to the city by two gates. Initially, it consisted simply of a part of the former foundry, the Ghetto Nuovo; only in 1543 did the area expand to include the Ghetto Vecchio, or "old foundry." The Ghetto Nuovissimo, erected in 1633, lay beyond the original foundry site and was called "ghetto" by association only. "Ghetto" thereafter rapidly became the established term for the Jewish residential area of Venice, which in official documents had never been referred to as the "Jewish quarter" or similar. By the end of the seventeenth century, the term had spread throughout almost all of Italy to refer to Jewish quarters enclosed by walls or similar structures and thus distinguished as separate urban entities.[15]

At the end of the eighteenth century, when Napoleon's armies came to Italy, they brought with them the ideals of freedom, equality, and brotherhood that had driven the French Revolution. In this context, the term "ghetto" became a verbal symbol of the ancien régime and its social boundaries, which the new ideals would overcome. For example, in 1797, the new town council of Padua, in declaring that the city's "Hebrews" were free to choose their place of residence, decried "the barbarous and meaningless name of Ghetto, which designates the street which they have been inhabiting hitherto" and renamed this street "Via Libera."[16]

New Connotations in the Nineteenth Century

Revolutionary as they were, these acts did not completely eliminate enclosed residential areas for Jews in the cities of Italy and the rest of Europe, because many governments reinstituted the former restrictions after Napoleon's defeat. Yet the symbolism of such emancipatory acts was not forgotten. During the nineteenth century, there was no clearly defined image of "ghetto" in Jewish public discourse. In pre-revolutionary times, as indicated above, it had been used almost exclusively to refer to specific localities in Italian cities, but popular travel diaries frequently mentioned it, spreading the term throughout Europe.

In the nineteenth century, "ghetto" became a pan-European trope, used in a range of contexts. In the restoration period after 1830, for example, Ludwig Börne used it to describe the situation of Germany. Whereas his critics frequently linked his sympathies for the ideals of the French Revolution to his upbringing in Frankfurt's *Judengasse* and alleged that he harbored a thirst for revenge against Germany because he was Jewish, Börne replied that such a thought was indeed far from his mind: "Haven't [the Germans] become my coreligionists and brothers in suffering? Isn't Germany nowadays the ghetto of Europe? Doesn't every German wear a yellow mark on his hat? Could I even harbor the slightest grudge against my hometown? Haven't all the Frankfurters, my former masters, become like the

Jews of old? Aren't the Austrians and the Prussians now their Christians?"[17] By contrast, in 1894, Bernard Lazare used the term "ghetto" figuratively to highlight the climate of suspicion and hatred around Jews that followed the arrest of Alfred Dreyfus. Although the Jews were "no longer cloistered, the streets in which they live are no longer cordoned off by chains," he said, the rising anti-Semitic wave generated "a ghetto more terrible than that from which [the Jews] could escape by revolt or exile."[18]

In the mid-nineteenth century, "ghetto literature" began to attract readers among the Jewish and non-Jewish population alike. Stories, novels, and dramas portrayed Jewish life in rural or small-town environments, with Bohemia and Galicia as popular backgrounds; these works transported images of traditional lifestyles, isolation, and backwardness among the Jewish population and juxtaposed "ghetto" and "emancipation." This implicitly contrasted the two and viewed emancipation not as overcoming exclusion from non-Jewish society, but rather as a transition from "tradition" to "modernity" that played out in international conflict.[19]

This genre of literature was a major purveyor of "ghetto" as a metaphor for "Jewish space." Great works on Jewish history by scholars such as Markus Isaak Jost and Heinrich Graetz utilized the term in a similar manner. Both historians extracted the term from its early modern Italian context and applied it to Jewish spaces in various historical settings, without subjecting it to systematic analysis. Literature and historiography created a detached and largely historicized topos of the ghetto: novelists regarded it as a phenomenon that, while still extant, belonged to the past, while historians used it as a motif to represent the conditions of Jewish life in the Diaspora in earlier times. Both regarded the ghetto as existing outside of time yet acknowledged that the term and the (imagined) space it represented held cultural significance for the Jewish experience.

The Ghetto and Jewish Modernity

One of the most important contexts for the term in the nineteenth century was the discourse that contrasted the "epoch of the ghetto" with the "era of emancipation."[20] This opposition became particularly popular in Central and Eastern Europe, where the road to Jewish emancipation proved especially long and onerous, even though the ghetto as an institution had not existed there in early modern times.[21]

The ghetto was a key motif in a late nineteenth-century journalistic debate in Galicia initiated by the biweekly journal *Ojczyzna* (Fatherland). This journal was published by the L'viv chapter of Agudas Achim (Union of Brothers), an organization devoted to the idea of Jewish integration into Polish society (assimilation). The periodical focused on the Jewish population in Galicia, and especially on its alleged resistance to modernization as promoted by the Assimilationists. This debate introduced "ghetto" as a metaphor for the situation of the Jewish masses, whose external circumstances, but also "unenlightened" leaders ("rabbis, Hasids, and miracle workers"), seemed to deprive them of their freedom and self-determination. Within this perspective, the ghetto is a dystopia, a place that needs to be overcome. Some years after this first Assimilationist debate, Zionist authors adopted the term, also using it in a dystopian sense but changing its emphasis. While they agreed that the Jewish masses were locked in a state of backwardness and isolation, they did not blame the Jews themselves for this. Rather, they believed that the non-Jewish environment, which had denied Jews the right to make decisions about their lives and continued to do so even in the present, was exclusively responsible.

The idea of the ghetto, in this discourse, embodied the past, but still it dominated the present. Its contrasting counterpart was emancipation, which stood for the ideals that defined the future but could be achieved in the present—if the ghetto were overcome. There is a distinct spatial dimension alongside this temporal aspect of the term: "ghetto" conveys a confined space, opposed to the unlimited space of modern society; it is narrow and static rather than broad and dynamic like the world envisioned in the emancipatory discourse.[22] The debate on the ghetto has to be contextualized against the backdrop of political developments then unfolding in Galicia between traditional close connections to Vienna and the increasing Polonization of the public sphere after the province gained autonomy in 1867.[23] The discourse likewise reflects the conflict between the older generation, devoted to the ideals of the *Haskalah,* and younger writers arguing for greater political and social integration into non-Jewish society.

Discussions of a "modern" Jewish society took place in German as well as in Polish or Hebrew, in a variety of publications. From its beginnings in 1868, the L'viv-based journal *Israelit. Organ des Vereins Schomer Israel,* which staunchly advocated Jewish assimilation into German culture, was printed in German. Between 1884 and 1886, however, a Polish version, *Izraelita. Organ Szomer Izrael,* appeared alongside it. The periodical's reason for exploring the Polish-language market was the previously mentioned journal *Ojczyzna,* which together with its Hebrew supplement *Ha-Mazkir* (The Recorder) openly propagated integration of Jews into the Polish nation. When the editors of *Ojczyzna* admitted defeat in their social and cultural struggle and discontinued the paper in 1892, the first Zionist journal to appear in Polish, *Przyszłość* (Future; publication ceased in 1897), was founded in the same publishing house; it published the work of several former contributors to *Ojczyzna.*[24] At the dawn of the twentieth century, *Wschód* (East; 1900–1912) continued the mission of *Przyszłość,* and in 1907 a new periodical entitled *Jedność. Organ Żydów Polskich* (Unity: Organ of Polish Jews; publication ceased in 1912) was launched, attempting to revive the traditions of the Assimilationist movement. All these publications shared principal aims: they sought to modernize Jewish society and show the "Jewish masses" a way to improve their social position. They disagreed vehemently, however, on the proper path to achieve that objective. Neither side felt able to compromise on the question of whether integration into non-Jewish society or the formation of a Jewish nation would lead to the overall improvement of Jewish society. Initially, both factions opposed traditional social structures, symbolized by the power of "rabbis and rebbes." Later, both focused on the problem of anti-Semitism; however, the social and political threat of anti-Semitism would not unite but only deepen the ideological divide between them. Zionists and Assimilationists accused each other of fostering anti-Semitism with their respective agendas and therefore of risking perpetuation of the ghetto.[25]

It was *Ojczyzna* that introduced the term "ghetto" into the discussion in 1882. In the years that followed, only Polish-speaking periodicals adopted the motif, with both the German and the Hebrew periodicals, including *Ha-Mazkir* in Galicia, refraining from entering the debate. Nonetheless, Alfred Nossig, Osias Thon, and other Galician Zionists, who worked closely with Theodor Herzl at the close of the nineteenth century, propelled the debate around the idea of the ghetto and Jewish society into German Zionist discourse.

The Ghetto as a Prison

The social, cultural, and emotional connotations of the concept of the ghetto for the Jewish population are manifested in the works we will now discuss. There is little specificity and certainty about what actually constitutes a ghetto. If such information is given at all, it appears very briefly and laconically, indicating that authors assumed that their readers already shared an idea of the ghetto as part of the Jewish experience. An unusually explicit description of a ghetto as a prison-like community can be found in a seminal article by Osias Thon on the historiosophical justification of Zionism, published in 1896: "Excluded from contact with anybody else, locked up in small groups within narrow and dim ghettos, the life of the Jews was rather simple. In a way, living conditions took on a very uniform, schematic character. At most now and again, or maybe a bit more frequently, persecutions, lootings and expulsions provided a little distraction."[26] In this piece, the ghetto appears as a constitutive factor of Jewish existence in the Diaspora during the period before emancipation, when Jews did not yet have the opportunity to develop their culture, and with it, their nationality, in the same way as other peoples. Up until this time, Thon explained, Jewish culture had been confined to religion, whose ceremonies defined the entirety of Jewish life: "In the same way as religious ceremony dominated the [Jewish] way of life, religious perception[s] filled the whole consciousness of the Jews."[27]

While Thon focused on the isolation and physical threat to Jewish existence in the ghetto, an unnamed author in *Przyszłość* expressed fears that Jewish culture might have become tainted by non-Jewish influence in this oppressive situation: "Jewish culture, which for long centuries had flowed not in its natural course, but in an artificially drained riverbed, [which was] moreover, mixed up with foreign ingredients, [was] contaminated by the stifling influences of unhealthy airs."[28] These fragments depict the ghetto as a pre-existing and immutable phenomenon, an integral part of Jewish life, itself beyond analysis even if its consequences were not—there was no need to differentiate between "Diaspora" and "ghetto." Salomon Schiller commented trenchantly in *Przyszłość* in the same year: "They imprisoned us in ghettos, but it was especially this keeping-us-away-from-them, this prison autonomy, that greatly stimulated the development of [our] own national and socio-cultural characteristics."[29]

Contributions to the Assimilationist press generally shared the view that Jewish culture had been sidetracked by the existence of the ghetto, but they differentiated between its external (i.e., non-Jewish) and internal foundations. External factors were said to prevail, for instance, "in the 'pale of settlement' of the [Russian] empire," where the Jewish masses were confined to a ghetto that was "spatially ... vast, though morally constricted." By contrast, they held that "in Galicia ... the persistence of the ghetto depends rather on internal than external causes."[30] Thus, one's interpretation of the causes of the ghetto's persistence depended on the political framework within which they could be found. In an unenlightened and backward state, such as Russia, external pressure confined the Jews to an excluded ghetto existence. In the enlightened and modern Habsburg Empire, however, the ghetto's persistence was said to derive from internal factors, so that Jews themselves were to blame for not joining modern society.

Between Past and Present

A travelogue published in the Assimilationist paper *Jedność* in 1908 states that that the ghetto of Rome dated back to ancient times. Founded under the reign of Emperor Titus after the destruction of Jerusalem and Jewish statehood in 70 CE, it housed a large number of prisoners of war he had brought to settle in Rome:

> In the course of time a great number of narrow, dirty, and unhealthy streets and corners developed, where narrow and small houses piled up one on another. The grounds are boggy; therefore the quarter is haunted by various epidemics, especially malaria, every so often. Titus probably for this very reason settled the subjugated there, after the incredibly brave resistance of the Jews.[31]

In dating this Roman ghetto back to ancient times, the author paradoxically underlines the timelessness of the phenomenon. Associating the ghetto with the destruction of the Second Temple, he links it directly to Jewish existence in the Diaspora. In this account, therefore, the ghetto acquires a metaphysical dimension that places it within the realm of theology rather than of history.

Accounts of the end of ghettos, by contrast, were usually firmly embedded in history. In an article on the history of the Jews under the reign of Emperor Franz Joseph, which appeared in *Jedność* on the occasion of the emperor's sixtieth jubilee, historian Majer Bałaban named a specific year as marking the end of the ghetto: 1867, when the Jews achieved legal equality, after which every citizen was treated equally before the law, could apply for any office, change his place of residence, and so on. Bałaban happily concluded that "the ghetto has been broken; nowadays many a Salomon Goldbaum owns a shop in the market square."[32] The end of the ghetto had changed Jewish topography, enabling Jews to move from the margins into the center (physically as well as metaphorically). In his view, it was important that the non-Jewish authorities, and not the Jews themselves, had paved the way to overcoming the ghetto.

Usually writers avoided such clear classifications. Those who viewed the ghetto as belonging to a past age praised "emancipation" without detailing what this actually meant.[33] Even so, many disputed that the ghetto had indeed been overcome and belonged to the past. Authors writing for the new Zionist weekly *Wschód*, founded in L'viv in 1900, in particular, acknowledged that the ghetto remained part of social reality. They associated the ghetto very closely with social deprivation and despair. The editorial of one of the first issues of *Wschód* lamented that the Jewish poor still overwhelmingly lived under medieval conditions. While everything around them had changed, their way of thinking and their way of earning a living remained old-fashioned. The author lamented that Galician Jews had lost their proverbial ability to adapt to new conditions and stayed within the narrow confines of the ghetto, while even Galician peasants had adjusted to the new circumstances.[34]

Both Zionist and Assimilationist authors frequently located the ghetto in the present. Travelogues from Rome, Venice, and Florence emphasized that Jews still inhabited the dilapidated streets formerly called "the ghetto." Further, they applied the term to other places where Jews lived in a relatively concentrated area, such as Warsaw in Poland, L'viv in Galicia, Baku in Azerbaijan, and even Chicago or New York, whose East Side was said to hold "the biggest Jewish ghetto in the world."[35] Common to all these "ghettos" was the thoroughly negative associations they evoked: the authors described the Italian ghettos and their buildings as extremely dilapidated (with the one exception of the synagogue in Florence, about which one author could not hide his admiration), they viewed Warsaw and L'viv as exemplifying abject poverty, and they characterized Jews in the province of Baku as being "religious to the point of fanaticism, conservative in all regards, not excluding their clothing. ... The masses are unimaginably poor."[36] Moreover, one author presented the development of the New York "ghetto" as an atypical conquest: "The Jews have conquered this quarter by slowly driving out the erstwhile immigrants—the Irish and the Germans." At the same time, it was clear that the Jews could be displaced by another wave of immigrants, the Italians. "If this process is to last for the next ten years, many Jewish institutions will find themselves located in the heart of the Italian quarter."[37] This narrative alluded to the early modern Jewish ghettos in Italy. Jews were being driven out of their living quarters as they had been in early modern times; once again a Diaspora situation had arisen that left Jewish institutions in a non-Jewish environment and dispersed the Jews themselves. Nonetheless, the author glossed over the reasons for this development, which lay not in legal or religious discrimination but in upward social mobility among the Jews.

It is striking here that the article on the "New York ghetto" appeared in *Jedność*, an outspokenly Assimilationist paper. Instead of expressing satisfaction about the slow yet ongoing dissolution of the ghetto, as the Galician Assimilationists of a generation before had done, the author bewailed the loss of Jewish space. He continued to associated the ghetto with poverty, overcrowding, and despair without any apparent sense of pleasure in seeing the demographic change.

The "Outer" and "Inner" Ghetto: Between Dystopia and Utopia

If we look at the ghetto as a spatial construction, the public debates in Galicia feature both dystopian and utopian connotations. Dystopian visions, in which the ghetto is cast as a kind of prison, can be divided into those that picture it as a phenomenon of the past or of the present (or as a past that obstinately remains). Conversely, some Zionist authors generated a utopian vision of the ghetto as a Jewish space within which a largely autonomous Jewish society could thrive, a refuge with the potential to counter some of the negative developments associated with Diaspora life.

Dystopian visions often attributed the existence of the ghetto explicitly to external factors, although they tended to be remarkably vague about precisely what these were—that is, who ordered the ghetto to be built and decreed that the Jews had to live there and nowhere else—regardless of their political orientation. Some alluded to non-Jewish authorities, such as the above-mentioned article by Salomon Schiller ("They imprisoned us in ghettos …"), but usually neither precise dates nor specific actors, such as political rulers or ecclesiastical authorities, were mentioned. It seems that this was not due to coincidence or readers' supposed familiarity with these contexts. Evidently, the "ghetto" as a concept did not need a clear starting point, cause, or initiator.

This was because the restricting and confining character of the ghetto was presumed to have arisen not only from pressure from without. A major reason the ghetto narrative was still important in the late nineteenth century was that it represented not just an external setting but also a state of mind: the "moral ghetto" was cast as a space that deprived Jews of the determination and power to meet the challenges of the modern world. In 1882, in one of the first articles to appear in *Ojczyzna*, Samuel Hirsch Margulies protested against Orthodox and Hasidic influences in Galicia and especially, in this context, against the continuing existence of the *cheder* schools run by the Jewish communities:

> Did we fight so hard for the human rights that had been denied to us for so long—only to voluntarily renounce them now? Should we wall ourselves up in the moral ghetto, now that the world at last has unlocked the rusted bolts of our age-long prison and pushed the doors wide open, because the world expects from us the active participation in works and labors for the fulfillment of the great task of mankind?[38]

Ignacy Schipper, in a 1911 essay on the "ghetto stories" of Mendele Mojcher Sforim, characterized the "ghetto psyche" as a state of having "lost control over life, apathetically suffering the collective process of life, unable to *consciously create* life ... the ghetto psyche will not strive to understand its own role in the process of life, but conceives of it instead as unchangeable and absolute. The content of the ghetto is tradition."[39]

The spiritual world of the ghetto was generally perceived as narrow and static; nevertheless, some authors, mostly with Zionist leanings, thought it problematic to stress only the negative aspects of the ghetto, as it also constituted the realm within which the vast majority of Galician Jews still lived. In 1895, in his inaugural rabbinical sermon, Osias Thon showered praise on the "ghetto seen from within" for the sense of community and quality of life it made possible:

> From the depths of the Jewish people comes bursting and tearing a ray of sunlight, clear and warm, evoking in our hearts a feeling of joyous contentment and great pride. There, within the walls of the ghetto, the spirit of God is alive and working. In the houses of prayer we hear the sound of sincere and heartfelt prayers sent unto God, that He may save His people just as has been foreseen. You can see in the bright eyes of the impoverished Jew, who is bent under the weight of his worries and suffering, that he firmly and unshakably believes in the perpetual vitality of his nation. ... All the time there is singing in this ghetto, and the base note of all these melodies is the love of God and of his great, chosen nation, of Israel. ... Everywhere we find unity and heartfelt brotherhood; everywhere we hear solemn hymns and wisdom. Such is the picture of Jewish history seen from within.[40]

While Thon still conceived of the ghetto as a prison and a place of suffering, he also saw it as a genuine Jewish place, where the Jewish community could survive in the Diaspora and God could reveal his deep connection to the People of Israel.

If the ghetto was not just a place of lethargic suffering and misery, but also a place of unity, where the People of Israel and their culture could persevere despite life in the Diaspora, then the space could be understood as a sort of preliminary stage to the renewed existence of the Jewish people in the Land of Israel. In 1901, the Zionist weekly *Wschód* printed an article about a previously delivered speech by Max Nordau, one of Theodor Herzl's closest collaborators and the editor-in-chief of the German paper *Die Welt,* that promoted this new understanding of the ghetto:

> Let us discard the old ghetto, let us free ourselves from its atmosphere, which is characterized by servant-like submissiveness and lack of self-esteem, and let us build a new ghetto, a ghetto of mutual support, of mutual refinement, a hotbed of Jewish science, a cradle of Jewish arts. ... "Hej, shoulder to shoulder"— onwards to our common work, and by the time when this ghetto of spirit, of Jewish thinking becomes reality, then the Jews will cease to be inhabitants of the ghetto.[41]

Nordau's vision added a new dimension to the concept of ghetto. Instead of the dystopian ghettos of the past and the present, the future could bring a utopia of Jewish community to life—a new "ghetto" as a step toward the Zionist dream.

Conclusion: Different Ideologies, Similar Imaginations

The writings we analyzed in this chapter were published by authors of radically divergent political persuasions during a period spanning over a quarter of a century; yet they show some remarkable similarities in how they imagined the ghetto, most prominently in the understanding of the concept as a confined space as opposed to an open one that could be reached only by leaving or overcoming the confinement. Again and again the authors emphasized the limitations and constriction of the ghetto, as well as the overwhelming stasis and isolation of life within it. Development and interaction with the space outside were prompted from the outside, or they were linked to the demise of the ghetto, which would occur when the Jews embraced modern society in the course of emancipation. Regardless of whether these authors thought of the ghetto as a past or ongoing phenomenon, none of them ever located himself inside the ghetto; all always depicted it as a foreign place whose inhabitants possessed no agency but were rather passive objects moved only by external incentives.

Culturally, the ghetto was, above all else, isolated—from non-Jewish society and from any influence that might introduce change or lead to self-awareness. It thus acquired an aura of timelessness: it was tied to Jewish life in the Diaspora, but it had no real beginning there, nor was there any development within it. This meant it had no organic end; the demise of the ghetto had to be brought from the outside: by emancipation, by individuals leaving the ghetto, or by its transformation into a new community. The biggest obstacle to overcoming the ghetto, as Osias Thon remarked in 1902, was "the masses—we [i.e., the Zionists] are still looking for them. We will win them over—there is no doubt about it. But they have not come by themselves, because they dwell in the 'ghetto.'"[42]

When he spoke of the masses that had yet to leave the ghetto, Thon also alluded to the social dimension of the problem. The ghetto was a prison, an isolated space, parallel to and separate from the world outside, conducive to backwardness and poverty. In this respect, again, there was little difference between Assimilationist and Zionist authors. For both, the ghetto was not simply a metaphor for the social problems of Jewish society in Galicia; it was, in a way, also the key to their solution, since these problems, too, would cease to exist when the ghetto was overcome, they believed.

Despite the different ways these two orientations envisioned a future without the ghetto, they had more than a superficial resemblance in how they viewed its origins. Neither the geography nor the initiators of the ghetto received more than a passing mention. Assimilationist authors tried very hard not to name non-Jewish authorities when writing about the ghetto because doing so could undermine their ultimate goal of integrating the Jews into non-Jewish society. They therefore tended instead to emphasize the internal constitutive factors of the ghetto and blame the Jews themselves, be they traditional, Orthodox, or Hasidic. Their Zionist counterparts had no such qualms, but they did not think much about the ghetto's origins. It was not the past that mattered to them, but the future; additionally, the ghetto was intrinsically linked to Jewish existence in the Diaspora, which made it a universal phenomenon.

The public debate in turn-of-the-century Galicia, then, was not fundamentally about the essence of the ghetto, but rather about its meaning and metaphorical significance within a broader analysis of Jewish society in Galicia (and beyond) and the political conclusions to be drawn therefrom. The ghetto as a concept was ripe for use as an argument in this debate because at the end of the nineteenth century it had long ceased to be a topographical demarcation. Instead, it had become a powerful spatial trope that evoked certain associations among readers while remaining sufficiently open to serve a range of differing interpretations.[43]

Jürgen Heyde has worked as project leader and research associate at the Leibniz Institute for the History and Culture of Eastern Europe (GWZO) since 2014; he is also extracurricular professor at Martin Luther Universität Halle-Wittenberg.

Notes

1. Gaisbauer, *Davidstern und Doppeladler,* 11.
2. Ravid, "From Geographical Realia"; Wacquant, "Ghetto"; on the academic discussion around the term "ghetto," see Heyde, "Making Sense," esp. 38–43.
3. Cf., for example, alongside further literature, Backhaus et al., *Frühneuzeitliche Ghettos*; Ruderman, "Cultural Significance."
4. Michman, *Emergence of Jewish Ghettos*; Tych, "Ghettos in Polen."
5. Himmelmann, "Ghetto."
6. The classic and probably most influential example is Katz, *Out of the Ghetto.*
7. Wirth, *The Ghetto.*
8. Wacquant, "Ghetto"; Wacquant, "Janus-Faced Institution." On metaphorical uses of the term, cf. Wacquant, "Urban Desolation"; Anderson, "Iconic Ghetto."
9. Fuchs and Krobb, *Ghetto Writing*; Ober, *Ghettogeschichte*; Glasenapp, *Aus der Judengasse.*
10. Hess, *Middlebrow Literature.*
11. Gerhardt, "Vom Raum zur Geschichte."
12. Shanes, *Diaspora Nationalism.*
13. Hoffmann, "Das Ghetto"; Heyde, "Making Sense."
14. On the etymology of the term "ghetto," see Roth, "Origins of Ghetto"; Toaff, "Getto-Ghetto"; Ravid, "Religious, Economic and Social Background," esp. 218–19.
15. Ravid, "On the Diffusion."
16. Mendes-Flohr and Reinharz, *Jew in the Modern World,* 122 (Padua, 28 August 1797).
17. Börne, "Menzel der Franzosenfresser," 889.
18. Lazare, "Le nouveau ghetto," *La Justice,* 17 November 1894, cited in Goldfarb, *Emancipation,* 290.
19. Hess, *Middlebrow Literature*; Ober, *Die Ghettogeschichte*; Glasenapp, *Aus der Judengasse.*
20. Baron, "Ghetto and Emancipation."
21. Tych, "Ghettos in Polen"; Bergman, "Rewir."
22. Tuan, *Space and Place,* 3–6.

23. Röskau-Rydel, *Galizien*, 145–52; Wandycz, *Lands of Partitioned Poland*, 214–38.
24. Mendelsohn, "From Assimilation to Zionism."
25. Gąsowski, *Między gettem a światem*, 123–35.
26. Thon, "Zur geschichtsphilosophischen Begründung."
27. Ibid., 13; Heyde, "Ghetto and Emancipation."
28. *Przyszłość*, no. 13, 5 April 1894, 146, "Za waszą i naszą wolność."
29. *Przyszłość*, no. 18, 20 June 1896, 137–40: Salomon Schiller, "Nasz byt narodowy." Schiller (1862–1925) was born into a Hasidic family. From 1890 onward, he studied in L'viv and later joined the group of editors of *Przyszłość*. He took part in the first Zionist Congress in Basel and emigrated in 1910 to Jerusalem (Österreichisches Biographisches Lexikon 1815–1950, vol. 10 [Vienna, 1994], 136), accessed 26 February 2015, www.biographien .ac.at/oebl/oebl_S/Schiller_Salomon_1862_1925.xml).
30. *Jedność*, no. 26, 23 June 1909, 4: "Reforma judaizmu."
31. *Jedność*, no. 14, 3 April 1908, 6: "Dzielnica żydowska w Rzymie" (author: Honor).
32. *Jedność*, no. 49, 4 December 1908, 5: Prof. Dr. Majer Bałaban, "Żydzi w Austryi za panowania Cesarza Franciszka Józefa I-ego (1848–1908) z szczególnem uwzględnieniem Galicyi."
33. Cf. *Ojczyzna*, no. 3, 1 February 1884, 10–11, "Którędy?," etc.
34. *Wschód*, no. 9, 30 November 1900, 1–4 [editorial].
35. *Jedność*, no. 27, 8 July 1910, 5: (L.), "Amerykańskie ghetta"; cf. *Jedność*, no. 14, 2 April 1909, 8: "Kronika: Kobieca straż bezpieczeństwa w Nowojorskim ghetto'; *Wschód*, no. 2 (og. zb. 298), 10 January 1908, 7, "Mark Twain a Szalom Alejchem."
36. *Wschód*, no. 9 (360), 26 February 1909, 5: Z Golusu.
37. *Jedność*, no. 27, 8 July 1910, 5: (L.), "Amerykańskie ghetta."
38. *Ojczyzna*, no. 9, 1 May 1882, 35–36, "O naszych stosunkach wyznaniowych (S. Margulies, sł. semin. rabin. W Wrocławiu)."
39. *Wschód*, no. 1, 6 January 1911 (po konfiskacie nakład drugi), 1–2, here 1: Ignacy Schipper, "Jubileusz S. J. Abramowicza (Mendla Mocher Sforim)."
40. Thon, "Ghetto—emancypacja," 32.
41. *Wschód*, no. 37, 18 June 1901, 10–11, here 10: "Listy z Wiednia."
42. Thon, "O emancypacyę," 51–52.
43. Mann, *Space and Place*, 123.

Bibliography

Anderson, Elijah. "The Iconic Ghetto." *Annals of the American Academy of Political and Social Science* 64, no. 2 (2012): 8–24.
Backhaus, Fritz, Gisela Engel, Gundula Grebner, and Robert Liberles, eds. *Frühneuzeitliche Ghettos in Europa im Vergleich*. Berlin: Trafo, 2012.
Baron, Salo W. "Ghetto and Emancipation: Shall We Revise the Traditional View?" *Menorah Journal* 14 (1928): 515–26.
Bergman, Eleonora. "The Rewir or Jewish District and the Eyruv." *Studia Judaica* 5, no. 9 (2002): 85–97.
Börne, Ludwig. "Menzel der Franzosenfresser." In *Sämtliche Schriften*, edited by I. Rippman and P. Rippmann, vol. 3, 871–984. Düsseldorf: Melzer, 1964.

Erb, Rainer, and Werner Bergmann. *Die Nachtseite der Judenemanzipation. Der Widerstand gegen die Integration der Juden in Deutschland 1780–1860.* Berlin: Metropol, 1989.

Fuchs, Anne, and Florian Krobb, eds. *Ghetto Writing: Traditional and Eastern Jewry in German-Jewish Literature from Heine to Hilsenrath.* Columbia, SC: Camden House, 1999.

Gaisbauer, Adolf. *Davidstern und Doppeladler. Zionismus und jüdischer Nationalismus in Österreich 1882–1918.* Vienna: Böhlau, 1988.

Gąsowski, Tomasz. *Między gettem a światem. dylematy ideowe żydów galicyjskich na przełomie XIX i XX wieku.* Kraków: Instytut Historii UJ, 1996.

Gerhardt, Kristiane. "Vom Raum zur Geschichte. 'Ghetto' als Metapher historischer Grenzziehung im 19. Jahrhundert." *PaRDeS. Zeitschrift der Vereinigung für Jüdische Studien* 17 (2011): 19–34.

Glasenapp, Gabriele von. *Aus der Judengasse. Zur Entstehung und Ausprägung deutschsprachiger Ghettoliteratur im 19. Jahrhundert.* Tübingen: Niemeyer, 1996.

Goldfarb, Michael. *Emancipation: How Liberating Europe's Jews from the Ghetto Led to Revolution and Renaissance.* New York: Scribe, 2009.

Hess, Jonathan M. *Middlebrow Literature and the Making of German-Jewish Identity.* Stanford, CA: Stanford University Press, 2010.

Heyde, Jürgen. "Ghetto and Emancipation: Reflections on Jewish Identity in Early Works of Ozjasz Thon." In *A Romantic Polish-Jew: Rabbi Ozjasz Thon from Various Perspectives,* edited by Michał Galas and Shoshana Ronen, 47–59. Kraków: Uniwersytet Jagielloński, 2015.

———. "Making Sense of 'the Ghetto': Conceptualizing a Jewish Space from Early Modern Times to the Present." In *Jewish and Non-Jewish Spaces in the Urban Context,* edited by Alina Gromova, Felix Heinert, and Sebastian Voigt, 37–61. Berlin: Neofelis, 2015.

Himmelmann, Werner. "'Ghetto'. Die juristische Definition." *PaRDeS. Zeitschrift der Vereinigung für Jüdische Studien* 18 (2012): 139–46.

Hoffmann, Christhard. "Das Ghetto – eine Begriffs- und Diskursgeschichte." In *Frühneuzeitliche Ghettos in Europa im Vergleich,* edited by Fritz Backhaus, Gisela Engel, Gundula Grebner, and Robert Liberles, 53–78. Berlin: Trafo, 2012.

Hsia, R. Po-chia, and Hartmut Lehmann, eds. *In and Out of the Ghetto: Jewish-Gentile Relations in Late Medieval and Early Modern Germany.* Cambridge: Cambridge University Press, 1995.

Jedność. Organ Żydów Polskich. Lwów, 1907–13.

Katz, Jacob. *Out of the Ghetto: The Social Background of Jewish Emancipation, 1770–1870.* Cambridge, MA: Harvard University Press, 1973.

Mann, Barbara E. *Space and Place in Jewish Studies.* New Brunswick, NJ: Rutgers University Press, 2012.

Mendelsohn, Ezra. "From Assimilation to Zionism in Lvov: The Case of Alfred Nossig." *Slavonic and East European Review* 49, no. 117 (1971): 521–34.

Mendes-Flohr, Paul, and Jehuda Reinharz. *The Jew in the Modern World: A Documentary History.* New York: Oxford University Press, 1995.

Michman, Dan. *The Emergence of Jewish Ghettos during the Holocaust.* Cambridge: Cambridge University Press, 2011.

Ober, Kenneth H. *Die Ghettogeschichte. Entstehung und Entwicklung einer Gattung.* Göttingen: Wallstein, 2001.

Ojczyzna. Lwów, 1881–92.

Przyszłość, Organ narodowej partyi żydowskiej. Kraków, 1892–97.

Ravid, Benjamin. "From Geographical Realia to Historiographical Symbol: The Odyssey of the Word 'Ghetto.'" In *Essential Papers on Jewish Culture in Renaissance and Baroque Italy,* edited by D. B. Ruderman, 373–85. New York: New York University Press, 1992.

———. "On the Diffusion of the Word 'Ghetto' and Its Ambiguous Usages, and a Suggested Definition." In *Frühneuzeitliche Ghettos in Europa im Vergleich,* edited by Fritz Backhaus, Gisela Engel, Gundula Grebner, and Robert Liberles, 15–38. Berlin: Trafo, 2012.

———. "The Religious, Economic and Social Background of the Establishment of the Ghetti of Venice." In *Gli Ebrei e Venezia,* edited by Gaetano Cozzi, 211–59. Milan: Ed. Comunità, 1987.

Röskau-Rydel, Isabel. *Galizien, Bukowina, Moldau.* Berlin: Siedler, 1999.

Roth, Cecil. "The Origins of Ghetto: A Final Word." *Romania* 60 (1934): 67–76.

Ruderman, David. "The Cultural Significance of the Ghetto in Jewish History." In *From Ghetto to Emancipation: Historical and Contemporary Reconsiderations of the Jewish Community,* edited by D. N. Myers and W. V. Rowe, 1–16. Scranton, PA: University of Scranton Press, 1997.

Shanes, Joshua. *Diaspora Nationalism and Jewish Identity in Habsburg Galicia.* Cambridge: Cambridge University Press, 2012.

Thon, Osias. "Zur geschichtsphilosophischen Begründung des Zionismus." In *Essays zur zionistischen Ideologie,* 3–28. Berlin: Kedem, 1930. First published in *Zion* 2, no. 11 (1896); *Zion* 3, nos. 1–3 (1896).

Thon, Ozjasz. "Ghetto—emancypacja (1895)." In *Kazania* (1895–1906), 31–34. Kraków: Austeria, 2010.

———. "O emancypacyę." *Rocznik Żydowski* 2 (1902/5662–63): 47–52.

Toaff, Ariel. "Getto-Ghetto." *American Sephardi* 6 (1973): 70–77.

Tuan, Yi-Fu. *Space and Place: The Perspective of Experience.* Minneapolis: University of Minnesota Press, 1977.

Tych, Feliks. "Ghettos in Polen." *Wiener Jahrbuch für jüdische Geschichte, Kultur und Museumswesen* 1 (2001): 69–75.

Viterbo, Ariel. "Da Napoleone all'Unità d'Italia." In *Gli Ebrei e Padova,* vol. 2, *Il cammino della speranza,* edited by C. De Benedetti, 1–51. Padua: Ed. Papergraf, 2000.

Wacquant, Loïc. "Ghetto." In *International Encyclopedia of the Social and Behavioral Sciences,* edited by Neil J. Smelser and Paul B. Baltes, 1–7. Amsterdam: Elsevier, 2004.

———. "A Janus-Faced Institution of Ethnoracial Closure: A Sociological Specification of the Ghetto." In *The Ghetto: Contemporary Global Issues and Controversies,* edited by R. Hutchison and B. D. Haynes, 1–31. Boulder: Westview Press, 2011.

———. "Urban Desolation and Symbolic Denigration in the Hyperghetto." *Social Psychology Quarterly* 20, no. 2 (2010): 1–5.

Wandycz, Piotr S. *The Lands of Partitioned Poland 1795–1918.* Seattle: University of Washington Press, 1993.

Wirth, Louis. *The Ghetto.* New Brunswick, NJ: Transaction Publishers, 1998 (original edition: Chicago, 1928).

Wschód. Tygodnik poświęcony politycznym, ekonomicznym i umysłowym sprawom żydostwa. Lwów, 1900–12.

3

Modernization and Memory in German-Jewish History

Nils Roemer

Religious, cultural, and economic modernization in the German-speaking regions coincided with increased urbanization in the second half of the nineteenth century. Moving out of rural communities into larger urban environments, Jews modernized their culture as they integrated into the emerging German national community. In the face of this transformation, continuities, symbolized by historical ruins and places such as synagogues and cemeteries, acquired profound new importance. Venerating and preserving these remnants of the past forged new relationships between past and present; preserving physical and textual objects was essentially an act of saving the past for the future. Moreover, spaces served to literally inscribe and map Jewish traditions into the wider national and urban German landscape.

Modern historical studies initially mapped the spaces circumscribing various national histories. Researching the nation quite literally involved traveling the country. Leopold von Ranke, like other nineteenth-century historians, journeyed across German lands to explore archival and library collections,[1] laying historical narrative over national space. Enveloping individual histories of princes and principalities, centers and peripheries, and the learned classes with a concept of space allowed historians to generate a unified national historical process; fashioning a space of shared historical experiences created a common past. The concept of national space also permeates the work of French historian Pierre Nora, who traced French national memory in monuments, street names, archaeological excavation sites, museums, exhibitions, tourist guidebooks, catalogues, photographs, paintings, and postcards.[2] Utilizing these varied sites of memory, Nora united multiple spaces into a map of French memory. Similar existing collaborative and multivolume works on German realms of memory operate within the arena of concepts of the nation based on territorial or cultural notions. This national perspective, constantly assumed, homogenizes memory without questioning the ambiguity of the central categories of German spaces of memory, such as Jewish ones. Yet despite a widespread Jewish propensity toward textual as opposed to spatial remembrance, the German landscape of the modern age remained dotted with venerable Jewish spaces. These deserve more attention than they have heretofore received in a few fleeting references, notwithstanding the growth of German-Jewish studies over the last decades.[3]

Ruins and Preservation

Nora's work contextualized the shifting meaning of memory spaces within the paradigm of modernity. Nora's *Realms of Memory* contends that there are no longer any "environments of memory" (*milieux de memoire*), but only "sites of memory" (*lieux de mémoire*).[4] A place of remembrance is hence a spatial remnant of an otherwise absent past. Remembrance is motivated by a nostalgic desire to counter modernity's "acceleration of history," which prompted the disappearance of environments of memory. Indeed, fascination with historical sites and ruins became a marker of a new modern Jewish spatialized culture from around 1850.[5]

Legal equality, religious reform, and cultural changes created a sense of historical discontinuity. Historical synagogues and cemeteries became spaces that seemed frozen in time, reminding communities of a distant past and an alternate future.[6] For Heinrich Heine, ruins represented something that we understood only if "we had ourselves become ruins."[7] Put differently, historical ruins acquired meaning precisely because they appealed to a profound sense of historical rupture. For the sociologist Georg Simmel, ruins were the opposite of the perfect moment, pregnant with unfulfilled potentialities.[8] Similarly, Walter Benjamin saw ruins as a form of allegorical thinking communicating "unstoppable decline" yet, in miraculously surviving, also representing a form of inheritance from a distant past.[9]

Not unlike Nora, Jan Assmann and John Czaplicka argue that the "objectification and crystallization of communicated meaning in different forms like texts, images, buildings, monuments, cities, or even landscapes and collectively shared knowledge is a prerequisite of its transmission in the culturally institutionalized heritage of a society." Cultural memory, they claim, "preserves the store of knowledge from which a group derives an awareness of its unity and peculiarity."[10] Their underlying assumptions about a spatialized culture of remembrance presuppose that culture is implicitly mapped onto series of discrete spaces. Spaces embody the culture of a particular group; they not only intertwine with but also overlap with social identities.

Jewish spaces, culture, and identity are not always coherent or identical with one another, on account of the history of Jewish expulsion, migration, and urbanization. Simultaneous dislocation and modernization generated a powerful culture of nostalgia. Having moved out of the *Judengassen* and away from small and rural regions, Jews recalled a past that was quickly vanishing.[11]

Heine developed an account of Jewish spaces in his *Rabbi of Bacharach,* portraying Abraham and Sarah escaping a pogrom by boat on the Rhine from Bacharach to Frankfurt, "the world-famous, free, imperial, and commercial city of Frankfurt-on-the-Main." Abraham instructs his wife to keep her eyes shut to the sights of the "magnificent wares" on display at the Easter fair as they journey along the city's historical landscape until they finally enter the Jewish space, the *Judengasse.*[12] Their flight constitutes a temporal and geographical move from smaller provincial communities to larger urban centers and a transition from an age of persecution and expulsion to a new age of civic emancipation with cultural and religious transformation. Thus, Heine mapped out a historical territory that encapsulated Bacharach, Frankfurt, and Worms—decisively different spaces that each represent different aspects of the German-Jewish past.

Heine relates the memory of Jewish spaces to the modernizing forces of Frankfurt. Indeed, historical transformation and urbanization function as the backdrop to the production of spaces and their remembrance. Jewish spaces were fashioned through cultural practices of engagement, preservation, and restoration. Their existence resulted from historical circumstance. What were once cultural and religious spaces became transformed into historical objects in text, archives, or the structure of museums whose purpose was to recall and map the German-Jewish past.

Ritualized Remembrance between Continuity and Change

The spatialization of history and culture entails a certain concealment of temporality.[13] In the nineteenth century, local spaces of memory such as cemeteries symbolically linked communities' past and present. Synagogues and cemeteries appeared to be timeless spaces. Burial grounds provided a sense of belonging and sacred realms of memory as modern communities developed along sharp denominational lines. In this context, such spaces became both contested realms and places of reconciliation, as the following historical example of a squabble at a cemetery illustrates. In the 1840s, religious differences erupted when Abraham Geiger and Salomon Tiktin both spoke at the funeral of a Breslau community member in the 1840s. Tiktin's supporters interrupted Geiger's eulogy with shouts, whereupon Geiger's followers did the same, leading to fisticuffs eventually halted by the police. The feud was not over the next day, and the community suspended Rabbi Tiktin from discharging his rabbinical duties.[14] Following the initial police investigation, several individuals were charged with disturbing a Jewish religious service. With the support of several rabbinical reports, the lawyer for the accused argued that a cemetery was not a consecrated space. At issue, then, was not only the religious opposition between Tiktin and Geiger but also the status of Jewish cemeteries. To refute the notion that cemeteries were non-consecrated spaces, the notable rabbis and scholars Zacharias Frankel and Samuel Holdheim reasserted the elevated religious status of cemeteries based on biblical and rabbinical textual evidence. Frankel's rabbinical response cited the special dignity of cemeteries considering their age, as evidenced by travelers' visits to burial grounds in Worms, Kraków, and Prague.[15]

In Frankel's view, religious tradition and modern sensibilities converged. Leopold Zunz, in his *Zur Geschichte und Literatur,* provided a similar intersection of religious piety and modern historical recollection by detailing the ways Jews commemorated their righteous dead and elaborating on the history of tombstones and epitaphs. He pointed out the historical value of these sources and called upon contemporaries to preserve the memory of Jewish heroes threatened by the ravages of time and persecution.[16] In response to Zunz's call, Jewish community leaders in Worms formed the Committee for the Renovation of the Jewish Monuments in Worms, which comprised neo-Orthodox representatives like Rabbi Bamberger and followers of the Reform movement such as Ludwig Lewysohn and Moses Mannheimer. In July 1853, the committee called for donations to renovate local Jewish monuments in German-Jewish, French-Jewish, and Anglo-Jewish periodicals; the ad emphasized that Worms still had a chapel dating back to Rashi, a tombstone of martyrs from the First Crusade, and the graves of several rabbinic luminaries. With the funds thus raised, the cemetery and the synagogue were renovated, leading to the publication of Ludwig Lewysohn's volume of tombstone inscriptions from the Worms cemetery.[17]

Worms was not the only community that attracted scholarly attention. A plethora of local community studies that appeared from the 1860s onward utilized historical evidence from cemeteries to document the age-old presence of Jews in Germany.[18] Jewish communities competed with one another to demonstrate that their roots reached back to antiquity and the much-celebrated Middle Ages, widely seen as the birthplace of modern Germany.[19] Serendipitous discoveries of tombstones sometimes promised to substantiate attempts to locate Jews in the German landscape as early as antiquity.[20]

During this period, local spaces of memory continued to ameliorate religious polarization in communities. In Worms, the committee that renovated the historical sites had partially overcome its internal differences, but differences between the religious groups continued to challenge the community's cohesion. In July 1876 a law, the *Austrittsgesetz,* was passed enabling German Jews to remain members of the community or leave. This new voluntary community seemed to threaten the fabric of Jewish spatial culture. The law was passed at a time of increasing Jewish urbanization and upward social mobility, which could potentially disperse them more broadly across the urban landscape. Occasioned by the new law, Samson Raphael Hirsch led "separationists" to form a distinct organizational framework for neo-Orthodox Jews in Frankfurt. Yet the large majority of neo-Orthodox Jews remained in general congregations and refrained from joining Hirsch, because of their age-old attachment to the traditional community.

The later leader of Frankfurt neo-Orthodox Jewry Jacob Rosenheim attributed at least some of this reluctance to follow Hirsch to the fact that the old *kehilla,* or community, although dominated by Reform Jews, represented the time-honored local historical tradition for many of its members. The local cemetery contained the graves of famous rabbinic leaders and righteous individuals from its past. Although this old Jewish community existed only in the imagination of individuals who had refrained from enlisting in Hirsch's separatist community, these local memories nevertheless held a powerful grip over them, Rosenheim believed.[21]

Even so, the Jewish communities in larger urban centers dispersed and fragmented, whereas, in smaller communities, historical memory during the period of the Kaiserreich remained attached to spaces and rituals that had themselves informed traditional forms of commemoration, thereby fashioning new continuities. Up to the 1880s, well over one thousand articles on the histories of individual communities and regions had been published in various Jewish and non-Jewish periodicals.[22]

Preserving Memory in Urban Centers

Urban growth and renewal contributed more often to the demise of existing structures than to the establishment of splendid new ones, engendering an ambivalent response of remembrance and forgetting in the second half of the nineteenth century. For example, when the fading Frankfurt ghetto's last buildings were demolished in 1884, all but one central feature of the *Judengasse* had disappeared.[23] Only the façade of the Rothschild house was restored following large-scale alterations carried out "with piety."[24] At first, members of the Frankfurt historical association welcomed the passing of the *Judengasse,* because it contradicted the city's self-image as progressive and tolerant.[25] Similarly, the *Allgemeine Zeitung des Judenthums* celebrated its razing, calling it a "monument of disgrace." Yet the *Allgemeine* also pointed out that works by Goethe, Heine, and Börne, as well as George Eliot's novel *Daniel Deronda,* which had become very popular in Germany, would preserve the *Judengasse*'s memory.[26]

Historical preservation and abandonment of the past thus commingled in the remembrance of Frankfurt's Jewish history at a time when the advance of urbanization had catapulted Berlin into being a new center of Jewish life. In 1866, a few days before Berlin celebrated the return of the victorious Prussian troops from the battles against Austria, the impressive Oranienburger Strasse synagogue was dedicated in the presence of Bismarck and various Berlin notables.[27] Designed by noted German architects Eduard Knoblauch and Friedrich August Stühler, the imposing Moorish-style house of worship featured a gilded dome and a brick façade with granite and sandstone inlays. This building inscribed a Jewish historical tradition into the metropolitan space that Ludwig Geiger, son of the Reform scholar Abraham Geiger, characterized as a "new city" that did "not know the splendor and disgrace nor the dignity and lowness of old German

cities."[28] To root the Jewish community more firmly in this new city, Ludwig
Geiger, at the request of local leaders, composed a history of Berlin's Jews to mark
the community's two hundredth anniversary.[29] Yet a decade later, in 1880, a new
cemetery opened in Weissensee, further multiplying and fragmenting spaces that
represented historical memory and continuity. Significantly, several noteworthy
cemeteries predate Weissensee. Founded in 1827, the Schönhauser Allee cemetery contained a row of honorary graves for individuals who had variously contributed to the community's reputation and well-being. Politicians Eduard Lasker
and Ludwig Bamberger and scholars Abraham and Ludwig Geiger and Leopold
Zunz are buried there.[30] In addition, the Grosse Hamburger Strasse cemetery, in
use from 1672 to 1827, enshrined the earliest community leaders, members of
the Jewish elite, and Enlightenment figures such as Moses Mendelssohn.[31] Nonetheless, Weissensee became a powerful site of commemoration of Berlin Jewish
history without eclipsing pre-existing cemeteries. Designed by the architect Hugo
Licht, the Weissensee cemetery featured honorary graves for luminaries like rabbi
and scholar Pinkus Friedrich Frankl, who succeeded Abraham Geiger as rabbi of
the Berlin Jewish Reform community and died in 1887; scholars David Cassel,
Gustav Karpeles, and Moritz Steinschneider; composer Louis Lewandowski; and
writers such as Karl Emil Franzos.[32]

Coinciding with the dispersal of these sites of memory, urban planning during
the 1880s transformed the old residential area of Berlin Jewry. With the construction of an east-west thoroughfare, Kaiser-Wilhelm-Strasse, Berlin metamorphosed
into a noisy capital city, and many historical buildings were destroyed.[33] By 1900,
Berlin had become even more of an urban center, constantly reinventing itself
and destroying its older traditions.

Often bereft of powerful historical sites, German Jews in the larger urban
centers nevertheless strove to recall the German-Jewish past by forging links between historical narratives, readers, and places of Jewish history. Eugen Täubler,
spurred by a desire to professionalize historical scholarship and preservation and
a longing to preserve the histories of rural and small-town communities in the
face of accelerated urbanization, founded the Gesamtarchiv der deutschen Juden
(Central Archive of the German Jews) in Berlin in October 1905.[34] Noting that
concepts such as "state," "nation," and "church" did not apply to the Jews in Germany, Täubler aimed to incorporate local histories into the German and Jewish
pasts; he proposed to spatialize and analyze Jewish history in Germany around
the concept of settlement to describe the particularities in these Jewish processes
of assimilation.[35]

In 1910, the archive reopened in Berlin's Oranienburger Strasse, close to
other important Jewish institutions of cultural remembrance. Its new interior
visually manifested its intent, containing modern features that fulfilled contemporary standards of archive design, putting historical remnants into identical-looking
folders placed on metal shelves. The design visibly captured the Gesamtarchiv's
key remit of preservation, aided by modern cement and iron construction.[36] Placed
beside the towering presence of the synagogue, the Gesamtarchiv helped trans-

form Oranienburger Strasse into the central location for Jewish historical memory in the capital. Within roughly ten minutes' walking distance from the Oranienburger Strasse synagogue was the Reform temple, the Hochschule, the Rabbinerseminar, and, opened in 1894, the substantial library of the Verein jüdische Lesehalle und Bibliothek (Association of Jewish Reading Rooms and Libraries).[37] A few buildings further along Oranienburger Strasse was the community library, opened in 1902. As early as 1861, Abraham Geiger had noted that "the most eloquent witness to the respect of spiritual work is the foundation and maintenance of a library. A library provides not only nutrition for the spirit, but is also a monument to the spirit where our ancestors are gathered. ... A library pictorially represents for us the connections between the ages, where gray antiquity is intertwined with the bright present." [38] At the end of the nineteenth century, these words were cited in the report of the Jewish community library in Berlin.

The Gesamtarchiv's incorporation of archival documents of German Jewry forged new continuities with the past and contributed, along with the other institutions and structures on Oranienburger Strasse, to fashioning Jewish spaces of memory. Nonetheless, many contemporaries felt little sense of Jewish space in Berlin as the Jewish community continued to grow and disperse. During the Weimar Republic, Berlin was largely perceived as a place of fragmentation and decline of Jewish life. This aggravated the sense of uncertainty that led writers to explore the paradoxes of modernity in the city's landscape and population. The perceived invisibility of the Jewish community and its tangible sites made visible Jewish spaces all the more important, such as the centrally located Scheunenviertel, where Eastern European Jews had settled, and Prenzlauer Berg.[39] Berlin's Jewish intellectual life flourished as about seventy thousand Eastern European Jews flocked to Germany. Many belonged to highly mobile groups that stopped in Weimar Germany, and Berlin in particular, only in transit to other destinations like London and New York; the celebrated Eastern European journalist in Germany Joseph Roth commented pointedly, "No Eastern Jew goes to Berlin voluntarily. Who in all the whole world goes voluntarily to Berlin?"[40] However, many Jews from within Germany and abroad did voluntarily go to Berlin, attracted by the bustling metropolis, and indeed Roth himself acknowledged that in Berlin "it is possible for a hawker to make a career."[41] What was arguably missing from Berlin for Roth was a distinct Jewish culture: "Berlin levels out differences and kills particularities. Hence the lack of a Jewish ghetto there."[42]

Despite this criticism, Berlin had become a hub for Russian Hebrew and Yiddish writers who had left their homes in Eastern Europe to escape political oppression and violent anti-Semitism. Many literary luminaries met in the newly emerging social spaces such as the Café Monopol and the Romanisches Café. Here Shalom Asch talked to Stefan Zweig, and Else Lasker-Schüler met Abraham Stencl, who portrayed the café as a colony and parliament, underlining the alleged cultural power of émigré intellectuals.[43]

Others sought out remnants of the visible Jewish presence in Berlin, like the converted Lutheran Franz Hessel, who described the changing cityscape in his book *Spazieren in Berlin* (Walking in Berlin).[44] Editor of *Vers and Prosa* and friends with Walter Benjamin, who published several translations of poems by Baudelaire in the monthly, Hessel aimed to make the city visible again in the face of its profound urban transformation. In his walks through the city, he perceived Berlin as restlessly caught between the vanishing old and the emergent new.[45] His awareness of Berlin's constantly changing appearance did not diminish his pleasure in strolling through it.[46] Rather, Hessel's *flânerie* united an appreciation of the modern city with an orientation toward the past. It provided him with aesthetic pleasure to "slowly walk through animated streets ... it is like bathing in the wave."[47] Hessel felt that recalling the past in a rapidly changing present made it possible to reconstruct fixed navigation points that acted to stabilize individuals in their interaction with and navigation through the city's space. As Benjamin, who otherwise projected onto Hessel his own theoretical historical memory, noted in a review of Hessel's book, Hessel was a native of Berlin, destined to narrate the stories the city had told him since he was a child instead of describing only its presence.[48] And his childhood memories allowed him to seek the vanishing old amid the new.[49]

Hessel, in his descriptions of his city strolls, mentions only the old Berlin quarters with the Jewish cemetery and the Scheunenviertel as representations of Jewish life in Berlin. Their location in one of the oldest parts of the city where Jews had originally settled intensified the sense of time travel in Hessel's accounts.[50] After leaving the Judenstrasse and the entrance to the great Jewish cemetery, Hessel's narrating *flâneur* pauses to remember the days when Jews were locked behind the gates of ghettos. During the day, their hats marked them out as prohibited from permanently residing in the city and as paying special taxes.[51] Thus, these strolls through the city brought an awareness of the legal, social, and economic advancement of Jews since those times, but also contrasted Jews' circumscribed existence in that past indirectly with their invisibility in modern times.

Hessel's narrator, venturing to the Scheunenviertel northeast of Alexanderplatz, senses an urgency, a need to approach the immigrant quarter quickly before it vanishes. There, he finds men with "ancient beards and sidelocks." "These streets are still a world unto themselves and home to the eternal foreigner until they ... have acclimatized so far that they are tempted to advance deeper into the West to discard the all-too-visible markers of their character."[52] He regrets that the contemporary Eastern European Jewish immigrants quickly adapted and assimilated to Berlin and finds Jews more attractive in the Scheunenviertel than in "their ready-to-wear clothes and at the stock exchange."[53]

Alfred Döblin, the accomplished writer and neurologist who left the Jewish community in 1912, similarly explored the Jewish areas of Berlin during the Weimar Republic.[54] He devoted his novel *Berlin Alexanderplatz* (1929) to the fictional exploration of this specific milieu, which many viewed as exotic. His medical practice was close to the Scheunenviertel, making visits quite easy for him; he attended guest performances by Eastern European theater groups from Vilnius with great interest. Before publishing his novel, he explored the city's

Jewish quarters, encountering the "spontaneous artistic productions of a living tribe."[55] He captured the city's vibrancy and constant change as well as its increasing ethnic and religious diversity. For Döblin, Eastern European Jews not only added to this diversity but came to symbolize it: "These people, who understand Yiddish, this mix of languages which is a naturally grown Esperanto, live at Alexanderplatz and have no money."[56]

Döblin's writing about Eastern European Jews expressed his view of Berlin as a cosmopolitan center. Although he depicted the "Galician settlement" of the Scheunenviertel, he said nothing about German Jews. The Scheunenviertel description formed a crucial part of his "Berlin Miniatures," which highlighted foreign newspapers, Russian cafés, as well as Chinese, Japanese, and Korean people as testaments to the capital's ethnic and cultural diversity.[57] Experiencing and writing about it asserted his right to the city—a right that seemed threatened when anti-Semitic Scheunenviertel riots erupted in 1923, inspiring him to scrutinize the area and its culture:[58] "I went again and again into the street of the pogroms, sat in the small Krakauer Cafe, observed the people, spoke to this and that one."[59]

In Search of a Jewish Past in German Cities

Berlin's modern social, political, cultural, and topographical changes amplified a sense of fragmentation and often coincided with the disappearance of Jewish spaces. In 1926, Berlin's Jews witnessed the sale of the large private Sally Kirschstein collection to the Hebrew Union College in Cincinnati, prompting Eugen Caspary to lament the departure of another visible and tangible link to the past in the *Jüdisches Jahrbuch*.[60] Caspary underlined not only the by now well-established tradition of viewing Berlin as a city "without center or shape [*Gestalt*]," but also asked, "Where is the Jewish Berlin? Nowhere in this city can one find a neighborhood that can be called the center of Jewish Berlin." Only in the fragmented landscape of clubs and associations did he detect the presence of Berlin Jewry.[61] In response to this perceived absence of visible spaces, community newspapers, also in 1926, dedicated a special issue to the question of community to familiarize readers with existing community structures, institutions, and associations. In this issue, Leo Baeck emphasized the formative and lasting influence of stable communal and spatial structures on the individual and called upon readers to preserve them.[62] The old communities functioned here as the only existing spaces whose memory sustained the self-image of wholly urbanized Jews. The historical spaces and their memories formed "for most Jews who live in large cities the reservoir from which they again and again gain strength and courage for the trying material and spiritual struggles [of life]."[63]

Such discussions established a lasting link between modernity, the metropolis, amnesia, and older regional communities and Jewish spaces and memory. The Jews of the Rhineland, for example, were prominent in a 1925 exhibition in Cologne entitled "A Thousand-Year Exhibition of the Rhineland," a part of wide-ranging festivities chronicling the region's German cultural heritage.[64] Officially opened by Cologne's then mayor Konrad Adenauer, the public display of age-old German history associated with this region fulfilled obvious political ends at a time when France still occupied parts of the area. Despite the exhibition's open nationalism, a special section featured the Jewish history of the Rhineland communities.

The major Jewish periodicals of this period, the *C.V.-Zeitung* and *Israelitisches Familienblatt,* among others, devoted extensive space to reviewing the exhibition's account of Jews' rich history in the Rhineland. For the *C.V.*, the exhibition validated the "synthesis of *Deutschtum* and *Judentum*" that it believed would silence all anti-Semitic voices. The Rhineland, the papers opined, was the first place where Jews had settled on German soil and therefore illustrated German Jews' long-standing attachment to Germany.[65]

The act of traveling to Cologne and visiting the exhibition enabled Germans to highlight their solidarity with the occupied areas and support the local economy. Along with German-Jewish associations of rabbis, cantors, and teachers, the western German branches of the *C.V.-Verein* and the *Reichsbund jüdischer Frontsoldaten,* the national association of Jewish soldiers who had served on the front in World War I, responded to the exhibition by moving one of their meetings to Cologne.[66] "Who would miss this celebration?" the *C.V.* newspaper asked.[67] Ludwig Holländer, chairman of the *C.V.*, welcomed the 250 delegates and expressed his hope that the convention in Cologne would bear witness to "the intimate attachment of German Jews to their German *Heimat.*"[68]

The exhibition publicly displayed the importance of local historical heritage and prompted similar celebrations and commemorations of local Jewish heritage elsewhere.[69] The local remembrance of spaces thus appealed to a much wider audience. Berthold Rosenthal targeted his regional history published in 1927 at Jews in Baden, as well as to "those co-religionists who left Baden and found a new *Heimat* beyond the yellow and red border posts." He hoped that his history could provide a "welcome way" for both Baden Jews and those with no connections to Baden "to transport [themselves] back to the happy days of childhood and lovely pictures of the *Heimat.*"[70]

Similarly, in Bavaria, the *Bayerische Israelitische Gemeindezeitung* had featured a succession of articles on the history of Bavarian Jewry since 1926. When the annual conference of the Bavarian Landesverband (state association), which had been constituted in 1920, convened in 1928 in Bamberg, moreover, it occasioned a historical review to strengthen a sense of commonality among Bavarian communities.[71] From 1926 to 1932, Theodor Harburger took over a thousand photographs for the Verband Bayerischer Landgemeinden (Association of Bavarian Rural Communities) as more and more rural communities lost inhabitants to ongoing migration to larger urban centers and abandoned artifacts; some religious objects were even traded as old silver. Preserving the memories of Bavarian provincial Jewish historical spaces, if not necessarily its objects, extended beyond local patriotism. In the pages of the *C.V.-Zeitung*, Harburger reminded his readers of the importance of smaller locations to the history of Judaism and propagated the Bavarian provincial communities and their heritages as essential components of German Jews' *Heimat*.[72] As a Zionist, Harburger simultaneously advocated preserving remnants of Jewish culture regionally and nationally, as well as within Europe.[73]

Aware of the continued significance ascribed to smaller communities, Leo Baeck viewed Jews' urbanization as the cause behind their cultural and spiritual uprooting and disenfranchisement, which occasioned a shift from what he called *Milieufrömmigkeit* (spatial piety) to the problem of *Individualfrömmigkeit* (individual piety). Establishing this opposition, Baeck saw urbanization as a transition wherein Jews' lives went from being governed almost entirely by Judaism in small towns to being dominated by the non-Jewish world.[74] Sharing Baeck's preference for rural and small-town communities, Jacob Jacobson, who had become the new director of the Gesamtarchiv in 1920, appealed to German Jews to reconnect with their former places of residence. He invited others to follow him to the provinces to survey the historical landscape of several small communities, arguing that they were where Jewish cultures had been formed. In Jacobson's view, this legacy of Jewish spaces ensured the continuation of German Jewry.[75]

Nostalgia for Jewish spaces and their preservation and re-creation appears not in opposition to modernity, but co-constitutive thereof. In the nineteenth century, veneration of these spaces served to negotiate a path between cultural and religious change and adherence to the past. Preserving remnants of the past became a way to refashion continuities despite apparent ruptures. This development became even more marked as the pace of modernization accelerated among German Jews in the twentieth century. Nevertheless, Jewish communities continued to recall the legacies of regional historical sites, spaces, and histories. Travel increasingly provided a new arena within which Jewish spaces could be explored even for those who otherwise lived in new urban centers such as Munich, Frankfurt, Hamburg, and Berlin.[76]

Nils Roemer is the director of the Ackerman Center for Holocaust Studies and the Stan and Barbara Rabin Professor in Holocaust Studies at the University of Texas, Dallas.

Notes

1. Smith, "Gender and the Practices of Scientific History," 1166.
2. Kritzman, foreword to *Realms of Memory*, 28.
3. François and Schulze, *Deutsche Erinnerungsorte*; and Nora, *Les lieux de mémoire*. For succinct criticism, see Carrier, "Places, Politics," 37; and Englund, "Ghost," 303, 318.
4. Nora, "Between Memory and History."

5. For works that have explored the spatial dimension of Jewish cultures, see Fonrobert and Mann, *Space and Place in Jewish Studies*; Brauch, Lipphardt, and Nocke, *Jewish Topographies*; Fonrobert and Shemtov, "Introduction"; Roemer, *German City*; Roemer, "The City of Worms"; Eshel, "Between Cosmos and Makom"; Silberstein, *Mapping Jewish Identities*.

6. On the meaning of ruins, see the excellent article by Stoler, "Imperial Debris."

7. Heine, "Aufzeichnungen," in *Sämtliche Schriften*, 6:607–65, here 651.

8. Simmel, "The Ruin."

9. Benjamin, "Ruins," 180.

10. Assmann and Czaplicka, "Collective Memory," 130.

11. Cohen, "Nostalgia and the Return to the Ghetto," in *Jewish Icons*, 154–85.

12. Heine, "Der Rabbi von Bacherach. Ein Fragment," in *Sämtliche Schriften*, 1:461–501. Here I use the English translation in Heine, "The Rabbi of Bacharach," in *Jewish Stories*, 21–80, here 43–46.

13. Massey, *Space, Place and Gender*; Massey, "Places and Their Past."

14. Meyer, *Response to Modernity*, 111.

15. "Die Heiligkeit jüdischer Kirchhöfe," 268.

16. Zunz, *Zur Geschichte und Literatur*, 304–458.

17. Lewysohn, *Nefashot Zadikim*, n.p.; and Roemer, "Jewish Provinciality."

18. See, e.g., "Zur Geschichte der Juden in Danzig"; Perles, "Geschichte der Juden in Posen"; Saalschütz, "Zur Geschichte der Synaogen-Gemeinde in Königsberg"; Wiener, "Geschichte der Juden in der Residenzstadt Hannover"; and Wiener, "Geschichte der Juden in der Stadt und Diöcese Speyer."

19. Moritz Güdemann, for example, documented residence of Jews in Magdeburg going back to the year 965. See his "Zur Geschichte der Juden in Magdeburg," 281; and Schorsch, "Moritz Güdemann," 46.

20. "Cöln"; "Ein jüdisch-antiquarischer Fund in Cöln."

21. Eisemann and Kruskal, *Jacob Rosenheim*, 32, 34.

22. Burkhardt and Stern, "Aus der Zeitschriftenliteratur."

23. "Demolition of the Judengasse."

24. "Der letzte Rest"; and "Noch einmal die Judengasse."

25. "Der Frankfurter Ghetto," 41.

26. "Das Ende der Frankfurter am Main Judengasse." See also chapter 6 (by Michael Meng) in this volume.

27. "Jewish Synagogue at Berlin"; "Berlin."

28. Geiger, *Geschichte der Juden in Berlin*.

29. *Festpredigten zur Säcularfeier*; and Geiger, "Die jüdische Gemeinde Berlin."

30. Etzold, Kirchner, and Knoblauch, *Historische Friedhöfe*, 13–19; "Der Jüdische Friedhof in Berlin"; and "Ein Grabdenkmal in Berlin."

31. Etzold, Kirchner, and Knoblauch, *Historische Friedhöfe*, 10–12. See also "Ein Gang durch den alten Berliner jüdischen Friedhof."

32. Etzold, Kirchner, and Knoblauch, *Historische Friedhöfe*, 30–36.

33. Julius Rodenberg described these before their demolition in "Im Herzen von Berlin."

34. "Bericht über die Tätigkeit des Gesamtarchivs der deutschen Juden."

35. Ibid., 63.

36. Ibid., 56.

37. Krüger and Rahmlow, "Das Leben im Umfeld der Neuen Synagoge."

38. Stern, *Bibliothek der jüdischen Gemeinde zu Berlin*, 3.

39. On the attempt to represent a Jewish Berlin, see Schlör, "Bilder Berlins als 'Jüdische

Stadt'"; and Schlör, *Das Ich der Stadt*. On the Scheunenviertel in particular, see also chapter 11 (by Anne-Christin Saß) in this volume.

40. Roth, *Wandering Jews*, 68.
41. Ibid., 71.
42. Ibid.
43. Pinsker, "Spaces of Hebrew and Yiddish Modernism."
44. Hessel, "Spazieren in Berlin," in *Städte und Porträts*, 7–192.
45. Ibid., 139.
46. Ibid., 9.
47. Ibid.
48. Benjamin, "The Return of the Flaneur," 262.
49. Ibid., 264.
50. Hessel, "Spazieren in Berlin," in, *Städte und Porträts*, 57–58.
51. Ibid.
52. Ibid.
53. Ibid.
54. Döblin, *Briefe II*, 259.
55. Döblin, "Deutsches und Jüdisches Theater," in *Kleine Schriften I*, 362–67, here 365.
56. Ibid., 367.
57. Döblin, "Berliner Miniaturen," in *Kleine Schriften II*, 87–92, here 87–88. See also Döblin, "Östlich um den Alexanderplatz," in *Kleine Schriften II*, 298–302. It originally appeared in *Berliner Tageblatt*, 29 September 1923.
58. Döblin, *Schriften zu Leben und Werk*, 64.
59. Ibid., 65.
60. Caspary, *Jüdisches Jahrbuch*, 159. On the Kirschstein collection, see Rauschenberger, *Jüdische Tradition*, 132–37; and Lowenstein, "Jewish Residential Concentration," 490.
61. Caspary, *Jüdisches Jahrbuch*, 9.
62. Baeck, "Unsere Gemeinde."
63. "Großstadt oder Kleinstadt," 369.
64. Haude, *"Kaiseridee" oder "Schicksalsgemeinschaft,"* 121–31.
65. Schweriner, "Köln—ein Markstein"; and Alexander, "Von der Wesensart." A special issue of *Der Schild* from 3 September 1925 was dedicated to covering the exhibition.
66. "Der 8. Lehrerverbandstag in Köln"; and Pierson, "Embattled Veterans," 151.
67. "Auf nach Köln."
68. Holländer, "Der westdeutschen Tagung zum Gruss!"
69. Kober, "Die Geschichte der deutschen Juden," 23. See, e.g., Hallo, "Jüdische Kult-und Kunstdenkmäler"; "Das Jüdische Museum in Kassel."
70. Rosenthal, *Heimatgeschichte*, vi.
71. Eckstein, "Der diesjährigen Verbandstagung." See also the special issue of *Menorah* 6, nos. 11–12 (1928): 651–720, devoted to the history of Bavarian Jewry.
72. Harburger, "Unsere deutsche Heimat."
73. Annette Weber, "Ein Dokument jüdischer Heimatkunde aus Bayern. Zur Entstehung und Bedeutung der Fotosammlung Theodor Harburgers," in Central Archives for the History of the Jewish People, *Theodor Harburger*, 23–41; Bernhard Purin, "Theodor Harburger und das jüdische Museumswesen in Bayern," ibid., 51–63.
74. Baeck, "Gemeinde in der Großstadt," originally published as Baeck, "Gemeinde in der Großstadt," *Der Morgen* 5 (1930): 583–90. See also Heitmann, "Die Krisis der Religion in der Großstadt."

75. Jacobson, "Fahrt in die Provinz."
76. See Roemer, "Traveling, Writing, and Seeing"; Roemer, "London and the East End"; Roemer, "Jewish Traveling Cultures"; and Roemer, "The City of Worms."

Bibliography

Anonymous articles:
"Auf nach Köln." *C.V.-Zeitung,* 12 June 1925, 414.
"Bericht über die Tätigkeit des Gesamtarchivs der deutschen Juden." *Mitteilungen des Gesamtarchivs der deutschen Juden* 3 (1911): 55–84.
"Berlin." *Allgemeine Zeitung des Judentums,* 18 September 1866, 604.
"Cöln." *AZJ,* April 24, 1854, 206–7.
"Das Ende der Frankfurter am Main Judengasse." *AZJ,* 10 June 1884, 386.
"Das Jüdische Museum in Kassel." *Der Schild,* 12 September 1927, 285–86.
"Demolition of the Judengasse." *The Times,* 8 December 1884, 6e.
"Der 8. Lehrerverbandstag in Köln." *C.V.-Zeitung,* 12 June 1925, 416.
"Der Frankfurter Ghetto." In *Frankfurt am Main. Seine Geschichte, Sehenswürdigkeiten und Verkehrsmittel. Den Theilnehmern an der zu Frankfurt vom 11.–15. September 1881 stattfindenen Generalversammlung des Gesamt-Vereins der deutschen Geschichts- und Alterthums-Vereins überreicht,* 39–42. Frankfurt am Main: Baumbach, 1881.
"Der Jüdische Friedhof in Berlin." *AZJ,* 21 September 1880, 603–5.
"Der letzte Rest der Judengasse in Frankfurt a. M." *AZJ,* 31 March 1885, 230.
"Die Heiligkeit jüdischer Kirchhöfe." *Zur Judenfrage in Deutschland* 1 (1843): 266–72.
"Ein Gang durch den alten Berliner jüdischen Friedhof." *Ost und West* 16 (October/November 1916): 387–98.
"Ein Grabdenkmal in Berlin." *AZJ,* 8 March 1881, 160–61.
"Ein jüdisch-antiquarischer Fund in Cöln." *AZJ,* 4 September 1854, 460–61.
"Großstadt oder Kleinstadt." *CV- Zeitung,* 1 July 1927, 369–70.
"Jewish Synagogue at Berlin, Consecration of Monday." *The Times,* 10 September 1866, 8c.
"Noch einmal die Judengasse in Frankfurt a. M." *AZJ,* 3 November 1885, 726–27.
"Zur Geschichte der Juden in Danzig." *Monatsschrift für die Geschichte und Wissenschaft des Judenthums* 6 (1857): 205–14, 241–50, 321–31, 401–41.
Alexander, Kurt. "Von der Wesensart des rheinischen Juden." *C.V.-Zeitung,* 26 June 1925, 453.
Assmann, Jan, and John Czaplicka. "Collective Memory and Cultural Identity." *New German Critique* 65 (1995): 125–33.
Baeck, Leo. "Gemeinde in der Großstadt." In Baeck, *Wege im Judentum. Aufsätze und Reden,* edited by Werner Licharz, 218–25. Gütersloh: Gütersloher Verlagshaus.
———. "Unsere Gemeinde." *Gemeindeblatt der Jüdischen Gemeinde zu Berlin* 16 (3 December 1926): 249–50.
Benjamin, Walter. "The Return of the Flaneur." In *Selected Writings II: 1927–1930,* translated by Rodney Livingstone et al., edited by Michael W. Jennings, Howard Eiland, and Gary Smith, 262–67. Cambridge, MA: Harvard University Press, 1999.
———. "Ruins." In *The Work of Art in the Age of Its Technological Reproducibility, and Other Writings on Media,* 180–86. Cambridge, MA: Harvard University Press, 2008.
Brauch, Julia, Anna Lipphardt, and Alexandra Nocke, eds. *Jewish Topographies: Visions of Space, Traditions of Place.* Aldershot: Ashgate, 2008.

Burkhardt, C. A. H., and Moritz Stern. "Aus der Zeitschriftenliteratur zur Geschichte der Juden in Deutschland." *Zeitschrift für die Geschichte der Juden in Deutschland* 2 (1888): 1–46, 109–49.

Carrier, Peter. "Places, Politics, and the Archiving of Contemporary Memory in Pierre Nora's *Les lieux de mémoire.*" In *Memory and Methodology,* edited by Susannah Radstone, 37–57. Oxford: Berg, 2000.

Caspary, Eugen. *Jüdisches Jahrbuch für Groß-Berlin auf das Jahr 1926.* Berlin-Grunewald: Fritz Scherbel u. Teilh. Verlag, 1926.

Central Archives for the History of the Jewish People and Jüdisches Museum Franken, and Jüdisches Museum Franken, eds. *Theodor Harburger, Die Inventarisation jüdischer Kunst und Kulturdenkmäler in Bayern.* Fürth: Jüdisches Museum Franken, 1998.

Cohen, Richard I. *Jewish Icons: Art and Society in Modern Europe.* Berkeley: University of California Press, 1998.

Döblin, Alfred. *Briefe II.* Edited by Helmut F. Pfanner. Düsseldorf and Zurich: Walter-Verlag, 2001.

———. *Kleine Schriften I.* Edited by A. W. Riley. Olten and Freiburg i. Br.: Walter-Verlag, 1985.

———. *Kleine Schriften II.* Edited by A. W. Riley. Olten and Freiburg i. Br.: Walter-Verlag, 1990.

———. *Schriften zu Leben und Werk.* Edited by Erich Kleinschmidt. Olten and Freiburg, i. Br.: Walter-Verlag, 1986.

Eckstein, Adolf. "Der diesjährigen Verbandstagung zum Willkommen. Ein kurzer Gang durch die Vergangenheit der israelitischen Kultusgemeinde Bamberg." *Bayerische Israelitische Gemeindezeitung,* 15 June 1928, 161–65.

Eisemann, Heinrich, and Herbert N. Kruskal, eds. *Jacob Rosenheim. Erinnerungen, 1870–1920.* Frankfurt am Main: Kramer, 1970.

Englund, Steven. "The Ghost of Nation Past." *Journal of Modern History* 64 (1992): 299–320.

Eshel, Amir. "Between Cosmos and Makom: Inhabiting the World and Searching for the Sacred Space in Jewish Literature." *Jewish Social Studies* 9, no. 3 (2003): 121–38.

Etzold, Alfred, Peter Kirchner, and Heinz Knoblauch. *Historische Friedhöfe in der Deutschen Demokratischen Republik: Jüdische Friedhöfe in Berlin.* Berlin: Institut für Denkmalpflege, 1979.

Festpredigten zur Säcularfeier der jüdischen Gemeinde in Berlin am 10. September 1871 von den Rabbiner Joseph Aub und Dr. Abr. Geiger. Berlin, 1871.

Fonrobert, Charlotte E., and Barbara Mann. *Space and Place in Jewish Studies.* New Brunswick, NJ: Rutgers University Press, 2012.

Fonrobert, Charlotte E., and Vered Shemtov. "Introduction: Jewish Conceptions and Practices of Space." *Jewish Social Studies* 11, no. 3 (2005): 1–8.

François, Etienne, and Hagen Schulze, eds. *Deutsche Erinnerungsorte.* 3 vols. Munich: C. H. Beck, 2001.

Geiger, Abraham. "Die jüdische Gemeinde Berlin." *Jüdische Zeitschrift für Wissenschaft und Leben* 9 (1871): 241–55.

Geiger, Ludwig. *Geschichte der Juden in Berlin.* Berlin: J. Guttentag, 1871.

Güdemann, Moritz. "Zur Geschichte der Juden in Magdeburg." *Monatsschrift für Geschichte und Wissenschaft des Judentums* 14 (1865): 241–56, 281–96, 321–35, 361–70.

Hallo, Rudolf. "Jüdische Kult-und Kunstdenkmäler im Hessischen Landesmuseum Kassel als eine Bild der Geschichte der Juden in Hessen." *Der Morgen* 4 (1928): 3–26.

Harburger, Theo. "Unsere deutsche Heimat." *C. V.-Zeitung,* 10 February 1928, 75–76.

Haude, Rüdiger. *"Kaiseridee" oder "Schicksalsgemeinschaft"*: *Geschichtspolitik beim Projekt "Aachener Krönungsausstellung 1915" und bei der Jahrtausendausstellung Aachen 1925*. Aachen: Aachener Geschichtsverein, 2000.

Heine, Heinrich. *Jewish Stories and Hebrew Melodies*. Edited by E. Petuchowski. New York: M. Wiener, 1988.

———. *Sämtliche Schriften*. Edited by Klaus Briegleb. Munich: Deutscher Taschenbuchverlag, 1997.

Heitmann, Ludwig. "Die Krisis der Religion in der Großstadt." *Der Morgen* 5 (1929): 421–33.

Hessel, Franz. *Städte und Porträts: Sämtliche Werke 3*. Edited by Bernhard Echte. Oldenburg: Igel Verlag, 1999.

Holländer, Ludwig. "Der westdeutschen Tagung zum Gruss!" *C.V.-Zeitung*, 26 June 1925, 447.

Jacobson, Jacob. "Fahrt in die Provinz." *Menorah* 9 (1931): 232–44.

Kober, Adolf. "Die Geschichte der deutschen Juden." *ZGJD* 1 (1929): 13–23.

Kritzman, Lawrence D. Foreword to *Realms of Memory: Rethinking the French Past*, edited by Pierre Nora and Lawrence D. Kritzman, ix–xiv. New York: Columbia University Press, 1992.

Krüger, Maren, and Regina Rahmlow. "Das Leben im Umfeld der Neuen Synagoge: Jüdische Einrichtungen 1826-1943." In *Neue Synagoge Berlin–Centrum Judaicum. "Tuet Auf Die Pforten": Die Neue Synagoge 1866–1995*, edited by Hermann Simon, 165–217. Berlin: Centrum Judaicum, 1995.

Lewysohn, Ludwig. *Nefashot Zadikim: Sechzig Epitaphien von Grabsteinen des israelitischen Friedhofes zu Worms regressiv bis zum Jahr 905 übl. Zeitr. nebst biographischen Skizzen*. Frankfurt am Main: Baer, 1855.

Lowenstein, Steven M. "Jewish Residential Concentration in Post-Emancipation Germany." *Leo Baeck Institute Yearbook* 28 (1983): 471–95.

Massey, Doreen. "Places and Their Past." *History Workshop Journal* 39 (1995): 182–92.

———. *Space, Place and Gender*. Minneapolis: University of Minnesota Press, 1994.

Meyer, Michael. *Response to Modernity: A History of the Reform Movement in Judaism*. New York: Oxford University Press, 1988.

Nora, Pierre. "Between Memory and History: Les Lieux de Mémoire." *Representations* 26 (1989): 7–25.

Nora, Pierre, ed. *Les lieux de mémoire*. 7 vols. Paris: Gallimard, 1992–98.

Perles, Joseph. "Geschichte der Juden in Posen." *MGWJ* 13 (1864): 281–95, 321–34, 361–73, 449–61.

Pierson, Ruth. "Embattled Veterans—The Reichsbund jüdischer Frontsoldaten." *Leo Baeck Institute Yearbook* 19 (1974): 139–54.

Pinsker, Shachar. "Spaces of Hebrew and Yiddish Modernism: The Urban Cafés of Berlin." In *Transit und Transformation: Osteuropäisch-jüdische Migration in Berlin, 1918–1939*, edited by Verena Dohm and Gertrud Pickhard, 56–76. Göttingen: Wallstein, 2010.

Rauschenberger, Katharina. *Jüdische Tradition im Kaiserreich und in der Weimarer Republik. Zur Geschichte des jüdischen Museumswesens in Deutschland*. Hanover: Hahnsche Buchhandlung, 2002.

Rodenberg, Julius. "Im Herzen von Berlin." *Deutsche Rundschau* 49 (1886): 81–101, 221–49.

Roemer, Nils. "The City of Worms in Modern Jewish Traveling Cultures of Remembrance." *Jewish Social Studies* 11 (2005): 67–91.

———. *German City—Jewish Memories: The Story of Worms*. Hanover, NH: University Press of New England, 2010.

————. "Jewish Provinciality and Cultural Heritage: The Production of Worms as a Destination in Modern Jewish History." *Jewish Studies Quarterly* 12 (2005): 80–100.

————. "Jewish Traveling Cultures and the Competing Visions of Modernity." *Central European History* 42, no. 2 (2009): 429–49.

————. "London and the East End as Spectacles of Urban Tourism." *Jewish Quarterly Review* 99, no. 3 (2009): 416–34.

————. "Traveling, Writing, and Seeing: Encountering and Representing Jewish Cultures." In *Jews and Journeys: Travel and the Performance of Jewish Identity*, edited by O. Bashkin and J. Levinson. Philadelphia: University of Pennsylvania Press, 2016.

Rosenthal, Berthold. *Heimatgeschichte der badischen Juden seit ihrem geschichtlichen Auftreten bis zur Gegenwart.* Bühl-Baden: H. Bissinger, 1927.

Roth, Joseph. *Wandering Jews.* Translated by Michael Hoffman. London: Granta Books, 2001.

Saalschütz, Jos. L. "Zur Geschichte der Synaogen-Gemeinde in Königsberg." *MGWJ* 6 (1857): 437–49.

Schlör, Joachim. "Bilder Berlins als 'Jüdische Stadt': Ein Beitrag zur Wahrnehmungsgeschichte der deutschen Metropole." *Archiv für Sozialgeschichte* 37 (1997): 207–29.

————. *Das Ich der Stadt: Debatten über Judentum und Urbanität, 1822–1938.* Göttingen: Vandenhoeck & Ruprecht, 2005.

Schorsch, Ismar. "Moritz Güdemann, Rabbi, Historian and Apologist." *Leo Baeck Institute Year Book* 11 (1966): 42–66.

Schweriner, Arthur. "Köln – ein Markstein in der Geschichte des C. V. Glänzender Verlauf der Kundgebung und Verbandstagung." *C.V.-Zeitung*, 3 July 1925, 469.

Silberstein, Laurence J., ed. *Mapping Jewish Identities.* New York and London: New York University Press, 2000.

Simmel, Georg. "The Ruin." In *Essays on Sociology, Philosophy and Aesthetics,* edited by Kurt H. Wolff, 262. New York: Harper and Row, 1965.

Smith. Bonnie G. "Gender and the Practices of Scientific History: The Seminar and Archival Research in the Nineteenth Century." *American Historical Review* 100, no. 4 (1995): 1150–76.

Stern, Moritz. *Bibliothek der jüdischen Gemeinde zu Berlin. Bericht über die Begründung der Bibliothek und die drei ersten Jahre ihres Bestehens, 3. Februar 1902 bis 31. März 1905. Nebst einer Beilage: Benutzungsordnung.* Berlin: Wertheim, 1906.

Stoler, Ann Laura. "Imperial Debris: Reflections on Ruins and Ruination." *Cultural Anthropology* 23, no. 2 (2008): 191–219.

Wiener, Max. "Geschichte der Juden in der Residenzstadt Hannover." *MGWJ* 10 (1861): 121–36, 161–75, 241–58, 281–97.

————. "Geschichte der Juden in der Stadt und Diöcese Speyer." *MGWJ* 12 (1863): 161–77, 255–68, 297–310, 417–31, 454–66.

Zunz, Leopold. *Zur Geschichte und Literatur.* Berlin: Veit und Comp., 1845.

4

From Place to Race and Back Again

The Jewishness of Psychoanalysis Revisited

Anthony D. Kauders

Psychoanalysis is not Jewish, but the idea that it might be has occupied the minds of many for over one hundred years. There have been several reasons for this preoccupation. Sigmund Freud himself feared that the early psychoanalytic "movement," composed primarily of Jews, might be interpreted along ethnic lines, which he believed would seriously jeopardize its prospects as a science. This concern was unfounded, as Gentiles came to join his camp and as Jews more often than not rejected his findings. Later, Freud's German-speaking opponents increasingly denounced psychoanalysis as a Jewish invention that originated in a specifically Jewish psyche. This obsession hardly made an impact elsewhere, as laypeople and professionals alike continued to ponder his system of thought rather than agonize over its pedigree. In the second half of the twentieth century, historians, literary theorists, and psychoanalysts returned to the subject, offering three distinct explanations for the emergence of Freudianism: a sociological one, frequently involving psychological reasoning; an intellectual one, usually examining the scientific, religious, and ideological progenitors of psychoanalysis; and, more recently, a racial one, controversially positing an unconscious ethnic bond that made psychoanalysis (and Judaism as such) possible. Although it is difficult to establish a chronology of these approaches, the sociological explanation, with its reliance on place, appears to carry less and less weight today, often being replaced by accounts that embrace Freudian speculation about transgenerational (epigenetic) inheritance, a perspective that ignores place altogether. We have moved, in short, from territory (the Habsburg Empire, Vienna, or the B'nai B'rith chapter in the Austrian capital) to a Rosenzweig-like Jewishness that lies beyond this world.

In what follows I shall take up these points in three distinct ways. First, I would like to indicate how, in the case of Germany, the reception of psychoanalysis rarely revolved around Freud's Jewish background, touching on this question only in the final years of the Weimar Republic and thereafter. Next, I will discuss the different ways, culminating in a problematic dependence on racial categories, in which scholars have addressed the purported "Jewishness" of psychoanalysis. Finally, and most importantly, I will hold that "old-style" sociological analysis could benefit from categories culled from the so-called spatial turn. The place(s) that helped produce psychoanalysis, I will maintain, did so because of particular resonances that engendered estrangement as well as engagement. Michel Foucault's concept of "other spaces" or "heterotopias" will prove helpful in understanding the appearance of this special Jewish place.[1]

The Fear of a Jewish Science

Sigmund Freud was the first to comment on the fact that so many Jews had joined his "movement."[2] He wondered why this was the case,[3] whether their prevalence was an advantage or disadvantage,[4] how Jews and Christians differed "psychologically" and "racially,"[5] and what it would take to ensure the survival of his school of thought in a hostile environment.[6] In a later paper on "resistance" to psychoanalysis, Freud speculated that his Jewish descent may have contributed to the antipathy he and his colleagues had encountered, though he admitted there was little evidence to support this claim.[7] An even stronger assertion—that psychoanalysis was inherently Jewish—has since been made, first and most forcefully in the final years of the Weimar Republic by Gentile opponents of psychoanalysis, and subsequently as part of Jewish attempts to strengthen Diaspora Jewry in the English-speaking world. For the one group, psychoanalysis figured as an example of what was wrong with the Jews; for the other, it provided Jews with a means of self-assertion and identity formation in the absence of (religious) tradition.

Even so, psychoanalysis was *not* regarded as a Jewish science in the first quarter of the last century. The shift in perception in the early 1930s owed much to an anti-Semitism that regarded Jews as rationalistic and law-bound, both of which conflicted with a growing desire to adopt an "emotional" and "holistic" stance in the sciences as well as in public life at large. In all of these discussions, however, place was never the real issue. Psychoanalysis may occasionally have been identified with Vienna, but Berlin soon emerged as a rival center, attracting aspiring analysts who believed that Germany's capital was more appealing politically and intellectually.

Freud never claimed that the science he had founded derived from ethnic particularity. While he would regularly comment on a shared "racial kinship" (*Rasseverwandtschaft*) with such analysts as Karl Abraham and Sándor Ferenczi, he nevertheless insisted that there was no "distinct Aryan or Jewish science. Their results should be identical; only their presentation may vary."[8] Either way, it would have been unusual, to say the least, for someone like Freud to assert such a difference. Not only had he studied with some of the most eminent neurologists and psychiatrists of the time, all of whom insisted on the universality and objectivity of science, but his ideas were also heavily indebted to sources—from Darwinism to Mesmerism—not commonly associated with Judaism or Jewishness. While historians of science have sought to trace these influences,[9] recently Jewish scholars have begun to challenge Freud's own position on the matter.[10] Let me first turn to his other intuition, however, namely, that the outside world repudiated psychoanalysis on the grounds that it was a Jewish affair.

If we look at responses to psychoanalysis in the first twenty-five years after the publication of *The Interpretation of Dreams* in 1900, nothing could be further from the truth, even among the prominent group of the new science's detractors. From 1900 to the mid-1920s, Freud's opponents rejected psychoanalysis on account of its purported "pansexualism" and lack of objectivity. Members of the scientific establishment voiced these criticisms with particular fervor. According to the elite of German-speaking psychiatrists and psychologists, Freud's modus operandi included so many objectionable elements that no serious scientist could ever utilize psychoanalytic methods. Psychoanalysis had discarded scientific methods that could be repeated and tested and universally applied. The critique exercised by the psychiatric and psychological establishment confronted psychoanalysis from what was deemed an objective, value-free vantage point, an ideal that culminated in the works of Gottlob Frege, Henri Poincaré, and Max Planck, for whom "science worthy of the name must be communicable to all" and capable of transcending "the privacy and individuality of representations and institutions."[11] Some of the same scholars who questioned Freud's methodology were incensed by the value he placed on sexuality in relation to both human development and the origins of neurosis.

As was the case for those who championed certain psychoanalytic ideas, the scholars who dismissed all or most of these did not belong to a recognizable ethnic or political community. In fact, if we were to retrospectively imagine such a community, we might mention that Freud's most influential critics during this period were Jews: the Breslau psychologist William Stern, the Berlin neurologist Hermann Oppenheim, the Heidelberg sexologist Arthur Kronfeld, and the Munich psychiatrist Max Isserlin. It would be difficult to find non-Jewish adversaries who were equally important at the time. Yet unless we wish to argue that these men (together with similarly disposed Jewish scientists) attacked Freud in order to disguise their own ethnic background or join a scientific culture dominated by non-Jews (for which there is no evidence), it might be best to conclude that well-wishers and foes alike did not classify psychoanalysis as Jewish.

This began to change toward the end of the Weimar Republic. Now Freud was no longer the failed psychologist who mistook hermeneutics for hard science or, even worse, the failed bourgeois who foolishly exaggerated the importance of sexuality. Instead, psychoanalysis came to be seen as a psychology that placed too much emphasis on reason at the expense of the unconscious. Initially, this critique was not anti-Semitic. Many Jews, among them Arthur Kronfeld, the theologian Max Wiener, and the economist Eduard Heimann, adopted a language that accentuated irrationalism over rationalism, community over society, and spiritualism over mechanization. In doing so, they were joining an ever-growing chorus of thinkers who challenged what they perceived to be the rampant "scientistic ideal of positivism," which would lead to a "cold" and "narrow" religion of reason.[12] In the early 1930s, this critique was transformed from a general call for emotionality (*Leidenschaftlichkeit*) against the domination of sober intellectualism into a struggle directed against the supposedly "Jewish" spirit of rationality. For the first time, the debate on psychoanalysis centered not on methodology, sexuality, or even the role of the unconscious, but on the belief that its "Jewishness" invariably begot a "one-sided" rationalism.[13] As much as Freud's earlier hunch that anti-Semitism was to blame for the hostility against psychoanalysis had been exaggerated, many Germans did come to believe that his ideas were the products of a Jewish psyche.

Spreading the Good News

One might imagine that this construction of psychoanalysis as "Jewish" would have been much less appealing after the Holocaust. And for the most part, it was. Historians, psychoanalysts, and literary theorists tended toward examining the social and intellectual origins of Freudianism, including spatial aspects that had contributed to the formation of a psychoanalytic community. More recently, however, these aspects have been ignored altogether, giving way to readings that come dangerously close to the anti-Semitic version mentioned above.

The sociological approach, regularly relying on psychological explanations, has been the most sensitive to place. It would be difficult to overstate the impact of Carl Schorske's *Fin-de-Siècle Vienna* in this respect.[14] His argument that the defeat of Austrian political liberalism (as embodied by the bourgeoisie) engendered apolitical alternatives to liberalism, from aesthetics to psychoanalysis, has prompted specialists to examine the social preconditions for the rise of the Freudian school. Steven Beller, to take one prominent example, noted that the city's bourgeoisie was much more diverse than *Fin-de-Siècle Vienna* made it out to be, the "core of the old Liberal intelligentsia, the officials and teachers" having actually voted political liberalism out of office. It was rather other members of the bourgeoisie, foremost among them Jews in journalism, medicine, and law, for whom Christian Socialism proved to be a real menace. Beller calculated that up to a third of grammar school (*Gymnasium*) graduates in Vienna were of Jewish descent and that most of these hailed from the Bohemian crown lands (Bohemia, Moravia, and Silesia). Sigmund Freud, together with Gustav Mahler and Karl Kraus, belonged to this group of students. The fact that Freud chose to do medicine was also typical, as 78 percent of *Maturanten* (holders of higher education entrance qualifications) with a "liberal bourgeois" background who attended medical school were Jewish, as were a "staggering 93 per cent ... of the merchants' sons in the *Gymnasium* sample who chose medicine." Beller concluded that Freud "could hardly have been anything else but Jewish, in the Viennese context."[15] But why were there so many Jewish *Gymnasium* pupils in the first place? Beller, perhaps not surprisingly, suggested that the Jewish religious tradition, albeit under the sway of secularism, continued to value education highly.[16]

Robert Wistrich's interpretation sounds no less familiar and is compatible with Beller's findings. Focusing more on the founder of psychoanalysis than on secondary-school leavers or medical students, Wistrich drew heavily on Freud's recollections to hypothesize a link between marginal social status on the one hand and innovative science on the other. The experience of anti-Semitism at university and elsewhere, we may recall, had led Freud into opposition against the so-called "compact majority," a move that he claimed had stimulated a "certain independence of mind."[17] Wistrich accepted this reading, only to add much more generally that exclusion tends to breed nonconformism and creativity. The mind of the secular Jew, furthermore, was "uncluttered by the vestiges of dogma and superstition, by the crushing burden of theological beliefs ingrown in Christian civilization."[18]

Both Beller and Wistrich assumed that, at least in the German-speaking context, Jews anticipated acculturation via education or *Bildung*. Others have expanded on this position, showing how Freud validated psychoanalytic knowledge by allying it with classical tradition (think of the association with Greek tragedy) as well as with philological methods of *Grammatik* and *Kritik*;[19] drawing parallels between the humanist concept of self-cultivation through *Bildung* and the ideal of character development inherent in psychoanalysis;[20] or, in a similar vein, comparing what took place in psychoanalytic therapy with either a *Bildungsroman* or Hegel's *Phenomenology of Spirit*.[21] Whatever the approach, however, Vienna, Austria, and German culture were found to have served as the background to the evolution of psychoanalysis, sometimes concretely (as in Beller's *Gymnasien*), at other times more broadly (as in references to a concept, *Bildung*, unique to Central Europe).

While the above works emphasized the social and geographical bases of psychoanalysis, other contributions have moved in the opposite direction, hypothesizing an intellectual or spiritual affinity between Jewishness and psychoanalysis. The best-known example here is David Bakan's *Sigmund Freud and the Jewish Mystical Tradition*. The book set out to uncover the esoteric origins of psychoanalysis, an objective that predictably required a rather heroic effort in reading between the lines, in putting forward connections no one else had noticed, and in convincing psychoanalysts that they were in fact practicing Judaism in disguise. Bakan was aware of these difficulties. It is for this reason that on at least two occasions, he resorted to place as the source of Freud's *Jewish* thought and action, albeit superficially and without any real interest in place as a concept at all. In the first instance, he established a causal nexus between the Galician homeland of Freud's parents, "a region whose atmosphere was saturated with Chassidism," and the "oral tradition" of psychoanalytic practice.[22] In the second, he invoked Leo Strauss, who had famously argued that authors contending with censorship on political or social grounds were forced to conceal their true intentions.[23] Freud belonged to these writers, Bakan argued, because "anti-Semitism … was so widespread and so intense at the time that the Jewish sources of his ideas would have dangerously exposed an intrinsically controversial theory to an unnecessary and possibly fatal opposition."[24] In other words, Vienna and the Austro-Hungarian Empire caused Freud to obscure the Jewish foundations of his enterprise.

All this remained conjecture, however, due to Bakan's inability to detail the exact transmission of one tradition to the next or fully disclose the secret ambitions of Freud the renegade Jew. And in any case, the author was much more concerned with textual analysis than with any kind of social history, explaining Freud's fantastic tale in which Moses is transformed into a Gentile with the no less fanciful fabrication of Freud as a modern-day Sabbatai Zevi bent on destroying the "restrictive hold of rabbinism on the Jews and on himself."[25] Bakan inspired a number of followers, who also focused almost exclusively on Jewish sources, as if Freud had never envisaged psychoanalysis to resemble a psychology of human nature, within the wider context of international medicine, neurology, and psychiatry.[26] Instead, these interpretations likened free association not to classical philology but to midrashic inquiry, with its interpretation of gaps, contradictions, repetitions, and anomalies.[27] Or the "importunate 'Yid,' released from the ghetto and *shtetl*," became the model for "Freud's coarse, importunate 'id'" that has to face the Western superego.[28] Or, even more nebulously, Judaism and Freudianism were deemed essentially the same, because both "defy all codification."[29] A notion consistent with this sort of thinking is the belief that analysts and modern Jews continually and restlessly examine the world.[30]

If these works insisted that psychoanalysis was a fundamentally Jewish science, irrespective of its geographical origins and international dissemination, another group of scholars used Freud's life and work to reflect on modern Jewish identity as such.[31] Yosef Yerushalmi must be singled out here, given the wide reception of his *Freud's Moses*.[32] Yerushalmi's main question revolved around the survival of post-emancipation Jewry, an issue he had already investigated in a previous work about Jewish historiography.[33] Devoid of all "but the most vestigial content," "psychological Jewishness" amounted to mere subjectivity, he remarked.[34] In other words, unless Jews continued to study classical texts and perform time-honored rituals, they would soon vanish. The New York–based historian therefore renounced Freud's proposition, most famously made in *Moses and Monotheism*, that Jewish identity could be attributed to unconscious transmission. "The basic modalities in the continuity of a religious tradition," Yerushalmi wrote, "are precept and example, narrative, gesture, ritual," all of which could not do without the active, conscious, self-aware involvement of human beings.[35]

Yerushalmi's main critic returned to Freud's text in order to spread the good news that Jewish identity was in fact communicable on an unconscious level, thereby ensuring the survival of the Jews without recourse to normative Judaism.

Eliza Slavet's plea for a "racial thinking" beyond racism is not only the latest attempt at ascertaining a relationship between psychoanalysis and Jewishness, it is also the work least mindful of place. At the same time, its allusion to inherent psychic characteristics comes very close to the anti-Semitic tracts of the 1930s.

Slavet's *Racial Fever* is an "attempt to explore race as a concept beyond the realm of physical variation and to consider racial thinking without reducing it to racism."[36] But Slavet was not principally interested in race and its permutations. Rather, she wanted to provide a definition of Jewishness that avoids the idea "that there are some people who are more Jewish than others" and "that there are certain individuals … who determine who is truly Jewish."[37] A routine attack on normative Judaism would not require an extended critique of the historian Yerushalmi, however, whose pessimistic outlook and skepticism toward "psychological" Jews paled in comparison to the much stronger reprobation emanating from within Orthodox Judaism itself. Slavet's strong sympathy for Freud's final book (whose most famous commentator required a commentary in turn) and equally strong distaste for any form of "essentialism" (which Freud promised to overcome) compelled her to describe Yerushalmi's *Freud's Moses* as a false prescription for Jewish flourishing.

But what made *Der Mann Moses* so appealing? The attraction for Slavet, I would argue, lies in the possibility of exchanging "bad" essentialism (religious laws, ethnic markers, authoritative strictures) for "good" essentialism (memory, history, trauma). Or, as she put it, "Rather than understanding difference as only a matter of external characteristics, or historical vagaries, Freud's emphasis on the inheritance of archaic memory suggests a model of understanding difference as something that is both internal to one's self or one's own community and as something beyond individual predilections and preferences."[38] Freud maintained that certain events in the distant past were so traumatic that the memories of these shocks were inherited from generation to generation. Rejecting heredity or history as the only path to (Jewish) survival, Freud, in Slavet's reading, amalgamated the two, mixing the biological and permanent with the psychological and cultural. The Jews, in this view, transmitted their archaic memories to future generations and did so both consciously and unconsciously. Although rituals, gestures, and texts provided one means of communication, they were not enough to guarantee the persistence of the Jewish people. Freud needed biology to explain the compulsion to be Jewish.

When Freud applied his notion of the "return of the repressed" to the Jewish people, Slavet's reading proposes, he was asserting that the Jewish people would survive "despite all reforms, rejections, repudiations, and repressions."[39] Neither grounded in authoritative texts nor predetermined along ethnic lines, Jewish identity therefore became the "very *process* of transmitting, awakening, and responding to the memory-traces of Moses."[40] Or, put differently, unconscious remembering was the condition for being Jewish, now and for all time. Slavet compared Freud's conception of the collective unconscious favorably to Jung's version of the theory, because although Jung searched for general symbols, archetypes, and structures common to humanity, Freud remained fixed to specific memories and historical events.[41]

This preference for Freud is not very convincing, however. It does not take into account that Jung in fact distinguished between a collective unconscious available to humankind and various forms of collective unconscious exclusive to nations or groups.[42] Freud may have been more specific about certain "events" (however mythic in character they happened to be), but Jung's alternative in the shape of stories and legends, such as that of Faust, or long-term historical experience, such as life in the Diaspora, was no less "concrete."[43] As a matter of fact, Jung was potentially less "essentialist" than Freud: the layers of the unconscious, the Swiss psychologist explained, were not passed on from one generation to the next; rather, they resembled the Kantian *a priori* in that they allowed for the *possibility* of coming into contact with previous images or perceptions.[44] The emotional investments of an earlier age would then reappear as specific archetypes in present-day dreams and fantasies.[45] More seriously still, Freud's Lamarckianism jeopardizes Slavet's project. As she herself admits, the belief that the Jewish people will survive no matter what lends credence to the idea that its identity is fixed once and for all.[46] For an anti-essentialist, this inescapability of Jewishness is hardly a happy prospect, and its proximity to racism, albeit in a rarefied, psychological form, cannot be played down. The parallel with Jung's later psychological juxtaposition of Germans and Jews is particularly glaring, for in subscribing to the (Freudian) view that Jewish continuity is the result of historical experience extended via unconscious memory, Slavet was not very far from agreeing with the (Jungian) view that in their historical antiquity (and consequent rationalism) Jews differ psychologically from Germans.[47]

Racial Fever is the latest attempt at defining Jewishness in the Diaspora. Yet by invoking Freud as a modern Jewish hero whose *Moses and Monotheism* offers an escape from "bad essentialism," Slavet has ended up racializing the Jewish psyche in a Jungian fashion. The good news that Freud's work can be used to embrace modern Jewish identity in a postcolonial, decentered, and anti-essentialistic fashion comes at a price: not only must we rely on dubious categories that rule out conversion or "going native," but in the end psychoanalysis, Judaism, and the Jews themselves are left without a particular place in this world.

The Return to Place

The closest work we have on the impact of place in the advance of psychoanalysis is Dennis Klein's seminal study *Jewish Origins of the Psychoanalytic Movement*. Unlike scholars who have emphasized hidden motivations (like Bakan) or unconscious connections (like Slavet), Klein was interested in the microsociology of the early movement over and above its religious content. The Jewishness he considered pertains to the fact that from 1902 to 1906, all seventeen members of the Psychological Wednesday Society (the predecessor of the Vienna Psychoanalytic

Society) had a Jewish background.[48] Klein rightly assumed that this detail is in need of an explanation. While Beller and Wistrich also examined the social history of Freudianism, Klein's was less a history of the Viennese Jewish community than of the peculiarity of psychoanalysis in the imperial city.

Klein's discussion of the local B'nai B'rith chapter is especially worth mentioning. Freud joined the lodge in September 1897. It was here that he found his first audience, which greeted, debated, and discussed psychoanalytic theory on a number of occasions. According to Klein, the Jewish society emerged as an "active intellectual forum for his metapsychological views during the productive five-year period 1897–1902," so much so that it became the "precursor of the movement of psychoanalysis."[49] In Klein's view, the association between Jews and psychoanalysis was not to be found in religious tradition or unconscious sympathy, but rather in the proximity of Jews dedicated to shared values and joining one another in a relatively closed setting: "Freud's accelerated recruitment, lecture, and leadership activity in the B'nai B'rith in 1901–2 points to the centrality of his development of a psychoanalytic school. By exchanging ideas with, and seeking advice from, fellow Jews, Freud promoted the view, commonly held in the brotherhood, that Jews played a primary role in affirming universal and humanitarian values."[50] Oskar Rie and Eduard Hitschmann, both members of the lodge, soon joined the psychoanalytic movement.

Concentrating on the B'nai B'rith or the Wednesday Society allows us to observe the combination of *Bildung* and minority status (or Enlightenment thinking and Jewish loyalty) at close hand, the importance of which other scholars have addressed more broadly.[51] What I will seek to do here, however, is to think about the emotions of place as they affected the special case of Jewish psychoanalysts. If poverty, homosexuality, and childhood, among others, are *felt* differently depending on the city or region in which they are experienced,[52] then the emergence of psychoanalysis can also be examined as an emotional response to constructions of place. When sixteen ethnic Jews joined Freud in Berggasse 19 from 1902 to 1906, they created a kind of heterotopia born out of estrangement and driven by engagement. Unlike the German Jews David Sorkin portrayed in his classic study, these men did not, in holding on to humanism and *Bildung,* create a Jewish subculture of which they themselves were unaware.[53] Likewise, their place was not the outcome of a pre-reflective habitus they had come to embody irrespective of their own objectives. Rather, they came together self-consciously (as Jews) to combine the ideals of humanism and science in a location that allowed this to happen without at the same time isolating them from the rest of the world.

Michel Foucault's concept of heterotopia is helpful here.[54] Though the French philosopher provided conflicting ideas in his short lecture on the subject, several points are worth noting in regard to our topic. First, heterotopias begin to "function at full capacity when men arrive at a sort of absolute break with their traditional time."[55] Second, they are related to other sites in that they "suspect, neutralize, or invent the set of relations they happen to designate, mirror, or reflect."[56] And third, they can lead to alternative or compensatory spaces that are "as perfect, as meticulous, as well arranged as ours is messy, ill constructed, and jumbled."[57]

Let me briefly address these points. While the Jews who met in Vienna's Berggasse 19 did not form a heterotopia that resembled or emerged out of a radical break in time, there was a real sense that a transition (from progress through *Bildung* within a liberal setting to the end of liberalism within an increasingly anti-Semitic environment) had occurred.[58] To be sure, the street belonged to the Alsergrund district already inhabited by a disproportionate number of middle-class Jews, for whom Jewishness rather than class was the "major criterion" by which they "selected a neighborhood."[59] Still, outrage over Jew hatred and other forms of injustice clearly influenced early members of the psychoanalytic movement, who were "disoriented and perplexed."[60] According to Louis Rose, "Freud set himself the task of molding his moralists into psychoanalysts," thus transforming "awareness gained from moral indignation into a new scientific consciousness."[61] Members of the Psychological Wednesday Society came together, in other words, to herald a new beginning—one that would allow Jews estranged from the medical establishment in particular and Austria in general to imagine change within the walls of a safe haven.[62] It is conceivable, furthermore, that the *ideal* of a psychoanalytic setting in which patient and analyst entered into a contractual relationship that guaranteed "the freedom of the patient from coercion by relatives and loved ones upon whom he or she is dependent"[63] owed something to the situation in Berggasse 19, where Jews assembled to escape the dependencies of Viennese society.[64] While hypnotherapists who reproduced experimental laboratories by eliminating all unpleasant sensual effects had heavily influenced the advancement of the notion of a specifically psychoanalytical site, the ideal of a protected space possibly restaged the experience of the early psychoanalytic community.[65]

This did not mean, however, that members of the Wednesday Society isolated themselves completely from the rest of society. Instead, the place they occupied "represented, contested, and inverted" other places.[66] Put differently, the early adherents of the psychoanalytic cause were entering a third space, in which they did not choose between one (humanism, science, universalism) or another (Jewishness, ethnicity, particularity), but came to challenge such choices altogether. Homi Bhabha's concept of third space describes this heterotopia quite well. Although developed to move beyond the binaries characteristic of (post)colonial thought, Bhabha's term can also be applied to the Wednesday Society (and the later Vienna Psychoanalytic Society): a space that "enables positions to emerge" by displacing the "histories that constitute it" and setting up "new structures of authority."[67] Older histories that underpinned psychoanalysis as psychotherapy and science (such as hypnosis, evolutionary theory and romantic psychology[68]) were combined with older histories that underpinned psychoanalysis as self-development (*Bildung,* morality, self-reflection) to produce something radically new. It is no accident, I would argue, that psychoanalysis matured into a hybrid,

Janus-faced enterprise that merged natural science (objectivity) with hermeneutics (subjectivity)—and that it was bitterly attacked for this very reason.

Yet what initially amounted to a gathering of like-minded Jewish men mutated into an alternative or compensatory space that would soon define an alternative order or, in the words of Robert Topinka, a site of "reordering."[69] The patterns of face-to-face interaction, network formation, and collective action initiated in Berggasse 19 generated a heterotopia with its own system of thought, hierarchies, practices, and rules. This is not to say that the product itself was Jewish. After all, the men who congregated there did so because of the alienation they were experiencing as Jews and Austrians, not in order to initiate a "Jewish" science. Predictably, the process of systematization led to the eventual demise of this Jewish heterotopia, as the new psychology attracted men and women from all walks of life. In Berlin, where in the 1920s psychoanalysis would become a profession rather than a calling,[70] the ethnic element was already less pertinent. Although the biographical data on Freudians within the German Psychoanalytic Society (of which most belonged to the Berlin chapter) is inconclusive, at least nine (and possibly twelve) of its twenty ordinary and extraordinary members in 1921 were not Jewish, with similar figures for subsequent years.[71] Psychoanalytical organizations elsewhere, especially but not exclusively in the English-speaking world, testify to the fact that psychoanalysis was transformed from an early Jewish heterotopia to a universal science whose appeal transcended ethnic boundaries.

These developments allow us, finally, to address what John Forrester has called the "historiographic tension between the need to find an explanation for the emergence of psychoanalysis in Vienna (rather than somewhere else) and the need for a historical characterization of psychoanalysis that will also explain its rapid dissemination and acceptance in Western culture in the early years of this century."[72] Once the outgrowth of a heterotopic Jewish place was standardized so as to be applicable anywhere (protests by anthropologists notwithstanding), Jewish Vienna no longer mattered. At the same time, however, place continued to make a difference, as the converging psychoanalytic traditions in London, Paris, and Buenos Aires show. Place, in sum, trumped race—and fortunately so.

Anthony D. Kauders is Reader in Modern European History at Keele University, England. He has published widely in the field of German-Jewish history, including *Democratization and the Jews* (2004), *Unmögliche Heimat* (2007), and *Der Freud-Komplex* (2014). Currently, he is researching the history of hypnosis in Germany.

Notes

1. As Jerram has noted, "space" and "place" have been used interchangeably, producing a "chaotic vocabulary" that does not distinguish between locations ("concrete places") and the "cultures, practices, values, and symbols that cluster there." Jerram himself suggests

that "space is material, location is relational or positional, place is meaningful." In this chapter I refer to the latter two. Jerram, "Space: A Useless Category," 404, 406.

2. To most of its adherents, early psychoanalysis felt like a movement. After World War I, the movement increasingly became a profession, with its own codified standards, rules, and regulations. Mühlleitner and Reichmayr, "Die Freudianer in Wien," 1052, 1062.

3. Freud to Karl Abraham, 20.7.1908, Freud and Abraham, *Briefe*, 57.

4. Freud to Abraham, 23.7.1908, in Freud and Abraham, *Briefe*, 57.

5. Freud to Abraham, 3.5.1908 and 11.10.1908, in Freud and Abraham, *Briefe*, 47, 64. Jones, *Life and Work*, 2:149, 153; Carotenuto, *A Secret Symmetry*, 116–21; Freud, "Vorrede," 569.

6. Freud to Abraham, 26.12.1908, in Freud and Abraham, *Briefe*, 75.

7. Freud, "Die Widerstände," 110.

8. Geller, *On Freud's Jewish Body*, 4.

9. See Ellenberger, *Die Entdeckung des Unbewussten*; Sulloway, *Freud: Biologist of the Mind*; McGrath, *Freud's Discovery*; Frankland, *Freud's Literary Culture*; Ffytche, *Foundation*.

10. See the next section.

11. Daston and Galison, *Objectivity*, 254, 273.

12. Kauders, *Freud-Komplex*, chap. 2; Kauders, "Weimar Jewry."

13. Kauders, *Freud-Komplex*, chap. 3.

14. Schorske, *Fin-de-Siècle Vienna*.

15. Beller, "Class," 57. See also Beller, *Vienna and the Jews*.

16. Beller, "Class," 55.

17. Freud, "Selbstdarstellung," 34–35.

18. Wistrich, *Jews of Vienna*, 564.

19. Winter, *Freud and the Institution of Psychoanalytic Language*.

20. Rose, *Freudian Calling*.

21. Kirschner, *Religious and Romantic Origins*, 196.

22. Bakan, *Sigmund Freud*, 37.

23. Strauss, *Persecution and the Art of Writing*.

24. Bakan, *Sigmund Freud*, 26.

25. Ibid., 166.

26. See also Brunner, "The (Ir)Relevance."

27. Meghnagi, "A Cultural Event," 63.

28. Cuddihy, *Ordeal of Civility*, 18–19.

29. Meghnagi, "A Cultural Event," 63–64.

30. Frosh, *Hate and the Jewish Science*, 31.

31. See, e.g., Santner, *On the Psychotheology*; Zaretsky, "Place of Psychoanalysis"; Gresser, *Dual Allegiance*.

32. See esp. Derrida, *Archive Fever*; Bernstein, *Freud and the Legacy of Moses*; Assmann, *Moses der Ägypter*.

33. Yerushalmi, *Zakhor*.

34. Yerushalmi, *Freud's Moses*, 10.

35. Ibid., 89.

36. Slavet, *Racial Fever*, 2.

37. Ibid., 5–6.

38. Ibid., 18.

39. Ibid., 191.

40. Ibid., 158.

41. Ibid., 64.
42. Kauders, *Freud-Komplex,* chap. 3.
43. Jung, *Wandlungen und Symbole,* 49.
44. Jung, "Über das Unbewusste," 23.
45. Ibid., 136, 144.
46. Slavet, *Racial Fever,* 191.
47. Jung, "Zur gegenwärtigen Lage der Psychotherapie."
48. Klein, *Jewish Origins,* xi. Mühlleitner and Reichmayr contest that all seventeen "were Jews" by pointing out that two of the original members, Hugo Heller and Edwin Hollerung, were Protestants whose parents had "probably" converted from Judaism; Mühlleitner and Reichmayr, *Freudianer,* 1080. However, Heller was born a Jew and converted to Protestantism in 1899. See Christine Dierks, "Hugo Heller – Chronologie," psyalpha: Wissenplattform für Psychoanalyse, http://www.psyalpha.net/biografien/hugo-heller/hugo-heller-chronologie (accessed 17 January 2017). Either way, the Viennese situation would have suggested to everyone involved that converts were often looked upon as "not quite belonging," and Heller and Hollerung would have been aware of the one-sided ethnic composition of the Psychological Wednesday Society.
49. Klein, *Jewish Origins,* 24.
50. Ibid., 93.
51. Bernstein, *Freud and the Legacy of Moses;* Winter, *Freud and the Institution of Psychoanalytic Language;* Rose, *Freudian Calling;* Kirschner, *Religious and Romantic Origins;* Gresser, *Dual Allegiance.*
52. Löw, *Soziologie der Städte,* 18.
53. Sorkin, *Transformation of German Jewry,* 173–78.
54. I refer to "Of Other Spaces: Utopias and Heterotopias" (based on a lecture given by Foucault in March 1967), as available online at http://web.mit.edu/allanmc/www/foucault1.pdf (accessed 10 July 2014), 1–9.
55. Foucault, "Spaces," 6.
56. Ibid., 3.
57. Ibid., 8.
58. On anti-Semitism's impact at this time, see Boyer, *Political Radicalism,* and Rechter, *Jews of Vienna,* 8.
59. Rozenblit, *Jews of Vienna,* 94.
60. Rose, *Freudian Calling,* 35.
61. Ibid., 26, 46.
62. On this alienation from the medical elite, see Gay, *Freud: A Life for Our Time,* 603.
63. Forrester, *Dispatches from the Freud Wars,* 85.
64. The situation would not improve after the war. During the late 1920s, there were two divisions in different wings of the medical school's anatomy building, one attended by Jews and Socialists and the other by Christian Socialists and anti-Semites. See Silverman, *Becoming Austrians,* 21.
65. Mayer, "Zur Genealogie des psychoanalytischen Settings," 23, 26; and Mayer, *Mikroskopie der Psyche,* 168–69.
66. Foucault, "Spaces," 3.
67. Rutherford, "Third Space," 211.
68. Mayer, *Mikroskopie der Psyche;* Sulloway, *Freud: Biologist of the Mind;* Ffytche, *Foundation.*
69. Topinka, "Foucault, Borges, Heterotopia," 56.
70. Schröter, "Zur Frühgeschichte der Laienanalyse," 1156, 1168–69.

71. See the *Korrespondenzblatt der Internationalen Psychoanalytischen Vereinigung 1910–1941*, http://www.psyalpha.net/files/docs/izp_korrespondenzblatt_1910-1941._michael_gie fer_2007_1.pdf (accessed 17 Janauary 2017).
72. Forrester, *Dispatches from the Freud Wars*, 190.

Bibliography

Assmann, Jan. *Moses der Ägypter. Entzifferung einer Gedächtnisspur.* Munich and Vienna: Hanser, 1998.

Bakan, David. *Sigmund Freud and the Jewish Mystical Tradition.* Mineola, NY: Dover, 2004.

Beller, Steven. "Class, Culture and the Jews of Vienna." In *Jews, Antisemitism and Culture in Vienna,* edited by I. Oxaal, M. Pollak, and G. Botz, 39–58. New York and London: Routledge, 1987.

———. *Vienna and the Jews, 1867–1938: A Cultural History.* Cambridge: Cambridge University Press, 1989.

Bernstein, Richard J. *Freud and the Legacy of Moses.* Cambridge: Cambridge University Press, 1998.

Boyer, John W. *Political Radicalism in Late Imperial Vienna: Origins of the Christian Social Movement 1848–1897.* Chicago and London: University of Chicago Press, 1981.

Brunner, Jerome. "The (Ir)Relevance of Freud's Jewish Identity to the Origins of Psychoanalysis." *Psychoanalysis and Contemporary Thought* 14 (1991): 655–84.

Carotenuto, Aldo. *A Secret Symmetry: Sabina Spielrein between Jung and Freud.* New York: Pantheon, 1982.

Cuddihy, J. Murray. *The Ordeal of Civility: Freud, Marx, Lévi-Strauss and the Jewish Struggle with Modernity.* New York: Beacon Press, 1974.

Daston, Lorraine, and Peter Galison. *Objectivity.* New York: Zone Books, 2007.

Derrida, Jacques. *Archive Fever: A Freudian Impression.* Chicago and London: University of Chicago Press, 1995.

Ellenberger, Henri. *Die Entdeckung des Unbewussten. Geschichte und Entwicklung der dynamischen Psychiatrie von den Anfängen bis zu Janet, Freud, Adler und Jung.* Zurich: Diogenes, 1973.

Ffytche, Matt. *The Foundation of the Unconscious: Schelling, Freud and the Birth of the Modern Psyche.* Cambridge: Cambridge University Press, 2012.

Forrester, John. *Dispatches from the Freud Wars: Psychoanalysis and Its Passions.* Cambridge, MA: Harvard University Press, 1997.

Frankland, Graham. *Freud's Literary Culture.* Cambridge: Cambridge University Press, 2000.

Freud, Sigmund. "Die Widerstände gegen den Psychoanalyse." *Gesammelte Werke* 14. Frankfurt am Main: S. Fischer, 1999.

———. "Selbstdarstellung." *Gesammelte Werke* 14. Frankfurt am Main: S. Fischer, 1999.

———. "Vorrede zur hebräischen Ausgabe von 'Totem und Tabu.'" *Gesammelte Werke* 14. Frankfurt am Main: S. Fischer, 1999.

Freud, Sigmund, and Karl Abraham. *Briefe 1907–1926.* Edited by H. C. Abraham and E. L. Freud. Frankfurt am Main: S. Fischer, 1965.

Frosh, Stephen. *Hate and the Jewish Science: Anti-Semitism, Nazism and Psychoanalysis.* London: Palgrave MacMillan, 2009.

Gay, Peter. *Freud: A Life for Our Time.* London: Papermac, 1988.

Geller, Jay. *On Freud's Jewish Body: Mitigating Cicumcisions.* Fordham, NY: Fordham University Press.

Gresser, Moshe. *Dual Allegiance: Freud as a Modern Jew.* Albany: State University of New York Press, 1994.

Jerram, Leif. "Space: A Useless Category for Historical Analysis?" *History and Theory* 52, no. 3 (2013): 400–19.

Jones, Ernest. *The Life and Work of Sigmund Freud.* Vol. 2, *Years of Maturity 1901–1919.* New York: Basic Books, 1955.

Jung, C. G. "Über das Unbewusste." *Gesammelte Werke* 10. Olten and Freiburg: Walter-Verlag, 1981.

———. *Wandlungen und Symbole der Libido. Beiträge zur Entwicklung des Denkens.* Munich: DTV, 1991.

———. "Zur gegenwärtigen Lage der Psychotherapie." *Zentralblatt für Psychotherapie und ihre Grenzgebiete* 7 (1934): 1–17.

Kauders, Anthony D. *Der Freud-Komplex. Eine Geschichte der Psychoanalyse in Deutschland.* Berlin: Berlin Verlag, 2014.

———. "Weimar Jewry." In *The Oxford Short History of Germany: The Weimar Republic,* edited by A. McElligott, 234–59. Oxford: Oxford University Press.

Kirschner, Suzanne R. *The Religious and Romantic Origins of Psychoanalysis: Individuation and Integration in Post-Freudian Theory.* Cambridge: Cambridge University Press, 1996.

Klein, Dennis B. *Jewish Origins of the Psychoanalytic Movement.* Chicago and London: Chicago University Press, 1984.

Löw, Martina. *Soziologie der Städte.* Frankfurt am Main: Suhrkamp, 2008.

Mayer, Andreas. *Mikroskopie der Psyche. Die Anfänge der Psychoanalyse im Hypnose-Labor.* Göttingen: Wallstein, 2002.

———. "Zur Genealogie des psychoanalytischen Settings." Österreichische Zeitschrift für Geschichtswissenschaften 14 (2013): 11–42.

McGrath, William J. *Freud's Discovery of the Unconscious: The Politics of Hysteria.* Ithaca and London: Cornell University Press, 1986.

Meghnagi, David. "A Cultural Event within Judaism." In *Freud and Judaism,* 57–70. London: Karnac, 1993.

Mühlleitner, Elke, and Johannes Reichmayr. "Die Freudianer in Wien. Die Psychologische Mittwochs-Gesellschaft und die Wiener Psychoanalytische Vereinigung 1902–1938." *Psyche* 51 (1997): 1051–1103.

Rechter, David. *The Jews of Vienna and the First World War.* Oxford: Littman Library, 2001.

Rose, Louis. *The Freudian Calling: Early Viennese Psychoanalysis and the Pursuit of Cultural Science.* Detroit: Wayne State University Press, 1998.

Rozenblit, Marsha. L. *The Jews of Vienna, 1867–1914: Assimilation and Identity.* Albany: State University of New York Press, 1983.

Rutherford, Jonathan. "The Third Space: Interview with Homi Bhabha." In *Identity: Community, Culture, Difference,* edited by Jonathan Rutherford, 207–21. London: Lawrence and Wishart, 1990.

Santner, Eric L. *On the Psychotheology of Everyday Life: Reflections on Freud and Rosenzweig.* Chicago: Chicago University Press, 2001.

Schorske, Carl E. *Fin-de-Siècle Vienna: Politics and Culture.* Cambridge: Cambridge University Press, 1980.

Schröter, Michael. "Zur Frühgeschichte der Laienanalyse. Strukturen eines Kernkonflikts der Freud-Schule." *Psyche* 50 (1996): 1127–75.

Silverman, Lisa. *Becoming Austrians: Jews and Culture between the World Wars.* Oxford: Oxford University Press, 2012.

Slavet, Eliza. *Racial Fever: Freud and the Jewish Question.* New York: Fordham University Press, 2009.

Sorkin, David. *The Transformation of German Jewry, 1780–1840.* Oxford: Oxford University Press, 1987.

Strauss, Leo. *Persecution and the Art of Writing.* Glencoe, IL: Free Press, 1952.

Sulloway, Frank. *Freud: Biologist of the Mind.* New York: Basic Books, 1979.

Topinka, Robert J. "Foucault, Borges, Heterotopia: Producing Knowledge in Other Spaces." *Foucault Studies* 9 (2010): 54–70.

Winter, Sarah. *Freud and the Institution of Psychoanalytic Language.* Stanford, CA: Stanford University Press, 1999.

Wistrich, Robert S. *The Jews of Vienna in the Age of Franz Joseph.* Oxford: Oxford University Press, 1989.

Yerushalmi, Y. Hayim. *Freud's Moses: Judaism Terminable and Interminable.* New Haven and London: Yale University Press, 1991.

———. *Zakhor: Jewish History and Jewish Memory.* Seattle: University of Washington Press, 1982.

Zaretsky, Eli. "The Place of Psychoanalysis in the History of the Jews." *Psychoanalysis and History* 8 (2006): 235–53.

5

Jewish Displacement and Simulation in the German Films of E. A. Dupont

Ofer Ashkenazi

In the years preceding the rise of Nazism, the German film industry employed hundreds of people of Jewish ancestry, among them some of the most innovative, popular, and studied filmmakers of the era.[1] Traditionally, however, scholars have tended to overlook the Jewish background of these filmmakers or deem it irrelevant to their analysis of Weimar cinema.[2] This chapter examines two films by the Jewish director and scriptwriter Ewald A. Dupont, *Das alte Gesetz* (*The Ancient Law*, 1923) and *Peter Voß, der Millionendieb* (*Peter Voß, Who Stole Millions*, 1931/1932). The former explicitly portrays a Jewish theme—the acculturation of Baruch Mayer, a Galician rabbi's son, in modern Vienna—and the latter is a crime adventure film, which allegedly has nothing to do with modern Jewish experience. I argue below that *Das alte Gesetz* exhibits a particular cinematic vocabulary that allowed Weimar Jewish filmmakers to contemplate the meanings and prospects of acculturation within the parameters of popular film genres. *Peter Voß* demonstrates how this vocabulary was utilized in different genres and without direct reference to Jewish circumstances.

The production of each film ended a few months after an alarming challenge to German-Jewish acculturation: *Das alte Gesetz* followed the assassination of Walther Rathenau and other eruptions of anti-Jewish violence; *Peter Voß* came after the phenomenal success of the Nazi Party in the 1930 election. The obvious differences between them notwithstanding, the two films shared a similar ambition to support Jewish acculturation in the face of the challenges it was presented with. Both films, I argue, responded to anti-Jewish sentiments with a similar emphasis on the vital role of displacement and simulation in the formation of modern identities.

The "authenticity" of belonging to a place had been a major theme in the discussion of Jewish identity in modern Germany, among Jews and anti-Semites alike.[3] Equally fundamental was the "genuine" relation between identity and its visual expressions.[4] The alleged Jewish "displacement" in Germany and the Jewish tendency to "disguise" one's true self among non-Jewish Germans were accordingly key arguments against Jewish acculturation. *Das alte Gesetz* and *Peter Voß* seem to address the calls—and actions—against Jewish acculturation by undermining the dichotomies between "here" and "there" and between "authentic" and "acted out." Both films depict a process in which individual identity is formed through spatial transitions and can, in a range of places, be genuinely expressed in various ways. There is, however, one essential disparity between Dupont's two films: while the 1923 melodrama presents an optimistic prospect for Jewish life within the German bourgeoisie, the 1931 comedy ends with a melancholic prediction of its end.

E. A. Dupont and *Das alte Gesetz* in Context

Ewald André (Andreas) Dupont was born in 1891 to an acculturated bourgeois family in the small town of Zeitz, Saxony.[5] Following in his parents' footsteps, he became a reporter and newspaper editor in the early 1910s. Dupont's biography, interests, and career were largely typical of his generation of Jewish filmmakers in Berlin.[6] Scholars tend to portray Dupont as a director exceptionally attuned to audience expectations.[7] In his oft-cited 1919 essay on filmmaking, Dupont indeed encouraged directors to comprehend and satisfy viewers' tastes and desires.[8] His best-known demonstrations of this attitude are the melodramas *Varieté* (1925) and *Piccadilly* (1928/1929), which depicted the popular entertainment milieu.[9] The craving for a broad audience and mastery of popular genres were not unusual among up-and-coming Jewish filmmakers of the era. Similarly unexceptional was Dupont's ability to write and direct in a variety of genres while appealing to various audiences across national borders.[10] Consequently, like many of his Jewish peers, Dupont's 1920s success led to opportunities outside of Germany. In 1928, after disappointing his employers at Universal Pictures, he moved from Hollywood to London, where the newly branded British International Pictures was endeavoring to extend its appeal beyond the British audience. Dupont's London-based films include some of the definitive productions of the era, such as *Moulin Rouge* (1928) and *Atlantis* (1929).[11]

What was supposed to be a triumphant homecoming in the early 1930s, however, was disastrous: The rise of Nazism led to his second (now forced) emigration, and attempts to resume his work in Hollywood were unsuccessful. *Das alte Gesetz* preceded Dupont's great blockbusters, and its topic, the transformation of the protagonist from a shtetl-born rabbi's son to a theater star, might appear at odds with his lasting interest in vaudeville melodramas. The film, also distributed under the title *Israel's Son*, stands out from Dupont's other works due to its clearly specified Jewish protagonists and its explicitly "Jewish" subject matter. A newspaper advertisement for the film features a drawing of Baruch, dressed as Hamlet, stepping over a Torah book, an iconic tablet of the Ten Commandments (figure 5.1). Such explicit reference to Jewish experience was exceptional for Jewish filmmaking in Weimar Germany. Dupont himself appears to have had limited interest in openly contemplating this issue. In more than seventy films he wrote and/or directed, he portrayed characters overtly defined as Jewish only once more, in the 1930 multilingual production *Two Worlds*.

At least at first glance, the small number of films of this period that refer to Jewish acculturation is a puzzle. The Weimar period marked unprecedented success for Jewish acculturation within German (bourgeois) society, alongside unprecedented anti-Jewish violence and the rise of racist ideologies.[12] For many prominent Jewish filmmakers of 1920s Berlin, as well as for Jewish film critics in the trade press, acculturation (and, presumably, anti-Semitism) represented intimate, firsthand experiences. The considerable scarcity of openly Jewish themes, protagonists, or experiences in films made by acculturated Jews might be related to the limited appeal of these subjects or to acculturated filmmakers' self-identification with the educated middle class rather than with "the Jews" (or, worse, with *Ostjuden*).

I would argue that in addition to such considerations and sentiments, overt Jewish motifs were scarce in Weimar film because they were not required for the cinematic contemplation of Jewish experience; Weimar Jewish filmmakers developed a cinematic vocabulary that facilitated the discussion of modern Jewish identities on the screen through a particular encoding of Jewish perspectives. "Jewish" here has little to do with traditional life in the Eastern European shtetl, which Weimar Jewish filmmakers had left behind or never experienced. The perspective they explored was of various designated "outsiders" endeavoring to integrate into modern middle-class society. Jewish filmmakers such as Ernst Lubitsch, Joe May, Henrik Galeen, Robert Siodmak, and others found ways to associate their protagonists' aspirations and behavior with both the stereotypical "rootless" Jew of the modern city and the "typical" young urbanite of the age of dramatic transitions in middle-class society. Understanding this mechanism of double-encoding sheds a different light on Weimar film and its popular genre films, in particular.

Dupont's 1923 *Das alte Gesetz* has a fundamental role within this context. It offers an open discussion of Jewish acculturation within the contemporary German-speaking culture, which provides us with a "dictionary" for the vocabulary Weimar Jewish filmmakers employed to reflect on Jewish embourgeoisement. Notably, Dupont constructed his display of Jewish experience around two main themes: simulation (acting, changing appearance) and displacement (alienation and migration). These themes were crucial—and familiar—pillars of Jewish modernization, and they appeared in numerous depictions of modern Jewish experience.[13] Tracing the cinematic vocabulary Dupont utilized to contemplate

Figure 5.1. Baruch's Hamlet in a newspaper advertisement for *Das alte Gesetz*. Furnished by Deutsche Kinematik

these themes yields insight into its use in many other films allegedly unrelated to Jewish experience. Dupont's *Peter Voß*, I argue, develops this vocabulary to provide a critical assessment of the project of Jewish acculturation in modern Germany.

Setting the Stage: Identity, Authenticity, and Simulation in *Das alte Gesetz*

Das alte Gesetz narrates the adventures of Baruch Meyer, a troubled son of a Galician rabbi, who leaves the Eastern European shtetl to become a revered actor in 1860s Vienna. His transformation from an *Ostjude* outsider to a champion of modern bourgeois culture is a slow process involving recurring humiliations and rejections by both Jews and Gentiles. Baruch's father, in particular, loathes his son's disregard for the "ancient law," whereas the non-Jewish theater audience ridicules the strange-looking actor's poor adjustment to the "laws" of modern social conventions. Baruch's luck changes when he performs in front of Archduchess Elizabeth Theresa, who falls in love with him and grants him the opportunity to impress the director of the Viennese Burgtheater. Baruch's outstanding talent soon makes him a star on this prestigious stage. Yet, while the city dwellers warmly embrace Baruch, he stubbornly retains his ties with his past. The Archduchess's affection is unrequited; Baruch chooses instead to marry his childhood sweetheart Esther, the daughter of the shtetl synagogue's sexton. In the final scenes, Baruch also makes peace with his formerly enraged father, who comes to the theater and joyously watches his son perform.

According to Irene Stratenwerth, *Das alte Gesetz* was the most successful Jewish-themed German silent film.[14] It generated mostly favorable reviews, which emphasized its effective juxtaposition of comic and sentimental elements, as well as the "authenticity" of the "ghetto" depictions.[15] Some critics applauded the "accurate" portrayal of the ghetto (the *Lichtbild-Bühne* reviewer identified it as the "Ghetto of Kraków").[16] Others added that while the film captures the atmosphere of the shtetl—"the darkness of the eastern ghetto"—viewers with "uncompromising zeal for the facts" were bound to be disappointed.[17] The question of historical "accuracy" was to be expected, since Paul Reno's script was "inspired" by historical reality: the (unsuccessful) struggle of a Jewish actor, Bogumil Dawison, to attain a permanent contract at the Burgtheater in the 1860s.[18] In their evaluation of the film, however, all reviewers demonstrated a similar understanding of Baruch's story as a metaphor, or a generic symbol, for the integration of Eastern European Jewry into the modern urban bourgeoisie of Central Europe. Berlin film critics consistently asserted that *Das alte Gesetz* was an effort to propagate Jewish acculturation and combat its adversaries. The *Film Kurier* wrote that Dupont's film was a response to those who "regard *Ostjuden* [with] superstitious horror" and a threat to the "wild, bellicose swastika people."[19] After the premiere, Fritz Olimsky warned in the *Berliner Börsen-Zeitung*, "The Kurfürstendamm audience smiled with pleasure, but don't show this film in Munich, lest they finally break away from us."[20]

In contrast to these assessments, Cynthia Walk and Tim Bergfelder recently argued that disguised as a celebration of Jewish assimilation, *Das alte Gesetz* describes its strict limits or even its "failure."[21] At the core of this reading lies Baruch's rejection of the archduchess and his subsequent marriage within the *Ostjuden* milieu, which in Walk's view works against the logic of the film and places an artificial barrier before the expected outcome, interracial intimacy. During the encounter between Baruch and the archduchess in her room, as her body leans toward him in expectation, she repeatedly asks him about his desires and fantasies. To her recurring question "Is there anything else you wish from me?" he absentmindedly mentions his aspiration to play Hamlet. Her final attempt to draw him into her arms includes a melodramatic toss of her handkerchief in a nocturnal garden scene, an act he conveniently overlooks. Thus symbolized, Walk argues, Baruch's life in the city "stops short of full integration,"[22] catering to the conservative non-Jewish audience, which feared the boundary crossing that Baruch embodied. The flattering reviews therefore did not reflect widespread pro-assimilation views, but rather approval—or acceptance—of segregation shared by bourgeois critics and (Jewish) filmmakers alike.

Yet, while the film does stop short of effacing past identities and differences of class, religion, and ethnicity, it does not depict "limited" acculturation. Acculturation, as opposed to conversion, presupposes a duality of identifications.[23] As several scholars have shown, in integrating into the educated bourgeoisie, Jews had adopted their environment's modes of experience, including the constant interplay, and tensions, between different notions of identity.[24] Baruch demonstrates the dualities and complexities intrinsic to successful acculturation. In marrying his childhood sweetheart, he intertwines his Jewish roots with his current middle-class environment. Dupont underlines Baruch's accomplishment by depicting the idyllic domestic routines of the newlywed couple in Vienna. Within a clichéd middle-class setting, they sit at the breakfast table surrounded by standard bourgeois status symbols (and indications of conventional "good taste"), from the tall ceiling, tapestry, random knickknacks, and kitschy pictures on the walls to the morning newspapers and servants. Baruch and Esther thus represent a model of Jewish embourgeoisement. This advocacy of acculturation, according to the *Film Kurier,* probably enraged the "wild, bellicose swastika people."

The concluding part of the film, in which the old rabbi comes to the theater and makes peace with his son, shows the modern city setting as an ideal arena for the continuation of Jewish life, a continuation that the shtetl—with its "old law" and traditional expectations—jeopardized. This scene encompasses two of Dupont's key themes in his model integration story. The first is the interaction between identity and simulation, which undermines the distinction between authenticity and acting. The film's fundamental metaphor equates successful acculturation with successful acting. Baruch's aspiration to become part of the modern bourgeoisie is phrased as a desire to be an actor. His fulfillment of the middle-class fantasy in the aforementioned breakfast scene with Esther follows an outstanding onstage performance.

Moreover, the film depicts acculturation as a process in which Baruch not only learns how to advance his stage acting, but also how to "play" outside the theater. Upon arriving in the city, he discards his sidelocks and assumes the appearance of his middle-class peers, including clothing, gestures, and shaving style. As the film advances, Baruch's offstage emulation of his peers mirrors his onstage success. While he still fails to simulate his companions' appearance and body language, he plays marginal roles in an insignificant group of traveling comedians (his ignorance of the kind of performing group he has joined is central to the comic effect at this point). When he finally looks like a "typical" bourgeois urbanite and behaves accordingly, he starts playing major protagonists of Western culture on the most respectable stage. Dupont's narrative is thus a story of the transformation of the Jew to an actor, who excels in simulating the "bourgeois character."

This metaphorical depiction fits well with a recurring motif in the discussion of Jewish integration into the *Bildungsbürgertum* in pre-1933 Germany. Numerous studies, essays, and works of fiction had associated Jewish acculturation with adopting the bourgeois *habitus,* that is, the values, tastes, and behavior of the non-Jewish middle class.[25] Such adaptation had inevitably involved emulation of the appearance, language, and gestures—as well as other public and private modes of expression—of that social environment. Various intellectuals who addressed this issue claimed that Jewish social experience therefore shaped the Jew as a "natural-born actor" (which, in turn, supposedly explained the noticeable Jewish presence in German theater).[26] Consequently, the identification of modern Jewish experience with empty simulation or even deception was fundamental to anti-Semitic writings from the later nineteenth century; anti-Semitic commentators identified Jewish acting either with an attempt to conceal a destructive Jewish "essence" or with Jews' complete lack of any "essence."[27] Some Jewish commentators, most notably Hannah Arendt, reiterated this emphasis on Jewish "aping" of Gentiles and acting out an identity they did not share.[28] Others associated Jewish "enthusiasm" for acting out with self-hatred, with attempted dissociation from their Jewish essence.[29] Still other Jewish intellectuals embraced the association between Jewish identity and bourgeois performance. Writers and critics such as Robert Weltsch, Arnold Zweig, and Willy Haas promoted a perception of simulation and acting as a revelation of a new, modern authenticity, of which the Jew was a product and a vanguard.[30] According to Zweig's view, the actor (and hence "the Jew") excels at existing in between different cultures and their moral and behavioral codes and can thus form a bridge between cultures and effectively criticize cultural conventions. The modern city, where performance is the (only) authentic way of life, becomes, accordingly, an ideal place—an ideal stage—for the modern actor.

I will demonstrate below how Dupont's film further constructs the modern city as a "Jewish" place. Its portrayal of identity formation in the city closely recalls Zweig's association between Jewish acting, authenticity, and urban displacement. First, *Das alte Gesetz* rejects the binary opposition of "authenticity" (of modern society) and acting (the Eastern Jew seeking integration in the city). The role Baruch learns to play is itself a product of social conventions. His patron, Elizabeth, explains that she lives by a set of social codes, "the new law," which she cannot escape, although they contradict her "authentic" feelings toward him. Rather than a natural development from the artificiality (or irrelevance) of the "old law" to the "new law," Dupont's viewer witnesses the Jewish protagonist adapt from one set of conventional codes to another. In the modern environment, the film suggests, acting does not challenge authenticity, since everyone is an actor playing according to, and in co-creation of, varying social norms.

Notably, the film also undermines the binary opposition of authenticity and acting in the traditional Jewish world. Baruch's acting is not a break with "authentic" Judaism, since the film's shtetl scenes identify (authentic) Jewish life with role-playing. The initial scene in the shtetl portrays the local Purim celebration, in which Jews put on costumes and play various characters from the ancient story of the persecution and salvation of the nation. Even after the stage act is over, Baruch's lover, Esther—the name of the Jewish heroine in the Purim story—is a living reminder of the play that initiated Baruch's longing to leave the shtetl. Put simply, if the "new law" refers to a modern style of acting, then the "ancient law" seems here merely to designate a different, old style of acting.

In the shtetl scenes, Dupont used actors who grew up in Eastern Europe and were identified by the press as "authentic" *Ostjuden,* in contrast to the acculturated Dupont. Accordingly, reviewers argued that they "authentically" represented shtetl characters and "mentality."[31] Such confusing identification of actors and acting styles with authenticity echoes an intriguing facet of the encounter between Baruch and Elizabeth. Their initial meeting takes place during his odd first performance as Romeo, in a quasi-parody version staged by the group he joined on his way to Vienna. At that point Baruch still maintains his "Jewish" look (symbolically, his apparent sidelocks), and his Romeo elicits humiliating laughs. Elizabeth is shown as being able to see beyond the surface and detect Baruch's remarkable essence; yet this "essence" consists in his talent for acting. In blurring the distinctions between acting and authenticity, Dupont's film counters efforts to dismiss acculturation as nothing but an act of "aping" the *habitus* of the educated bourgeoisie and hence an assault on the authentic identity of both Jews and non-Jewish Germans. In other words, in *Das alte Gesetz* Dupont transforms the cliché equating Jewish modernization with acting into a manifesto in favor of acculturation.

Heimat and Home in *Das alte Gesetz*

The second principle of Dupont's narration of acculturation is the spatial encoding of identity, which highlights its multilayered nature in modernity. At first glance, the title of Dupont's film *The Ancient Law* suggests that his discussion of Jewish identity will be more concerned with its relations to memory (or the commandment to remember, *zachor*) and source (*makor*), rather than its interactions with place (*makom*).[32] Yet, instead of contrasting the "ancient law" with the "modern" place, Dupont subordinates laws—or social norms—to places. In this film, each place has its familiar set of laws, which are incompatible (only) as long as the places are detached.

Das alte Gesetz begins, indeed, as a story about two distinct places—the shtetl and the city—and about the distinct parcel of values, meanings, memories, and social practices embedded within each. The Jewish realm, the shtetl, is structured as a nexus of narrow alleyways, crowded and dark, tightly framed by small, mostly wooden, run-down buildings. While they lack the skewed lines and diagonal shades of the ghetto Paul Wegener imagined in his classic *Der Golem, wie er in die Welt kam* (1920), the shtetl streets form an enclosed world, foreign and eerie to the modern viewer.[33] The indoor settings make a similar impression. Walls, furniture, and characters are shot from close ranges, often from a lower angle, highlighting the claustrophobic nature of "Jewish" living spaces and constantly locating the protagonists within small frames (formed, for instance, by ceiling and floor, corners of rooms, supporting columns, and shelves). The light comes down from tiny apertures in the upper part of the dark walls, in a way that often situates the protagonists between the viewers and the only source of light. In some scenes, this mise-en-scène fluctuates between indicating warm intimacy or stubborn backwardness; in other scenes—in particular, Baruch and Esther's dialogues—it underscores the passionate desire to escape.

By contrast, Vienna's cityscapes are spacious and bright, streets are wide, buildings are tall, and facades are shiny. The city also contains extensive parks, which bring (tamed) nature—and with it, as Stefan Zweig maintained, a sense of *Heimat,* of an organic relationship between humans and their habitat—into the modern urban landscape.[34] The park, the countryside within a city, is where the archduchess first sees Baruch, so it is a crucial place in his transition from foreigner, "a Jew," to successful member of the *Bildungsbürgertum* who happens to be Jewish. The park's open spaces stand in stark contrast to the shtetl's landscape. Similarly, the city's private spheres are roomy, with high ceilings, large windows letting in ample "natural" sunlight, and decorative artifacts. Unlike the ghetto, the city is not an alienated place, enclosed and detached from its surroundings, but rather intertwined with nature.

At first glance, therefore, it appears that *Das alte Gesetz* narrates the replacement of one identity with another by transitioning from the Jewish place to the non-Jewish (Central European), bourgeois place. Several studies dedicated to the intermingled formation of place and identity in modern German and Jewish cultures have particularly underlined cultural representations of the encounter between the "Jewish place"—the ways in which Jewish identity is imagined as spatial and expressed in space—and modernity.[35] The formation of the modern city as a "place" in the bourgeois cultural imagination and its association with the constitution of bourgeois identities are equally familiar.[36] In modern German culture, this matrix of modern places and identities is supplemented by the concept of *Heimat,* an authentic national place, apolitical and atemporal, which shapes and reflects the essential qualities of the national community.[37] Dupont's imagery thus draws on well-known cultural symbolism in its blending of notions of the self originating in Jewish, bourgeois, and German national identity discourses.

The above-mentioned reading of Dupont's use of spatial metaphors would easily fit with the anti-Semitic stereotype of the treacherous Jew. The scenery of the ghetto resembles the conventional scenery of expressionist film, which connotes the irrational pathologies of psychiatric institution inmates (*Caligari*), murderers (*Raskulnikov, Hintertreppe*), and foreign invaders (*Nosferatu, Der Golem*).[38] Detached from its natural environment, it allegedly reflects Jewish rootlessness and inability—or unwillingness—to organically belong to a place. In a way that resembles Veit Harlan's infamous staging of the Jews' entry to Stuttgart in *Jud Süss* (or the vampire's coming to Bremen in Murnau's *Nosferatu*), Baruch comes from the ghetto of the East in order to stir up the social structures and values embedded in the city, "infiltrating" it with the aid of the aristocratic archduchess, and never fully becomes an urbanite; his "nature" sees him remaining always a stranger in the city.

Such a reading, however, overlooks some of the complexities in Dupont's spatial symbolism. First, *Das alte Gesetz* dedicates a long sequence to the space in between the shtetl and the city.[39] Wandering around with the group of comedians, Baruch is educated in the ways of modern entertainment, learns the canonical texts of modern culture (*Hamlet*), and acquires the basics of bourgeois etiquette. It is largely the long journey toward Vienna, and not the city itself, which teaches Baruch how to be both a Jew and a modern entertainer. The extensive scenes dedicated to the journey undermine the dichotomy of identities and instead characterize acculturation as a fluid concept, a scale of degrees of disassociation from the ghetto.

The movement between shtetl and city has more than one direction. After his initial success, Baruch returns to the shtetl in hopes of making peace with his parents. It is not a homecoming of a prodigal son, since he has no intention to stay; yet it is not merely tourism: he feels at home there, and when his father refuses to embrace his new way of life, he takes Esther, a living reminder of the shtetl, with him to the city. He likewise opens his city house to visitors from the shtetl and makes his private sphere an in-between place, where city and shtetl are not separated or contradicting ideals.

Dupont further undermines the dichotomy between the shtetl and the city by introducing the shtetl during the masquerade of Purim, with everyone acting according to a prescribed role. Baruch originates, therefore, from a large, all-encompassing stage. He eventually completes his journey by finding his place, by feeling at home, on the stage in the city. He does not move from one "world" to another, but rather between two versions, two reflections of the same place, the stage. In this regard, the city (and the space in between, where the comedians perform) mirrors the shtetl; the disparities, the differences between the provincial shtetl and the modern imperial center, arise solely from cultural conventions, not from the fixed, foreign "nature" of the shtetl Jew.

Moreover, the shtetl in the film includes two distinct landscapes. The above-mentioned ghetto landscape is complemented by another location, apparently on the outskirts of the Jewish town. Several fundamental shtetl scenes occur in this setting, including the intimate encounters between Baruch and Esther, his contemplations of the future and realization that he must leave, and the arrival of Ruben Pick, a modern "wandering Jew," who disturbs the balance between the rabbi and his son. The houses shown in this location, overlooking vast open fields, are far less crowded and decrepit than in the other shtetl scenes; the walls are brighter and the sky is visible between them. The scenery here resembles the model German village Dupont created in his pioneering *Heimatfilm, Die Geier-Wally* (1921). The Jewish ghetto, which other scenes depict as detached from its environment, appears here as rooted in the *Heimat*.[40] Consequently, Vienna, with its parks that blend urban life and "authentic" nature, does not appear as fundamentally different from the uprooted shtetl.

Das alte Gesetz demonstrates a sophisticated utilization of familiar spatial symbolism. Dupont employs clichés that normally accentuate well-defined, exclusive notions of identity and makes them the foundations for a new notion of fluid, multilayered German-Jewish identity. His imagery principally refers to two different, albeit interrelated, spatial distinctions, each of them complicating the association between identity and authenticity. The first distinction is between the *Heimat*—the authentic, timeless and apolitical landscape that forms and manifests the homogeneous nation—and the city, the ever-changing landscape that breeds alienation, rootlessness, and "inorganic" diversification of society. This distinction has been a vital component of modern Central European (bourgeois) culture since the later nineteenth century.[41] The notion of authentic, inherent belonging to a place embedded in the imaginary national *Heimat* posed a potential threat to the full integration of Jews, particularly those recently immigrated, in German middle-class society.[42] In *Das alte Gesetz*, Dupont uses some *Heimat*-like scenery but confuses its conventional meanings. The idyllic rural scenery is here part of the city and its modern leisure culture; the uprooted Jewish ghetto replaces the supposedly authentic, premodern environment. Yet Dupont does not romanticize the shtetl and its alleged authenticity. His Vienna, like Stefan Zweig's imperial city, indeed offers (to bourgeois urbanites) an ideal combination of nature and city, of progress and authenticity. The city's ability to possess this duality is associated with its ability to incorporate, acculturate, and express some aspects of the shtetl within it.

In addition to the cultural expectations related to the notion of *Heimat*, Dupont exploits here another fundamental spatial distinction recurrent in the discourse about acculturated Jewish identity, the differentiation between the private and public spheres. German-Jewish acculturation was, to a great extent, an endeavor of Jews to be "invisible in public," that is, to be inconspicuous in the bourgeois cityscapes;[43] this endeavor allowed for and encouraged expressions of individual heritage and background in private. As in the simplified dictum of Moses Mendelssohn and Juda Leib Gordon "Be a man in the street and a Jew at home," an acculturated bourgeois Jew could maintain certain distinctive practices in private, while sharing "appropriate" practices in public.[44]

In *Das alte Gesetz*, elements of the shtetl are apparent in the private apartment—most notably the presence of "shtetl people" such as Esther, Pick, and Baruch's parents—but are absent in public. We might read this as an indication of the success of Baruch's acculturation: the diverse cityscapes allow multiple expressions of identity, which can simultaneously—albeit not at the same place—contain Jewish heritage and (non-Jewish) bourgeois *habitus*. This latter reading fits with Dupont's characterization of the theater, a public sphere in which Baruch can be various characters of the Western canon while maintaining the symbols of his Jewish tradition. If Jews aspired to be "invisible" as Jews in public, then the stage performance enables Baruch to bear a distinguished symbol (to carry his prayer book) without standing out as a foreigner. Modern performance is not only a symbol of successful acculturation, but also provides cultural spheres in which acculturation can be perfected—in which a Jew can maintain his difference and still be invisible.

A Dictionary of Acculturation: From *Das alte Gesetz* to *Peter Voß*

In blurring the distinction between acting and authenticity and complicating the conventional associations between displacement and identity, Dupont's *Das alte Gesetz* develops a cinematic vocabulary that fits a favorable interpretation of Jewish acculturation. Performance, disguise and emulation of norms, and movement between "homes" became major themes in the films of Dupont and many other Jewish filmmakers in Germany of the 1920s and early 1930s.

Dupont's 1931 film *Peter Voß, Who Stole Millions* is a peculiar version of a popular story, which was filmed in Germany four more times between 1920 and 1960.[45] Based on Ewgar Seeliger's 1913 novel of the same name,[46] the plots of all cinematic versions maintain a similar framework: resourceful detective Bobby Dodd chases around the world after Peter Voß, who has allegedly robbed millions from a bank but in fact has only staged a robbery of an already empty safe. While in all *Peter Voß* films the protagonist fakes a robbery to aid his bankrupt patron, the identity of the patrons and the reasons for the bankruptcy vary (from a criminal scheme to stock-market speculations). Dupont's script for the first "talkie" adaptation met with mixed, or confused, reviews. As Fritz Olimsky wrote in the *Berliner-Börsen-Zeitung*, it was disappointingly limited in its scope and deviated from Seeliger's plot promptly after the opening scenes.[47] While Dupont thought of this comic adventure as a combination of "art and sensations," Berlin's *Morgenpost* described it as "an erotic quasi-documentary film [*Kulturfilm*]."[48] Yet, both the above-mentioned critics highlighted Dupont's exceptional "attention to details" and ability to satisfy what audiences crave.

Dupont's film was indeed remarkably different in its plot, scope, and emphases from the six-episode adaptation that preceded it, made shortly after World War I, and from the succeeding adaptation, made in Goebbels's studios during World War II. Dupont's Voß is a clerk in a small bank that had lost its capital through sloppy investments; even though he invested in accordance with his employer's advice, he is held responsible for the loss. When a rich client, Pitt, wishes to withdraw his entire deposit, Voß stages a break-in to the bank safe and escapes on a ship toward Marseilles, supposedly with Pitt's millions. Detective Bobby Dodd and Polly, Pitt's daughter, trail him. In order to escape them, Voß must repeatedly disguise himself as various characters. Dodd and Polly realize they must hide their identity as well to trap him. Yet Voß is always better at pretending to be someone else; Dodd and Polly are always one step behind. From Marseilles Voß escapes to Morocco, where he and Polly become lovers and he stages his death in a car accident to elude Dodd a last time. Then, a telegram arrives from Germany and clears Voß's name (and, with a miraculous stock-market turn, the bank regains the lost funds and is able to pay back Pitt's deposit).

The film is structured around two themes: simulation and displacement. At the core of almost every scene lies deception, disguise, and role-plays, incessantly conducted by the three main characters. Voß is the unchallenged master of disguise; Dupont dedicates long shots to his acting preparations. For instance, we see him trying on various moustaches (including a narrow, Hitler-style moustache, accompanied by a short parody of Hitler's facial expression). He constantly exploits people's expectations of a correlation between appearance, behavior, and authenticity. Only in the Moroccan desert, where Voß finally manages to evade Dodd, is his true self revealed—in both his (reciprocal) true love for Polly and the telegram that clears his name and allows him to state his identity. The second major theme is of a journey from the heart of modern Germany, Hamburg's banking district, to the Orient. In Morocco, Voß both stands out as a European and becomes invisible in traditional local clothing. It is only in a European fantasy of the ideal Orient that Voß ultimately finds his place.

Peter Voß has generally been overlooked by scholars of Weimar film and, indeed, in studies dedicated to Dupont's filmography. Like reviewers of the early 1930s, scholars have disregarded it as a cliché adventure film, not commensurate with Dupont's reputation as a brilliant observer of urban entertainment milieus. Surprising similarities between *Peter Voß* and *Das alte Gesetz*, however, suggest that we should read it as a late commentary on the themes Dupont developed in his acculturation film of 1923. The vital role of displacement and simulation indicates that Dupont regarded the later film as an opportunity to reconsider the prospect of Jewish acculturation in Germany during the final years of the Weimar Republic.

Peter Voß was made shortly after Dupont returned to Germany from England and a few months before he left the country for good. Despite the different circumstances and the different genre, its protagonist's ambitions and behavior closely resemble those of Baruch. Like Baruch, Voß is a natural-born actor who shifts between identities as new roles demand. Like Baruch's, Voß's freedom, his ability to be himself, is dependent on superb acting skills. In a manner comparable to Baruch's ability to use acting to reunite with his father, Voß's successful acting brings him closer to authenticity in the form of true love, the genuine feelings he shares with Polly, which would have not been discovered otherwise. We might characterize Voß as a reincarnation of Baruch, the acculturated Jew. In fact, Voß appears to be an exaggerated version of Baruch, changing his disguise much more rapidly, adapting to new (national) environments faster, and thriving in cosmopolitan spheres. Embodied in Voß, the tension between authenticity and identity also reaches an overstated climax: only after he successfully stages his own death is Voß able to show the telegram that clears his name and allows him to declare his identity.

The spatial duality in *Peter Voß*, the modern urban (financial) center and the premodern periphery of the Orient, also resembles that of *Das alte Gesetz*. As in the earlier film, and arguably to a greater degree, the 1931 film emphasizes the territories in between these places and the journey itself. Yet *Peter Voß* handles these spatial categories differently. First, the direction of the movement is exactly the opposite of Baruch's. Instead of toward the modern European urban centers from the eastern periphery, Voß moves away from those centers toward the peripheral, premodern "east." If Voß is a variation of Baruch, the context of acculturation is a principal subtext: while it is not explicitly stated here, the parallels between the Middle Eastern terrain and the eastern ghetto, both relating to Jewish (national) isolationism, should not be overlooked; similarly, the Hamburg banking district resembles 1860s Vienna as a symbolic sphere of Jewish acculturation. Within this metaphorical framework, Voß starts as a model acculturated Jew and moves away from, is forced to escape, the society to which he belonged. The second difference between Voß and Baruch is the former's reluctance to go back. Baruch travels between two worlds and is at home in both (or, at least, contains two worlds in his bourgeois home). Voß ends the film in the Orient. The final scene shows the donkey that, presumably, was to take him back to a ship and on to Germany. But Voß and Polly are not on the donkey; they have found intimacy, happiness, amid the desert hills.

The utilization of the same symbolism as in *Das alte Gesetz* makes *Peter Voß* an intriguing commentary on the former's themes. As in the 1923 film, the likable protagonist is identified with the fundamentals of acculturation (the emulation of the social environment, transformation of identities, quick adaptation to "foreign" places). His talent for acculturation enables him to successfully overcome injustice and bias and eventually leads him to his true love, Polly. In the two films, acculturation comes with displacement, with departure from the "home" and settlement in a foreign terrain where the characters were not "supposed" to be. As in the case of Baruch, Peter Voß's soulmate comes from the world he had left behind. There are two obvious differences between these cases of displacement. First, Baruch finds his acculturated Jewish place within the urban bourgeoisie, whereas Voß drifts away from this environment. Second, and more importantly, Baruch incorporates the shtetl into the bourgeois environment of the modern city; the modern city develops from a location of displacement to a complex sphere that interacts with the shtetl, integrates some elements of it, and allows movement to and from the shtetl. *Peter Voß*, by contrast, ends with a spatial dichotomy, with the European bourgeois landscape, on the one hand, and the oriental scenery, on the other. Voß has to choose between them, and despite his patent sense of displacement, he chooses to stay in North Africa. The comic tenor of *Peter Voß* notwithstanding, its version of displacement insinuates a much less optimistic view of the prospect of Jewish acculturation in middle-class Germany. *Peter Voß* is not a prophetic vision of the demise of Jewish life in Germany. Nevertheless, it provides a candid insight into the Jewish discourse of acculturation in the months preceding Hitler's rise to power.

Ofer Ashkenazi is a senior lecturer and the director of the Koebner-Minerva Center for German History at the Hebrew University of Jerusalem. He is the author of the books *Weimar Film and Modern Jewish Identity* (2012) and *A Walk into the Night: Reason and Subjectivity in Weimar Film* (2010).

Notes

1. Ashkenazi, *Weimar Film*, 1–16; Isenberg, *Weimar Cinema*, 9.
2. Eisner, *Haunted*; and Kracauer, *From Caligari*. See the discussion in Rogowski, "Introduction."
3. Fisher and Mennel, *Spatial Turns*; Brauch, Lipphardt, and Nock, *Jewish Topographies*; Lowenstein, "Urbanization."
4. Bayerdörfer, "Jewish Self-Presentation."
5. Bretschneider and Hampicke, "Biographie."
6. Dorner, "Dupont"; St. Pierre, *E. A. Dupont*, 11–26.
7. Bergfelder, *Life*, 26–29.
8. Dupont, *Wie ein Film*.
9. Walk, "31 January 1929."
10. St. Pierre, *E. A. Dupont*, 18–20.
11. His subsequent multilingual films include *Two Worlds* (1930); *Cape Forlorn* (1930/1931); and *Salto Mortale* (1931).
12. Mendes-Flohr, *"Kriegserlebnis"*; Niewyk, *Jews*, 43–81.
13. Cesarani, Kushner, and Shain, *Place and Displacement*; Aschheim, "Reflections."
14. Simon and Stratenwerth, "Auszug aus dem Ghetto," 230.
15. For instance: Effler, "Das alte Gesetz," 14; Ihering, "Das alte Gesetz."
16. "Purim," 29; Olimsky, "Filmentwürfe von Künstlerhand."
17. "Das alte Gesetz."
18. Prawer, *Between Two Worlds*, 22–23.
19. "Das alte Gesetz."

20. Olimsky, "Das alte Gesetz."
21. Walk, "Romeo"; Bergfelder, *Life*, 29.
22. Walk, "Romeo," 90–91.
23. Mendes-Flohr, *German Jews*; Kaplan, "Tradition"; Funkenstein, "Dialectic"; Boyarin, "Other Within."
24. Van Rahden, *Jews*; Kaplan, *Jewish Daily Life*, 173–269; Mosse, *German Jews*.
25. Lässig, *Jüdische Wege*.
26. Nietzsche, "Vom Probleme des Schauspielers," 235; Schach, "Das jüdische Theater."
27. Panizza, "Operated Jew"; Blüher, *Secessio Judaica*, 63.
28. Arendt, "Jew as Pariah," 99–100.
29. Lessing, "Jewish Self-Hatred."
30. A. Zweig, *Juden*, 23–25; Weltsch, "Theodor Herzl," 158; Haas, "Hugo von Hoffmannsthal."
31. "Purim"; Effler, "Das alte Gesetz," 14.
32. Mann, *Space and Place*, 11–25. See also Goldstein, "Memory."
33. Wildmann, "Desire."
34. S. Zweig, *World of Yesterday*, 13–14.
35. Roemer, *German Cities*; Rechter, "Geography"; Brinkmann, "From Hinterberlin."
36. Schütz, "Berlin."
37. Schumann, *Heimat denken*, 5–12; Boa and Palfreyman, *Heimat*, 1–10.
38. *Das Cabinet des Dr. Caligari* (1919/1920, dir. Robert Wiene); *Raskolnikow* (1922/1923, dir. Robert Wiene); *Hintertreppe* (1921, dir. Leopold Jessner, Paul Leni); *Nosferatu* (1922, dir. Friendrich Wilhelm Murnau).
39. A similar argument was made about Murnau's *Nosferatu*: Mayne, "Dracula."
40. Von Moltke, *No Place*; Kaes, *From Hitler*, 161–92
41. Confino, *Nation*; Lekan, *Imagining*; Umbach, *German Cities*, 64–68; Applegate, *Nation*.
42. Kayserling, *Juden als Patrioten*, 4; Wassermann, *Mein Weg*, 20–21; Rosenthal, *Heimat-geschichte*.
43. Sorkin, "Invisible Community."
44. Gordon, "Hakiza Ami," 17.
45. The others are *Der Mann ohne Namen* (1920/1921, dir. Georg Jacoby; a series of six sequels, the episode *Der Millionendieb* correlates to the plots of later versions); *Peter Voss, der Millionendieb* (1943–45, dir. Karl Anton); *Peter Voss, der Millionendieb*, (1958, dir. Wolfgang Becker); *Peter Voss—der Held des Tages* (1959, dir. Georg Marischka).
46. Seeliger, *Peter Voß*.
47. For instance, Olimsky, "Das alte Gesetz."
48. Dupont's quote in Gehler, "Peter Voß," 290; "*Peter Voß*," *Morgenpost* (Berlin).

Bibliography

Applegate, Celia. *A Nation of Provincials: The German Idea of* Heimat. Berkeley: University of California Press, 1990.
Arendt, Hannah. "The Jew as Pariah: A Hidden Tradition." *Jewish Social Studies* 6, no. 2 (1944): 99–122.
Aschheim, Steven E. "Reflections on Theatricality, Identity and the Modern Jewish Experience." In *Jews and the Making of Modern German Theatre*, edited by Jeanette R. Malkin and Freddie Rokem, 21–38. Iowa City: University of Iowa Press, 2010.

Ashkenazi, Ofer. *Weimar Film and Modern Jewish Identity.* New York: Palgrave Macmillan, 2012.

Bayerdörfer, Hans-Peter. "Jewish Self-Presentation and the 'Jewish Question' on the German Stage from 1900 to 1930." In *Jewish Theater: A Global View,* edited by Edna Nahshon, 153–74. Boston and Leiden: Brill, 2009.

Bergfelder, Tim. "Life Is a Variety Theater—E. A. Dupont's Career in German and British Cinema." In *Destination London: German-Speaking Emigrés and British Cinema, 1925–1950,* edited by Tim Bergfelder and Christian Cargnelli, 24–35. New York: Berghahn Books, 2008.

Blüher, Hans. *Secessio Judaica. Philosophische Grundlegung der historischen Situation des Judentumes und der antisemitischen Bewegungen.* Berlin: Der Weisse Ritter, 1921.

Boa, Elizabeth, and Rachel Palfreyman. *Heimat—A German Dream: Regional Loyalties and National Identity in German Culture 1890–1990.* Oxford: Oxford University Press, 2010.

Boyarin, Jonathan. "The Other Within, and the Other Without." In *The Other in Jewish Thought and History,* edited by Robert L. Cohn and Laurence J. Silberstein, 424–52. New York: New York University Press, 1994.

Brauch, Julia, Anna Lipphardt, and Alexandra Nocke, eds. *Jewish Topographies: Visions of Space, Traditions of Place.* Burlington, VT: Ashgate, 2012.

Bretschneider, Jürgen, and Evelyn Hampicke. "Biographie." In *Ewald Andre Dupont. Autor und Regisseur,* edited by Jürgen Bretschneider, 111–26. Munich: Text+Kritik, 1992.

Brinkmann, Tobias. "From Hinterberlin to Berlin: Jewish Migrants from Eastern Europe in Berlin Before and After 1918." *Journal of Modern Jewish Studies* 7, no. 3 (2008): 339–55.

Cesarani, David, Tony Kushner, and Milton Shain, eds. *Place and Displacement in Jewish History and Memory: Zakor v'Makor.* London: Vallentine Mitchell, 2009.

Confino, Alon. *The Nation as a Local Metaphor: Württemberg, Imperial Germany, and National Memory, 1871–1918.* Chapel Hill: University of North Carolina Press, 1997.

"Das alte Gesetz." *Film Kurier,* 30 October 1923.

Dorner, Maren. "Ewald A. Dupont." In *Pioniere in Celluloid: Juden in der frühen Filmwelt,* edited by Hermann Simon and Irene Stratenwerth, 289–91. Berlin: Henschel, 2004.

Dupont, Ewald André. *Wie ein Film geschrieben wird und wie man ihn verwertet.* Berlin: Reinhold Kühn, 1919.

Effler, Erich. "Das alte Gesetz." *Reichsfilmblatt* 44 (1923).

Eisner, Lotte Henriette. *The Haunted Screen: Expressionism in the German Cinema and the Influence of Max Reinhardt.* Berkeley: University of California Press, 2008 [1952].

Fisher, Jaimey, and Barbara Caroline Mennel, eds. *Spatial Turns: Space, Place, and Mobility in German Literary and Visual Culture.* Amsterdam: Rodopi, 2010.

Funkenstein, Amos. "The Dialectic of Assimilation." *Jewish Social Studies* 1, no. 2 (1995): 1–14.

Gehler, Fred. "Peter Voß, der Millionendieb." In *Deutsche Spielfilme von den Anfängen bis 1933,* edited by Günther Dahlke and Karl Günter, 290–91. Berlin: Henschel, 1988.

Goldstein, Jonathan. "Memory, Place and Displacement in the Formation of Jewish Identity in Rangoon and Surabaya." *Jewish Culture and History* 9, nos. —3 (2007): 101–13.

Gordon, Jehuda Leib. "Hakiza Ami" [Awake, My People] (1863). Reprinted in *Kitvei Yehuda Leib Gordon: Shira.* Tel Aviv: Dvir, 1959.

Haas, Willy. "Hugo von Hofmannstal." In *Juden in der Deutschen Literatur: Essays über Zeitgenossische Schriftsteller,* edited by Gustav Krojanker, 139–64. Berlin: Welt-Verlag, 1922.

Ihering, Herbert. "Das alte Gesetz." *Berlin Börsen-Courier,* 1 November 1923.

Isenberg, Noah William, ed. *Weimar Cinema: An Essential Guide to Classic Films of the Era.* New York: Columbia University Press, 2009.

Kaes, Anton. *From Hitler to* Heimat*: The Return of History as Film.* Cambridge, MA: Harvard University Press, 1989.

Kaplan, Marion A. *Jewish Daily Life in Germany, 1618–1945.* New York: Oxford University Press, 2005.

———. "Tradition and Transition: The Acculturation, Assimilation and Integration of Jews in Imperial Germany. A Gender Analysis." *Leo Baeck Yearbook* 27, no. 1 (1982): 3–35.

Kayserling, Meyer. *Die Juden als Patrioten.* Berlin: Albert Katz, 1898.

Kracauer, Siegfried. *From Caligari to Hitler: A Psychological History of the German Film.* Princeton, NJ: Princeton University Press, 1947.

Lässig, Simone. *Jüdische Wege ins Bürgertum. Kulturelles Kapital und sozialer Aufstieg im 19. Jahrhundert.* Göttingen: Vandenhoek & Ruprecht, 2004.

Lekan, Thomas M. *Imagining the Nation in Nature: Landscape Preservation and German Identity, 1885–1945.* Cambridge: Cambridge University Press, 2004.

Lessing, Theodor. "Jewish Self-Hatred." 1930. Reprinted in *The Jew in the Modern World: A Documented History,* edited by Paul Mendes-Flohr, and Jehuda Reinharz, 272–74. Oxford: Oxford University Press, 1995.

Lowenstein, Steven M. "Was Urbanization Harmful to Jewish Tradition and Identity in Germany?" *Studies in Contemporary Jewry* 15 (1999): 80–106.

Mann, Barbara. *Space and Place in Jewish Studies.* New Brunswick, NJ: Rutgers University Press, 2012.

Mayne, Judith. "Dracula in the Twilight: Murnau's *Nosferatu* (1922)." In *German Film and Literature,* edited by Eric Rentschler, 23–27. New York: Methuen, 1986.

Mendes-Flohr, Paul. *German Jews: A Dual Identity.* New Haven, CT: Yale University Press, 1999.

———. "The *Kriegserlebnis* and Jewish Consciousness." In *Judisches Leben in Der Weimarer Republik/Jews in the Weimar Republic,* edited by Wolfgang Benz, Arnold Paucker, and Peter Pulzer, 225–38. London: Leo Baeck Institute, 1998.

Mosse, George Lachmann. *German Jews beyond Judaism.* Bloomington: Indiana University Press, 1985.

Nietzsche, Friedrich. "Vom Probleme des Schauspielers." *Die fröhliche Wissenschaft, Werke.* Vol. 2. Munich: Carl Hanser Verlag, 1969.

Niewyk, Donald L. *The Jews in Weimar Germany.* Baton Rouge: Louisiana State University Press, 1980.

Olimsky, Fritz. "Das alte Gesetz." *Berliner Börsen-Zeitung,* 1 November 1923.

———. "Filmentwürfe von Künstlerhand." *Berliner Börsen-Zeitung,* 7 October 1923.

———. "*Peter Voß.*" *Berliner Börsen-Zeitung,* 26 March 1932.

Panizza, Oskar. "The Operated Jew." Translated by Jack Zipes. *New German Critique* 21 (1980) [1893]: 63–79.

"*Peter Voß.*" *Morgenpost* (Berlin), April 1932.

Prawer, Siegbert Salomon. *Between Two Worlds: The Jewish Presence in German and Austrian Film, 1910–1933.* New York: Berghahn Books, 2005.

"Purim im May-Atelier." *Lichtbild-Bühne,* 1923.

Rechter, David. "Geography Is Destiny: Region, Nation and Empire in Habsburg Jewish Bukovina." *Journal of Modern Jewish Studies* 7, no. 3 (2008): 325–37.

Roemer, Nils H. *German Cities, Jewish Memory: The Story of Worms.* Lebanon, NH: Brandeis University Press, 2010.

Rogowski, Christian. "Introduction: Images and Imaginaries." In *The Many Faces of Weimar Cinema,* edited by Christian Rogowski, 1–12. Woodbridge: Camden House, 2010.

Rosenthal, Berthold. *Heimatgeschichte der badischen Juden seit ihrem geschichtlichen Auftreten bis zur Gegenwart.* Bühl: Konkordia, 1927.

Schach, Fabius. "Das jüdische Theater." *Ost und West* 5 (1901): 347–58.

Schumann, Andreas. *Heimat denken.* Cologne: Böhlau, 2002.

Schütz, Erhard. "Berlin: A Jewish Heimat at the Turn of the Century?" In *Heimat, Nation, Fatherland: The German Sense of Belonging,* edited by Jost Hermand, and James D. Steakley, 57–86. New York: Peter Lang, 1996.

Seeliger, Ewald Gerhard. *Peter Voß, der Millionendieb.* Berlin: Ullstein, 1913.

Simon, Hermann, and Irene Stratenwerth. "Auszug aus dem Ghetto – das jüdische Drama im Film." In *Pioniere in Celluloid: Juden in der frühen Filmwelt,* edited by Hermann Simon and Irene Stratenwerth, 220–46. Berlin: Henschel, 2004.

Sorkin, David. "The Invisible Community: Emancipation, Secular Culture and Jewish Identity in the Writing of Berthold Auerbach." In *The Jewish Response to German Culture,* edited by Jehuda Reinharz and Walter Schatzberg, 195–211. Hanover, NH: University Press of New England, 1985.

St. Pierre, Paul Matthew. *E. A. Dupont and His Contribution to British Film: Varieté, Moulin Rouge, Piccadilly, Atlantic, Two Worlds, Cape Forlorn.* Cranbury, NJ: Fairleigh Dickinson University Press, 2010.

Umbach, Maiken. *German Cities and Bourgeois Modernism, 1890–1924.* Oxford: Oxford University Press, 2009.

Van Rahden, Till. *Jews and Other Germans: Civil Society, Religious Diversity, and Urban Politics in Breslau, 1860–1925.* Madison: University of Wisconsin Press, 2007.

Von Moltke, Johannes. *No Place Like Home: Locations of Heimat in German Cinema.* Berkeley: University of California Press, 2005.

Walk, Cynthia. "Romeo with Sidelocks: Jewish-Gentile Romance in E. A. Dupont *Das alte Gesetz* and Other Early Weimar Assimilation Films." In *The Many Faces of Weimar Cinema,* edited by Christian Rogowski, 84–101. Woodbridge: Camden House, 2010.

———. "31 January 1929: Limits on Racial Border-Crossing Exposed in Piccadilly." In *A New History of German Cinema,* edited by Jennifer M. Kapczynski and Michael D. Richardson, 185–95. Rochester, NY: Boydell & Brewer, 2014.

Wassermann, Jakob. *Mein Weg als Deutscher und Jude.* Berlin: G. Fischer, 1921.

Weltsch, Robert. "Theodor Herzl und wir." In *Vom Judentum. Ein Sammelbuch,* edited by H. Kohn et al., 155–64. Leipzig: Kurt Wolf Verlag, 1913.

Wildmann, Daniel. "Desire, Excess, and Integration: Orientalist Fantasies, Moral Sentiments and the Place of Jews in German Society as Portrayed in Films of the Weimar Republic." In *Orientalism, Gender, and the Jews Literary and Artistic Transformations of European National Discourses,* edited by Ulrike Brunotte, Anna-Dorothea Ludewig, and Axel Stähler, 137–55. Berlin: De Gruyter, 2014.

Zweig, Arnold. *Juden auf der deutschen Bühne.* Berlin: Welt-Verlag, 1928.

Zweig, Stefan. *The World of Yesterday: An Autobiography.* Lincoln: University of Nebraska Press, 1964 [1943].

6

Layered Pasts

The Judengasse *in Frankfurt and Narrating German-Jewish History after the Holocaust*

Michael Meng

This chapter discusses, on one level, the post-1945 history of the former space of the *Judengasse* (Jews' Lane) in Frankfurt am Main, while addressing, on another, a challenge central to historical narration that the example of the *Judengasse* illuminates and that the concept of "layered pasts" may be able to address, to some extent. The challenge presents itself to anyone engaged in narrating the past, including myself and those involved in my story here, the Frankfurters who, in the late 1980s, debated their city's Jewish past. What is the challenge? It might be expressed as the tension that exists between the fluidity and discontinuity of the past (*histoire*) and the order and continuity that narration (*récit*) imposes on the past from its position of hindsight.[1] As Raymond Aron put it, "Retrospect creates an illusion of fatality which contradicts the contemporaneous impression of contingency."[2] Historical narratives strive to capture the contingency of the past but do so from the vantage point of knowing the way history ends; *histoire* unfolds in a dynamic and unpredictable way, but *récit* organizes the past into a cohesive narrative.

The tension between *histoire* and *récit* is particularly evident in discussions about the place of the Holocaust in German-Jewish history: Can the complexity of German-Jewish history be told in a form that avoids the teleology of linear historical narration? Can the history of Jews in modern Germany be reconstructed in such a way that duly acknowledges the centrality of the Holocaust without reducing the complexity and richness of German-Jewish history to Nazism?

If not expressed precisely in this manner, such questions played a major role in a debate that erupted in Frankfurt am Main when, in 1987, construction workers stumbled upon remnants of a *mikveh* dating from the fifteenth century. The *mikveh* was a relic from Frankfurt's *Judengasse,* which had functioned as a compulsory, segregated, and enclosed area of Jewish residence for nearly 350 years.[3] The discovery of the *mikveh* sparked an intense debate about the city's Jewish past, known as the "Börneplatz Conflict." One dimension of this debate concerned the concrete issue of what physically should be done with the discovered ruins; another involved a protracted discussion about continuity and teleology, namely, how the early modern history of forcing Jews to live in an enclosed and segregated space relates to the Holocaust.

In a concrete way, the Börneplatz Conflict raised the tension between *histoire* and *récit.* And it did so in the politically divisive context of 1980s West Germany. While liberals and leftists in Frankfurt argued that the Holocaust represented a caesura in German history that ought to be remembered along with the long history of German persecution of Jews, conservatives resisted what they viewed as a leftist-liberal attempt to "shame" Germans into viewing their history from an excessively critical perspective.[4] Hence, several prominent politicians of the conservative Christian Democratic Party claimed that the history of the *Judengasse* had little connection to the Holocaust. In response, local politicians of the Social Democratic Party, among others, insisted that continuities did exist between the Nazi measures against the Jews and earlier periods of anti-Jewish persecution in German-speaking Central Europe.

This chapter analyzes this debate with the aim of examining both the political dynamics that shaped it and, more deeply, the theoretical issues that underpinned it. It discusses the postwar history of the former space of the *Judengasse* and the Börneplatz debate, before concluding with a brief theoretical discussion about narration. Inspired by this volume's focus on space, the final section will consider the extent to which the metaphor of "layered pasts" can capture, perhaps with greater flexibility, the complexity of the past than linear history can. If historians typically tend to narrate history according to a linear employment of time, perhaps the geological pattern of layers suggested by Sigmund Freud and developed by Reinhart Koselleck suggests a partial way out of the dilemmas of teleology.[5] This spatial imagination of history may help us see the multiple layers of the past that overlap in spaces such as Frankfurt's *Judengasse.* Before approaching the concept of layered pasts, however, let us turn first to the early postwar years, when urban reconstruction erased many markers of Germany's past in the rush to build new modern cities. It is within the dialectic of urban modernism—its negation of the old in the construction of the new—that the postwar history of Frankfurt's *Judengasse* must first be placed.

An Erased Past

Two prominent travelers to early postwar West Germany, Hannah Arendt and Amos Elon, arrived at similar observations about the country's urban reconstruction: it was unfolding at a strikingly quick pace and in a remarkably uniform style. Modernist buildings and planning projects were rapidly being implemented across West Germany. This rush to build in a style seemingly sanitized of the most immediate past troubled both Arendt and Elon.[6] Modernism seemed to provide an escape from the burdens of the past. In 1950, Arendt detected a boastful sense in West Germany that the country was becoming the "most modern" in Europe.[7] Traveling to Germany a decade later, Elon saw a country built up even more completely.[8]

In truth, West Germany was not quite as completely modern as Elon and Arendt made it out to be, for alongside the pedestrian precincts of capitalist modernity stood historically preserved or reconstructed buildings and streets. Efforts to keep some elements of the old city centers, the *Altstädte,* animated discussions about postwar reconstruction across West Germany. Historical preservationists contested contemporary processes of modernization, claiming that modernist architecture and planning would efface the particularity and individuality of the city unless some elements of the past were preserved. This critique carried some weight in West Germany, despite the dominance of urban modernization. While the position of historical preservation in West Germany's reconstruction was complex, one would be hard pressed to find a German town without at least one historical building in it that had been reconstructed in the historical style.[9]

Preservationists found themselves compelled to make choices regarding which buildings should be restored, preserved, or reconstructed. Although preservationists had to let many buildings go to the wrecking ball, they tended to show less interest in preserving one type of building in particular: the communal spaces of Germany's prewar Jewish community. Jewish sites were generally ignored and neglected across Germany. In the early decades after the war, Jewish spaces rarely appeared to urban planners, historic preservationists, or perhaps even to ordinary citizens as "valuable" elements of the urban landscape that should be preserved, reconstructed, or commemorated. Why this might have been so is a complex question involving many levels of explanation; one might think about it in terms of German silence about the Holocaust, lingering prejudices against Jews, the absence of Jewish space in postwar discourses of urban loss, exclusive notions of cultural heritage, or the destructive dialectic of urban modernization.[10]

In the case of Frankfurt, the erasure of the former space of the *Judengasse* and Börneplatz should be viewed in light of the massive changes that Arendt and Elon observed across West Germany, changes that affected Frankfurt in an especially acute manner. Frankfurt embraced urban modernization to an extraordinary degree. Frankfurt's medieval city center was rebuilt anew almost entirely, and its layout was changed substantially, including the area of the former *Judengasse.*[11] In 1945, nearly all the buildings on Börnestrasse constructed in the late nineteenth century had been completely destroyed.[12] The area was cleared for, and altered by, the construction in the mid-1950s of the four-lane Kurt-Schumacher-Strasse, a north-south thoroughfare that ran through the main part of the former *Judengasse* and through the area where the Hauptsynagoge had once stood, the city's principal synagogue, unveiled in 1860 and destroyed by the Nazis in 1938. According to a 1957 city map, only two areas of the former *Judengasse* still existed, a sliver of its northernmost section and Börneplatz.[13]

By the mid-1950s, nothing remained of Frankfurt's Jewish past in this area except two ruins. Although a Nazi report claimed that all traces of the Börneplatzsynagoge had been "completely cleared," it appeared to be wrong.[14] In 1957, a Frankfurt resident wrote to the city's mayor to request that the city properly clear the space of the former synagogue, since some remnants were being handled profanely:

> On Börneplatz in Frankfurt am Main there still exists today a remaining part of the formerly noble synagogue, which the Nazis damaged. Among this remaining part there is today a parking lot for old cars. Before the former *aron kodesh* (holy shrine) dirty junk is lying around. This is a scandalous desecration of a once holy Jewish site.[15]

The mayor's office responded that the area would be cleaned up in due course, to which the resident wrote back pleading that the city not continue to allow the desecration of "a remaining part of a former synagogue."[16] The space was cleared soon thereafter to make room for the construction of a flower market, which then existed until the 1970s.

The other ruin, which still stands today, is a piece of the Staufenmauer, a wall first erected in 1180, originally to protect the city of Frankfurt, which by the late fifteenth century was functioning as the western enclosure of the *Judengasse*. In 1711, the wall was damaged by fire and rebuilt. The ruin that survives is a small portion of the rebuilt wall, which was never torn down during Frankfurt's defortification in the early nineteenth century.[17] Hidden for decades among rows of newly built houses, the wall was only rediscovered after 1945, when workers clearing the rubble happened upon it.[18]

Postwar urban planners initially envisaged tearing down the Staufenmauer to make room for connecting Töngesgasse with Allerheiligenstrasse across Kurt-Schumacher-Strasse. But these plans were not realized, and the Staufenmauer was restored after the war as "a venerable and such an important ruin for the physical development of the city."[19] In press clippings about the wall, hardly any mention was made of its relationship to the *Judengasse*. The most extended journalistic discussion of the *Judengasse* during the early postwar years appeared in a three-paragraph article published in the *Frankfurter Neue Presse* in 1956, which confined itself to the basic facts: that the last remnants of the former *Judengasse* were "disappearing" as new streets, houses, and stores were being built in the area.[20]

A Reappearing Past

Cities are like palimpsests. Past layers of a city's history stubbornly remain, if in highly fragmentary and complex ways, as Sigmund Freud noted in *Civilization and Its Discontents*. In historical cities such as Rome, multiple layers of time overlap and interact in the urban landscape. While new layers of time settle over those of the past, some traces of the past always seem to re-emerge. Elements of the past that had been ignored for years, decades, or even centuries can come unexpectedly into view again.[21]

Frankfurt is, of course, very different from Rome; the massive destruction of World War II erased Frankfurt's past in a near total way that Freud may not have been able to imagine. Nevertheless, his metaphor seems apt here. Even in a city as profoundly destroyed as Frankfurt, layers of the past can suddenly reappear and provoke interest among residents. It was precisely this that happened regarding Dominikanerplatz/Börneplatz in the late 1970s as the square's history started to be recollected in the public sphere. Although commemorations of this space can be found earlier,[22] the late 1970s represent the beginning of a turning point in public efforts to remember Frankfurt's Jewish history. This emergence of interest in the city's Jewish past can be explained in several ways. Generally speaking, it emerged out of a growing sensitivity to the destructive effects of urban modernization and planning in Germany and elsewhere, a sensitivity that began to develop in the early 1970s as exemplified by the *Häuserkampf* in Frankfurt's Westend, a series of protests against gentrification and urban renewal in the city.

The effort to retell the *Judengasse*'s history began with Paul Arnsberg, a journalist and historian who escaped to Palestine in 1933 and returned to Frankfurt in 1958. Arnsberg wrote a three-volume history of the *Judengasse*, which appeared in 1983, after his death in 1978. A year before he died, Arnsberg sent a letter to Frankfurt's city councilor for cultural affairs, expressing his discomfort with the postwar handling of Dominikanerplatz.[23] He proposed converting the area into a green space and erecting a monument to chronicle the history of the Frankfurt Jewish community. Arnsberg also requested that the name of the square be changed to Börneplatz.

The square's name was changed to Börneplatz in November 1978, barely more than a year after Arnsberg's request. His other suggestion, however, went nowhere. Instead, city authorities decided to sell the parcel of land to its municipal utilities group, which planned to use the plot to build an administrative and service center. In July 1984, an architectural competition for the center was announced, without prior consultation of Frankfurt's Jewish community or the Kirchheimische Stiftung (an organization devoted to the preservation of Frankfurt's Jewish history). Salomon Korn, then one of the leading figures of Frankfurt's Jewish community, himself an architect, and currently the vice president of the Central Council of Jews in Germany, subsequently met with city administrators about their plans for Börneplatz. The assurances he received that the Jewish community would be more involved in the process were largely not met during the project's initial conceptualization. Frustrated by its marginalization, the board of the Frankfurt Jewish community issued a press release calling for greater sensitivity to the historical and symbolic importance of Börneplatz.[24]

By winter 1985, the CDU-led city government had heeded some of these concerns. It proposed constructing what it called the "new Börneplatz," on which a Jewish memorial would be erected.[25] Meanwhile, something rather remarkable happened: interest in Frankfurt's Jewish past increased among some segments of society as it did throughout West Germany more broadly for a complex set of reasons.[26] Throughout the mid-1980s, an alliance of Jewish community members, historians, Social Democrats, Greens, church leaders, and other residents interested in Frankfurt's Jewish history engaged in a series of discussions about the history of Börneplatz and the need to remember it. The discussion quickly became highly emotive. A number of Frankfurters expressed deep frustration with the postwar transformation of the area into a traffic intersection. "The former center of Jewish life in Frankfurt," wrote Eva Demski in an open letter signed by several dozen Frankfurters, "is today one of the ugliest, loudest sites in this city."[27] Another appeal, signed by 130 professors of Frankfurt's university, suggested that erecting on this space a "profane administrative building for the city's electricity, water, and gas utilities" would mean the effacement by "concrete and bureaucratic routine" of any possible attempts to remember and mourn the loss of Frankfurt's prewar Jewish community. "It would be a final victory of administration—and of gas, of all things—over the remembrance of Jewish life in this city."[28] The allusions to the bureaucratic processing of death by Nazi officials such as Adolf Eichmann and to the National Socialist method of killing through gas could hardly be missed.

Despite these appeals and internal attempts made by the SPD, the Greens, and Jewish community leaders to persuade city administrators to rethink the project, the CDU-led city government proceeded. Site excavation began in early 1987 before coming to a halt in mid-May when ruins from the *Judengasse* were discovered. This discovery injected tremendous energy into the movement opposing the project. A civic group, "Save Börneplatz" (*Aktionsbündnis – Rettet den Börneplatz*), was established to save the ruins and stop the construction of the administrative center. As city authorities showed no signs of listening to their demands, a group of about thirty Frankfurters occupied the construction site on 28 August 1987, and Save Börneplatz organized a demonstration attended by about eight hundred people. A handful of demonstrators then occupied the site for several more days until, on 2 September 1987, some two hundred police officers forced the protesters off the area. A metal fence was erected around the construction site. While the administrative center was in the end erected on Börneplatz, some of the archaeological ruins of the *Judengasse* were reconstructed and placed inside the building in an exhibition devoted to the history of the *Judengasse*. A memorial to the Holocaust was also erected on the southeastern part of Börneplatz.

What was at stake for Jewish and non-Jewish Frankfurters engaged in these protests? Both Jews and non-Jews viewed the *Judengasse* as a historically distinctive space in a city with very few historical markers. The city's postwar history of modernization was critiqued for producing a monotonous and dehistoricized urban landscape. Urban modernism had greatly contributed to effacing Frankfurt's particularity after the war, and those involved in the effort to preserve Börneplatz felt it was time to prevent it from doing so again.

In short, Jews and non-Jews were engaging in one of the fundamental tasks of collective memory, that is, nourishing bonds of continuity and permanence.[29] Jews and non-Jews sought to establish a form of continuity with the past that had been violently severed by Nazism and forgotten for years after the war. This was a fraught undertaking. Jews tended to emphasize both the importance of mourning the loss of those killed during the Holocaust and situating the post-1945 Jewish community within the long history of Jewish life in the city. When by the 1980s post-1945 Jewish life in West Germany seemed more secure than it perhaps ever had been, Jews sought to establish a sense of continuity with their predecessors. Moreover, Jews in Frankfurt were particularly sensitive to matters relating to the past after the heated discussion over an attempt to stage Rainer Werner Fassbinder's controversial play *Garbage, the City, and Death* (written in 1975) at Frankfurt's Schauspielhaus in October 1985. About thirty members of the Frankfurt Jewish community occupied the stage and prevented the play from being performed in light of its ambiguous portrayal of Jews and its handling of the topic of anti-Semitism.[30]

The motives of non-Jewish Frankfurters for nourishing ties of continuity with their city's Jewish past were different; non-Jewish Frankfurters expressed a desire to view Jewish history as part of "their" history, wishing to integrate Jewish history into their understanding of the past after its erasure during the National Socialist period and the early postwar eras. Also, non-Jewish Germans saw the preservation and memorialization of the *Judengasse* as providing a measure of justice for past wrongs for which they were partly responsible as Germans.

But these basic differences should not obscure two key similarities that tie together Jewish and non-Jewish participants in the Save Börneplatz initiative. First, both Jews and non-Jews critiqued their city's disregard for the past based on a common rejection of urban modernization, a rejection that had been growing throughout the 1970s and 1980s in Germany and other parts of the world. Frankfurt became, in fact, a major site of protest against urban renewal projects. Second, at stake for both Jewish and non-Jewish Frankfurters was the issue of how the past should be narrated. The publicist and writer Micha Brumlik put the issue concisely in 1987: "Should the community's writing of history or museum politics primarily serve enlightenment and solidarity with the victims of history, or should it aim at augmenting the positive feeling of self-worth of its citizens— even if such augmentation rests on omissions and misinterpretations?"[31] The choice, Brumlik suggested, was between two kinds of histories: a national history that reprises positive national narratives and traditional patterns of identifica-tion, on the one hand, and a critical national history along the lines proposed by Theodor Adorno and Jürgen Habermas, on the other—the latter interrogating conventional national narratives by unearthing subaltern and antiheroic histories that have long been marginalized or forgotten by the national community.

Almost as if he knew where the debate was going, the question Brumlik asked emerged at the center of the discussion just a few days later in the wake of several statements made by the city's leading CDU politicians. The CDU-led city administration had remained insistent on continuing the project as planned partly because, it argued, millions had already been spent on or committed to it. This financial argument found understanding from some opponents of the project. But the tone of the debate, and the issue at its heart, changed when, on 5 September 1987, Walter Wallmann, former mayor of Frankfurt and at the time minister-president of the German federal state of Hesse, gave a speech to the local branch of the Frankfurt CDU, in which he said that the history of the *Judengasse* should be "no cause for shame." "It is not correct," he claimed, "that Börneplatz and the Jewish ghetto have something to do with Auschwitz. And it is thus not correct that a direct path led from this ghetto to Auschwitz." Rather, he suggested, the *Judengasse* should be celebrated and remembered as a space of emancipation: the ghetto was, after all, torn down in "the spirit of the Enlightenment."[32] Two weeks later, Frankfurt's mayor, Wolfram Brück (CDU), supported his colleague's positive interpretation of the *Judengasse*'s history, expressing the view that the *Judengasse* had functioned as a "refuge" for Jews in Germany and, moreover, that it was "not Christian anti-Judaism, but the [errant] path of the German people since the Enlightenment that led to the German catastrophe."[33]

This interpretation of the past came at an especially crucial moment in West German debates about German history. It appeared as West Germany was engaged in intensely political discussions about the Nazi past, not least during the Bitburg controversy and the *Historikerstreit*. Even more importantly, it emerged in an era when the Christian Democrats, led by Kohl, had initiated an effort to stress the "positive" dimensions of German history. In so doing, the Christian Democrats sought more than merely to perpetuate a Whiggish interpretation of German history. They intended to supplant, or at least diminish, what they viewed as an overly negative portrayal of German history by internal and external revisionists: by West Germany's liberal-left intelligentsia that was advancing a "non-German German identity" around the memory of Nazi violence and by the purveyors of Holocaust memory in the United States who were purportedly solidifying an "anti-German" image of the Federal Republic through the creation of institutions such as the United States Holocaust Memorial Museum.[34] An angst around the memory of the Holocaust shaped the CDU's politics of history throughout the 1980s.[35]

The CDU's intent was to suggest that a positive history was possible for Germany even after Auschwitz. In Frankfurt, Wallmann and Brück went further, arguing that the history of Jews in their city could support such a positive reconstruction of the past. The response to this revisionist attempt on history was sharp. The board of the city's Jewish community said that the mayor's characterization of the *Judengasse* as a space of protection for Jews was "false and contradicts the historical facts"; "the Frankfurt ghetto," the board's statement continued, "served to oppress, disenfranchise, exclude and discriminate against

Jews."[36] The Börneplatz initiative wrote that the mayor was attempting to offer not just a "neoconservative, but a *völkisch* conception of history," an understanding of the past "without any feeling of guilt."[37]

Layered Pasts and Repetitions

The Börneplatz debate turned into a political conflict about history and identity. In this sense, it mirrored the other debates about the past that West Germany experienced in the 1980s. But one topic in the Frankfurt debate that did not appear as explicitly in the other discussions of the 1980s warrants attention for the important issue about narrating German-Jewish history it raises—namely, continuity, or, more precisely, the interrelationship among temporal periods.[38] This issue was brought into sharp relief in Frankfurt because of the multi-temporal nature of the Börneplatz debate. Indeed, the extent to which the ruins of the *Judengasse* and the history they represent should or could be connected to the Holocaust inescapably dominated the debate.

When the ruins of the *Judengasse* emerged into view, they were understood within the contemporary political concerns of 1980s West Germany. A layer of history far removed from the world of the late twentieth century quickly became assimilated into the concerns of that world. One example of this temporal assimilation was the circulation of the word "ghetto" throughout the debate. "Ghetto" originates from early sixteenth-century Venice, where, in 1516, the Venetian senate permitted Jews to stay in the city on the condition that they live on a single island known as Ghetto Nuovo (New Ghetto). The island gained that name because it served as the dumping ground of waste from the Ghetto Vecchio, the site of an old brass foundry.[39] The *Judengasse*, as far as we know, was not referred to as a "ghetto" until the mid- to late nineteenth century (it was called the *Judengasse*).[40]

Now, if by "ghetto" we mean a compulsory, segregated, and enclosed area of Jewish residence, we may technically call the *Judengasse* a ghetto. But this is no mere technical matter. For the word "ghetto" provokes emotions and memories that far outstrip the putatively neutral facts that one may wish to claim the term represents. As a term, "ghetto" has accrued multiple layers of meaning over the past five hundred years that cannot easily be divorced from the word; it is entangled with such diverse histories as Jewish emancipation, Eastern European Jewry, shtetl nostalgia, Nazism, racial segregation, apartheid, the *banlieues,* and *Parallelgesellschaften* (the term used in present-day Germany to refer to the de facto segregation of migrant groups). In the course of the Börneplatz debate, two historical contexts became associated with "ghetto": the emancipatory history of Jews moving "out from the ghetto" throughout the nineteenth century and the history of the Nazi creation of over one thousand ghettos in Eastern Europe. A third historical context, though rarely mentioned, was the contemporary perception that West Germany's Turkish migrant population had segregated itself into "ghettos."[41] The only mention of this postwar context was made by writer and publicist Alfred Grosser who invoked "ghetto" to criticize the exclusion of migrants by German society. The "ghetto" of the *Judengasse,* he said, "should also cause us to be aware of other ghettos," including the "ghettos of today [in which] Turks and Muslims [are enclosed] among us."[42]

The Nazi period was, however, the most significant context for the Börneplatz debate. The use of the word "ghetto" in 1980s West Germany suggested a possible relationship between the *Judengassen* of early modern German-speaking Central Europe and the ghettos of Nazi-occupied Eastern Europe. One either affirmed or denied a connection between these two historical periods. While Walmann and Brück asserted there was no relationship, activists from the Börneplatz initiative insisted there was one: "from the ghettos of the Middle Ages and the early modern period, which the Börneplatz represents, a path was led to Treblinka and Sobidor, if not a *direct* path yet a *path* nevertheless."[43] Another activist implied an even stronger link between the Holocaust and Frankfurt's *Judengasse,* writing that the ruins represented "no longer only physical witnesses of the oppression in the ghetto, but also—and perhaps above all—a symbol of the catastrophe of the Holocaust."[44] Occasionally, some participants were more hesitant about making such claims. After implying a connection between the *Judengasse* and the Nazi ghetto, Ignatz Bubis, chair of the Jewish Community in Frankfurt, expressed concern about viewing history in too teleological a manner: "But I do not want to understand this ghetto that existed for centuries, despite all the discrimination and oppression of Jewish life that grew from it, as a precursor to the ghetto of the twentieth century, [evidencing] a seamless connection to the medieval ghettos."[45]

Bubis's remark raises one of the key challenges in writing German-Jewish history in particular and history in general—the dilemma of establishing continuity while at the same time resisting teleology. This challenge exists for anyone involved in recollecting the past, because a central tension runs through the practice of historical reconstruction itself. Historians seek to reconstruct the unfolding of something into what it is not yet, with full knowledge of what it already is. How might historians resist forcing their subject matter to conform to an end that they identify and for which they then seek antecedent conditions? How might historians describe the openness and spontaneity of their subject matter when they have foreknowledge of its conclusion?

Perhaps the dilemma might be described in terms of a tension between content and form. While history deals with a fluid and indeterminate content, its governing form—the narrative—stabilizes the past by virtue of its very ordering of time in a permanent account of the past; narratives contain the past within a general form of explanation. The most common form has been the linear narrative, which orders the past based on the establishment of endpoints. Linear narratives prioritize endpoints. The end shapes the historical narrative.

Could history be told without ends? Imagine a piece of history as free as possible from linear narration. What would it be? Perhaps it would be something akin to the "novels" of Raymond Federman that seek to defy conventional narrative patterns through detours, digressions, and circumvolutions even as they return to the common theme of the Holocaust.[46] If historians would likely resist going as far as Federman, this resistance alone is revealing: it underscores the challenge of locating less teleological ways of thinking about the past.

In other words, the challenge is to identify a conceptual pattern that might capture the complexity and contingency of the past more subtly than the linear pattern seems capable of doing. One possibility might be suggested by spatial history. In *Zeitschichten,* Reinhart Koselleck proposes the geological notion of "layered time" as an alternative to linear history.[47] If Koselleck's metaphor of layered time does not seek to resolve the problem of teleology and the stabilization of the past into historical narratives—"the suspension of history [*Stilllegung der Geschichte*]" into narrative form, to borrow Martin Heidegger's concise formulation—it seeks to understand the past in terms of multiple temporalities.[48] As geologists study the history of the Earth preserved through the strata of its sedimentary rocks, so, too, historians might consider their subject matter as made up of multiple layers that at one and the same time interact with one another and exist on their own.

Yet, Koselleck's spatial metaphor can only go so far. Let us return to Freud, who recognized the epistemological challenge of the layered metaphor. Freud's analogy of human memory to ancient Rome reveals the impossibility of locating a "pictorial" representation that can adequately capture the complexity of historical reconstruction. His analogy suggests that human recollection may be partial and uncertain. If we know the layers of the past that appear to us, we may never know about the many other histories of human activity that have been washed away and built over. If it is true that suddenly, as in the case of Frankfurt in 1987, a layer of the past unknown for centuries can suddenly erupt into view, such discoveries only remind us of the many others pasts that are never to be recovered because they have been permanently erased or absorbed by another layer of time. We may not even know about their erasure; by definition we cannot know what remains unknown to us.

Our knowledge of the cultural, economic, religious, intellectual, and social history of Jews living in the *Judengasse* over some 350 years cannot be anything but fragmentary; hence, the caution Bubis showed toward drawing lines of continuity between the *Judengasse* and the Holocaust might be taken as a caution about reducing a barely known multilayered past to one of its more visible layers as well as a caution about asserting with certainty an understanding of the temporal relationship among those layers of the past. For if one cannot claim to know the whole history of the *Judengasse,* to see all the layers of its past in one total and unitary view, then how can one assert with certainty the existence of lines of either continuity or discontinuity? If one cannot see all of the layers, how can one claim to know how they relate to one another?

This does not mean to suggest that historians should set the issue of temporal relationships aside. Rather, it means that historians may need a more precise way of coming at these relationships than that offered by the concept of continuity, which presupposes a complete whole in which all parts fit together without any spatial or temporal gaps. Perhaps another concept, suggested by the common human experiences of time that partly inspired Koselleck's notion of layered temporalities, might prove more adequate. Our quotidian experience of time is repetition: birth, life, and death.

When faced therefore with the challenge of thinking about the *longue durée* relationship of the Holocaust to the history of the *Judengasse*, we might perceive a fundamental repetition in the recurrence of an idea across the layers of time visible to us: the idea that Jews represent a minority population viewed as distinct from and unequal to the majority. The repetition lies in the negative idea of the "Jew" as a figuration of anything that the majority marks as different, distasteful, and dangerous.[49] To say this intends neither to imply the inevitability of this notion nor to diminish the radically new character of Nazism and its hatred of Jews. Nazism emerged as a revolutionary movement that aimed to rescue Germany from what it saw as its apocalyptic end in modern life and to create a new civilization freed of the tedium and rootlessness that it associated with "modernity." The primary figure of the "modern" for the Nazis was the rational, liberal, Communist, and cosmopolitan Jew.[50] For the Nazis, the Jew had to be excluded, segregated, expelled, and ultimately killed as both a symbol and agent of modernity. Nazism enacted, then, a revolutionary repetition of anti-Judaism: it repeated negative ideas and stereotypes of Jews found in the Western tradition since Paul but radically rejected, ultimately, previous solutions to the "Jewish question"—not least ghettoization—in favor of eliminating once and for all what it perceived as the Jewish menace.

In sum, this brief theoretical discussion has, if anything, attempted to bring out the sheer complexity of thinking about temporal connections, a multifaceted issue that lay at the heart of the Börneplatz Conflict and discussions about the place of the Holocaust in modern German history more broadly. Although some participants in the Frankfurt discussion emphasized this complexity, most tended to make generalizations about continuity or discontinuity that are difficult to sustain when the issue of historical narration is considered more deeply. Yet to make this point is not to render impossible thinking about temporal connections. On the contrary, the spatial history of the *Judengasse* may provide a compelling temporal pattern for emphasizing both what separates and what unites multiple layers of time. While historians can mark the distinctiveness of one layer of time over another, they can at the same time identify the substratum—the repeated element—that courses through these layers.

Michael Meng is associate professor of history at Clemson University. His first book *Shattered Spaces: Encountering Jewish Ruins in Postwar Germany and Poland* was published in 2011 by Harvard University Press. He coedited with Erica Lehrer the volume *Jewish Space in Contemporary Poland*, which was published by Indiana University Press in 2015.

Notes

1. Genette, *Narrative Discourse.*
2. Quoted in Grethlein, *Experience and Teleology in Ancient Historiography,* 4.
3. From this perspective, the *Judengasse* can, retrospectively, be called a ghetto according to Ravid's definition in "All Ghettos."
4. Grossmann, "The 'Goldhagen Effect'"; Gassert and Steinweis, *Coping with the Nazi Past;* Herzog, *Sex after Fascism;* Moses, *German Intellectuals and the Nazi Past.*
5. It is uncontroversial to suggest that the dominant pattern by which *récit* organizes *histoire* is linear. The more interesting issue is to ask whence this pattern emerges historically. The importance of Judeo-Christian eschatological notions of time seems decisive. See Löwith, *Meaning in History;* Heidegger, *Überlegungen,* 4.
6. Diefendorf, *In the Wake of War.*
7. Arendt, "The Aftermath of Nazi Rule," 343.
8. Elon, *Journey through a Haunted Land,* 13.
9. Campbell, "Resurrecting the Ruins, Turning to the Past"; Koshar, *Germany's Transient Pasts;* Rosenfeld, *Munich and Memory.*
10. Meng, *Shattered Spaces.*
11. Müller-Raemisch, *Frankfurt am Main.*
12. Institut für Stadgeschichte (hereafter Ifs), *Frankfurt am Main Stadtplan, 1957,* no. 1c (Konstabler Wache).
13. Ibid.
14. IfS, Magistratsakten, 5.800, 97, Bauamt, Abbruch der Synagogen, XXVI. Meldung, 12 June 1939.
15. IfS, Magistratsakten 971, unpaginated, letter of 28 April 1957.
16. IfS, Magistratsakten 971, unpaginated, letter of 18 May 1957, and letter of 15 July 1957.
17. On the city's defortification, see Mintzker, *Defortification of the German City.*
18. IfS, Sammlung Ortgeschichte, S3/G, 18.551, "Die staufische Stadtmauer. Rücksicht in der Bebauung der Umgebung," Pressestelle der Stadt Frankfurt am Main, 15 June 1953.
19. Ibid.
20. IfS, Sammlung Ortgeschichte, S3/G, 18.551, "Bagger an der Staufenmauer. Frankfurts ältestes Stadtgebiet wird aufgebaut," *Frankfurter Neue Presse,* 17 October 1956.
21. Freud, *Civilization and Its Discontents.*
22. See, for example, the 1966 architectural sketch for a memorial to be located on Börneplatz, which is on display in the permanent exhibition of the Judengasse Museum, Frankfurt am Main.
23. Letter from Arnsberg of 19 July 1977, reprinted in Demski et al., *Frankfurter Börneplatz,* 21.
24. Statement by Jewish Community of January 1985, reprinted in Demski et al., *Frankfurter Börneplatz,* 22–23.
25. Salomon Korn, "Der neue Börneplatz," February 1989, reprinted in Demski et al., *Frankfurter Börneplatz,* 23–24.
26. See Meng, *Shattered Spaces.*
27. Open letter by Eva Demski, February 1986, reprinted in Demski et al., *Frankfurter Börneplatz,* 26–28.
28. Open letter by Valentin Senger, June 1986, reprinted in Demski et al., *Frankfurter Börneplatz,* 29–30.

29. The following remarks are based on my reading of the extensive material from the debate republished in Demski et al., *Frankfurter Börneplatz*, and conserved in the IfS.
30. Markovits, Benhabib, and Postone, "Rainer Werner Fassbinder's *Garbage, the City and Death.*"
31. Speech by Micha Brumlik, 29 August 1987, reprinted in Demski et al., *Frankfurter Börneplatz*, 79.
32. Speech by Walter Wallmann, 5 September 1987, reprinted in Demski et al., *Frankfurter Börneplatz*, 99–100.
33. Speech by Wolfram Brück, 17 September 1987, reprinted in Demski et al., *Frankfurter Börneplatz*, 113, 116.
34. The concept of the "non-German German identity" is from Moses, *German Intellectuals*, 229–45.
35. Eder, *Holocaust Angst*. My thanks go to Jacob Eder for providing me with a copy of his manuscript.
36. Statement by the board of the Frankfurt Jewish community, 21 September 1987, reprinted in Demski et al., *Frankfurter Börneplatz*, 119.
37. Statement by the Börneplatz Initiative, September 1987, reprinted in Demski et al., *Frankfurter Börneplatz*, 120.
38. The issue of continuity is of course hardly new; it has shaped discussions about German history for the past seventy years now, and more broadly, it stands at the center of writing history itself. Blackbourn and Eley, *Peculiarities of German History*; Sheehan, "Paradigm Lost?"; Smith, "*Sonderweg* Debate."
39. On the word "ghetto," see Michman, *Emergence of the Jewish Ghettos during the Holocaust*, 20–31; Ravid, "All Ghettos"; Ruderman, "The Cultural Significance of the Ghetto in Jewish History"; and the four essays by Cala, Heyde, Lipphardt, and Steffen in *Simon Dubnow Institute Yearbook* 4 (2005).
40. I base this on my research on the history of the Frankfurt *Judengasse* during the late eighteenth and nineteenth centuries conducted in the archive collection held by the IfS.
41. "Die Türken kommen—rette sich, wer kann," *Der Spiegel* 31 (1973): 24–34.
42. Speech by Alfred Grosser, 4 September 1987, reprinted in Demski et al., *Frankfurter Börneplatz*, 92.
43. Brumlik, "Erinnern und Erklären," 13. Original emphasis.
44. Ness, "Wenn Leben Erinnern heißt—Gedanken zu Walter Benjamins Gleichnis vom Engel der Geschichte," 14.
45. Speech by Ignatz Bubis, 9 November 1987, reprinted in Demski et al., *Frankfurter Börneplatz*, 125.
46. See, e.g., Federman, *Double or Nothing*, 2.
47. Koselleck, *Zeitschichten*.
48. Heidegger, *Die Frage nach dem Ding*, 33.
49. Nirenberg, *Anti-Judaism*.
50. Traverso, *Origins of Nazi Violence*.

Bibliography

Arendt, Hannah. "The Aftermath of Nazi Rule: Report from Germany." *Commentary* 10, no. 4 (1950): 342–53.
Backhaus, Fritz, Gisela Engel, Robert Liberles, and Margarete Schlüter, eds. *The Frankfurt Ju-*

dengasse: Jewish Life in an Early Modern German City. Portland, OR: Vallentine Mitchell, 2010.

Blackbourn, David, and Geoff Eley. *The Peculiarities of German History: Bourgeois Society and Politics in Nineteenth-Century Germany.* Oxford: Oxford University Press, 1984.

Brumlik, Micha. "Erinnern und Erklären. Unsystematische Überlegungen eines Beteiligten zum Börneplatz-Konflikt." *Babylon. Beiträge zur jüdischen Gegenwart* 3 (1988): 9–17.

Cala, Alina. "'The Discourse of 'Ghettoization'—Non-Jews on Jews in 19th- and 20th-Century Poland." *Simon Dubnow Institute Yearbook* 4 (2005): 445–58.

Campbell, Brian. "Resurrecting the Ruins, Turning to the Past: Historic Preservation in the SBZ/GDR 1945–1990." PhD diss., University of Rochester, 2005.

Demski, Eva, Bettina Mähler, Hans Sarkowicz, and Michael Best. *Der Frankfurter Börneplatz.* Frankfurt am Main: Fischer Taschenbuch Verlag, 1988.

Diefendorf, Jeffry M. *In the Wake of War: The Reconstruction of German Cities after World War II.* Oxford: Oxford University Press, 1993.

Eder, Jacob. *Holocaust Angst: The Federal Republic of Germany and Holocaust Memory in the United States since the Late 1970s.* Oxford: Oxford University Press, 2016.

Eley, Geoff. *The "Goldhagen Effect": History, Memory, Nazism—Facing the German Past.* Ann Arbor: University of Michigan Press, 2000.

Elon, Amos. *Journey through a Haunted Land: The New Germany.* Translated by Michael Roloff. New York: Holt, Rinehart and Winston, 1967.

Federman, Raymond. *Double or Nothing.* 3rd ed. Boulder, CO: Fiction Collective 2, 1998. First published by Swallow Press, 1971.

Freud, Sigmund. *Civilization and Its Discontents.* Translated by James Strachey. New York: Norton, 2005 [1930].

Gassert, Philipp, and Alan E. Steinweis. *Coping with the Nazi Past: West German Debates on Nazism and Generational Conflict, 1955–1975.* New York: Berghahn Books, 2007.

Genette, Gérard. *Narrative Discourse: An Essay in Method.* Translated by Jane E. Lewin. Ithaca, NY: Cornell University Press, 1983.

Grethlein, Jonas. *Experience and Teleology in Ancient Historiography: Futures Past from Herodotus to Augustine.* Cambridge: Cambridge University Press, 2013.

Grossmann, Atina. "The 'Goldhagen Effect': Memory, Repetition, and Responsibility in the New Germany." In *The "Goldhagen Effect": History, Memory, Nazism—Facing the German Past,* edited by Geoff Eley, 89–129. Ann Arbor: University of Michigan Press, 2000.

Heidegger, Martin. *Die Frage nach dem Ding.* Tübingen: M. Niemeyer, 1962.

———. *Überlegungen VII–XI (Schwarze Hefte 1938/1939)* GA 95. Frankfurt: Vittorio Klostermann, 2014.

Herzog, Dagmar. *Sex after Fascism: Memory and Morality in Twentieth-Century Germany.* Princeton: Princeton University Press, 2007.

Heyde, Jürgen. "The 'Ghetto' as a Spatial and Historical Construction—Discourses of Emancipation in France, Germany, and Poland." *Simon Dubnow Institute Yearbook* 4 (2005): 431–43.

Hutton, James, ed. *Aristotle's Poetics.* New York: Norton, 1982.

Koselleck, Reinhart. *Zeitschichten: Studien zur Historik.* Frankfurt: Suhrkamp, 2000.

Koshar, Rudy. *Germany's Transient Pasts: Preservation and National Memory in the Twentieth Century.* Chapel Hill: University of North Carolina Press, 1998.

Lipphardt, Anna. "'Dos amolike yidishe geto'—Blick auf das jüdische Viertel in Vilne." *Simon Dubnow Institute Yearbook* 4 (2005): 481–505.

Löwith, Karl. *Meaning in History.* Chicago: University of Chicago Press, 1949.

Markovits, Andrei S., Seyla Benhabib, and Moishe Postone. "Rainer Werner Fassbinder's *Garbage, the City and Death*: Renewed Antagonisms in the Complex Relationship between Jews and Germans in the Federal Republic of Germany." *New German Critique* (1986): 3–27.

Meng, Michael. *Shattered Spaces*. Cambridge, MA: Harvard University Press, 2011.

Michman, Dan. *The Emergence of Jewish Ghettos during the Holocaust*. Cambridge: Cambridge University Press, 2011.

Mintzker, Yair. *The Defortification of the German City, 1689–1866*. Cambridge: Cambridge University Press, 2012.

Moses, A. Dirk. *German Intellectuals and the Nazi Past*. Cambridge: Cambridge University Press, 2007.

Müller-Raemisch, Hans-Reiner. *Frankfurt am Main: Stadtentwicklung und Planungsgeschichte seit 1945*. Frankfurt am Main: Campus-Verlag, 2001.

Ness. "Wenn Leben Erinnern heißt—Gedanken zu Walter Benjamins Gleichnis vom Engel der Geschichte." In *Stationen des Vergessens—der Börneplatzkonflikt,* edited by Georg Heuberger, 14. Frankfurt am Main: Jüdisches Museum, 1992.

Nirenberg, David. *Anti-Judaism: The Western Tradition*. New York: Norton, 2013.

Ravid, Ben. "All Ghettos Were Jewish Quarters, but Not All Jewish Quarters Were Ghettos." *Jewish Culture and History* 10, nos. 2–3 (2008): 5–24.

Rosenfeld, Gavriel D. *Munich and Memory: Architecture, Monuments, and the Legacy of the Third Reich*. Berkeley: University of California Press, 2000.

Ruderman, David B. "The Cultural Significance of the Ghetto in Jewish History." In *From Ghetto to Emancipation: Historical and Contemporary Reconsiderations of the Jewish Community,* edited by David N. Myers and William V. Rowe, 1–16. Scranton: University of Scranton Press, 1997.

Sheehan, James. "Paradigm Lost? The 'Sonderweg' Revisited." In *Transnationale Geschichte: Themen, Tendenzen und Theorien,* edited by Gunilla Budde, Sebastian Conrad, and Oliver Janz, 150–60. Göttingen: Vandenhoeck & Ruprecht, 2010.

Smith, Helmut Walser. "When the *Sonderweg* Debate Left Us." *German Studies Review* (2008): 225–40.

Steffen, Katrin. "Connotations of Exclusion——'Ostjuden,' 'Ghettos,' and Other Markings." *Simon Dubnow Institute Yearbook* 4 (2005): 459–79.

Traverso, Enzo. *The Origins of Nazi Violence*. New York: New Press, 2003.

PART II

Transformations

*Emergences, Shifts, and Dissolutions
in Spaces and Boundaries*

7

The Representation and Creation of Spaces through Print Media

Some Insights from the History of the Jewish Press

Kerstin von der Krone

The Jewish press, which emerged in the seventeenth century, is an essential part of modern Jewish history. Like the printing press and print media in general, it significantly changed patterns of communication, as well as of the production and distribution of knowledge. As such it created a new space of social interaction and negotiation of interests. We might thus define the Jewish press as all forms of periodical literature by and for Jews, which often but not necessarily dealt with Jewish affairs.

This chapter explores the Jewish press from a spatial historical perspective while drawing on paradigmatic incidents and developments from Central European history. This approach conceives of the Jewish press as both a place and space of information and discussion that gained momentum as a communicative and discursive practice in modernity due to its periodicity, currency, and efficiency. The concepts of "place" and "space," as Barbara Mann has pointed out, can both be rendered with one Hebrew word, *makom*.[1] In her study *Space and Place in Jewish Studies*, Mann cites Yi-Fu Tuan's distinction between place and space, the first corresponding to "pause" and the second to "movement."[2] We might go further and consider "space" to be an extension of place characterized by any form of motion; with regard to printing and print media, motion has a literal and metaphorical dimension, with the circulation and distribution of books and periodicals giving rise to the establishment of a public arena within which information and ideas could be exchanged and disseminated.

The Jewish press has continuously played a significant role in capturing and constituting Jewish places and spaces by reporting and reflecting on Jewish affairs at the local, national, and transnational levels and thereby transmitting an understanding of the political, cultural, and geographical dimensions of Jewish life. From the eighteenth century onward, the Jewish press served as an arena for the negotiation of what constituted modern Judaism; it played a part in transforming Jewish reading cultures[3] and complemented other forms of public interaction such as socializing in clubs and societies or attending public events including synagogue services.[4]

The historical Jewish press contributed greatly to the constitution of Jewish public spheres, providing information and a space for discussion for schools and tendencies within modern Judaism such as the *Maskilim* (followers of the Jewish Enlightenment), Reformers, Orthodox, and Zionists. Thus, the Jewish press became a crucial instrument for defining the boundaries of such groups, shedding light on the dynamic processes of diversification, fragmentation, and spatialization that shaped modern Jewish history.

Finally, the Jewish press connected Jewish places and spaces, captured the realities of Jewish life, and mediated the various ways in which Jews experienced modernity and its challenges. In general, the press relied significantly on two characteristics of modernity: acceleration and standardization. The introduction of standardized time and new means of transportation for information, goods, and persons, such as the postal service and the railway system, changed people's understanding of temporality and space, prompting unprecedented mobility, and created a new experience of acceleration.[5] Many scholars have highlighted the importance of the printing press and print media in this process, usually defining them as *agents of change,* tools with which reality could be captured and transformed.[6] As Todd Presner and others have indicated, mobility was even more central in the Jewish case, due to the historical experience of exile and the dynamics of migration and urbanization it entailed.[7]

This spatial analysis of the Jewish press focuses on three intertwined dimensions: first, location and locality as represented in, through, and by the Jewish press; second, spatiality and processes of spatialization in modern Jewish history, on which the Jewish press reflected and to which it simultaneously contributed; and third, mobility and temporality, which manifested themselves dramatically in the press as a modern practice of communication with a capacity to connect Jewish places and create Jewish spaces.[8]

The Beginnings of the Jewish Press

The Jewish press came into being in the late seventeenth century in Amsterdam, one of the major Jewish communities of the time, a center of Hebrew typography that was also an early center of print media in general due to its economic wealth and liberal politics.[9] The Spanish-language *Gazeta de Amsterdam,* presumably first published in 1675, may have been the first Jewish paper. Providing information on international events and economic developments, it largely relied on translations from the Dutch press. Yet it is unclear whether the *Gazeta* can be classified as Jewish—since the only indication of this is its editor and printer David de Castro Tartas (1630–98)—and whether it targeted Spanish-speaking Sephardic Jews in Amsterdam or an audience beyond the city.[10] Amsterdam was also the home of the *Dinstagishe un Fraytagishe Kurant*, a Yiddish paper appearing from 1686 to 1687, which likewise relied substantially on translations, reported mainly on international news, and only occasionally referenced Jewish affairs.[11] Little is known about its distribution and therefore about its audience, though it may have extended beyond the Yiddish-speaking Ashkenazi community of Amsterdam.[12]

Regardless of the open questions surrounding them, both these papers illuminate the first attempts to establish a Jewish press and highlight how significant vital cultural centers, appropriate political and economic conditions, and intellectual curiosity were to its emergence. Centers of commerce with a flourishing printing industry were always important to the Jewish press and Jewish printing. Jewish periodicals represented these cultural centers to the broader Jewish world but also served the particular interests of the local community. The first Hebrew periodical benefited not only from Amsterdam's vitality but also from a supportive religious and cultural environment: *Pri Ets Haim* (1721–61), issued by the Sephardic *beit midrash* Ets Haim, was the first press organ to periodically publish rabbinical *sh'elot ut'shuvot* (responsa). This central instrument of rabbinical decision-making, formerly composed of handwritten manuscripts and a form of official communication, thus became accessible to the broader public. The example of *Pri Ets Haim* may fit into what Daniel Gutwein once characterized as modern patterns of traditional Jewish communication. Traditional elites adopted the periodical as a modern manifestation of printing culture to maintain the status quo in Jewish politics.[13]

Like the non-Jewish press of the time, the early Jewish press largely was a product of the educated elites, merchants and scholars, who were targeted as the main audience as well. However, the Jewish press served traditional as well as newly emerging elites, becoming a significant instrument of agency for the latter, in particular. About a century after the first Jewish papers appeared in Amsterdam, the *Haskalah*, the Jewish Enlightenment movement, extensively made use of the new opportunities the printing press presented. Its most important periodical, *HaMeassef* (1784–1811), challenged rabbinical authorities by promoting Enlightenment ideas, calling for a reorganization of Jewish life, including Jewish education, and advocating a new attitude toward Hebrew.[14] The *Maskilim* aimed to modernize the holy language, Hebrew, and widen its use beyond religious purposes, with the ultimate intention of establishing it as a cultural language alongside German. The promotion of Hebrew and German by the Berlin *Haskalah* accompanied a devaluation of Yiddish, which helps explain why, after the *Dinstagishe un Fraytagishe Kurant*, only one other Yiddish paper—the *Dirnfurter prifilegirte Tzaitung* (1771), printed in Breslau—appeared in Central Europe for more than a century.

Despite these efforts on the part of the Berlin *Haskalah,* Hebrew gradually lost favor as a Jewish publishing language in Central Europe, giving way to vernacular languages from the late eighteenth century onward. Yet it continued to be used in a small number of periodicals such as *Bikkurei HaIttim* (1820–31) and *Keren Hemed* (1833–56), which were both connected to the Galician *Hokhmat Israel* (Wisdom of Israel).[15] In addition, Orthodox periodicals were either bilingual or would include Hebrew supplements. By the mid-nineteenth century, Eastern Europe became the center of a Hebrew culture and of a flourishing Hebrew press that not only reached a much larger audience than its Western European counterpart but established a tremendous variety of formats.[16] Although Eastern European Jews had a higher literacy rate in Hebrew than Western European ones at the time, Yiddish was their mother tongue. This formed the basis for a Yiddish press that began to emerge in the 1860s under much more difficult circumstances.[17] For example in Tsarist Russia, the Yiddish press, like Yiddish printing in general, was hindered by state policies that suppressed the use of the language up to 1905.[18]

Nonetheless, it would be overly simplistic to divide the spaces of the Jewish press along linguistic lines. The languages Jewish periodicals used were related to their target audiences, the circumstances of their production, and their agenda. The adoption of vernacular languages, such as German, French, English, Hungarian, Polish, or Russian, reflected both social and cultural transformations—including linguistic changes—and the degree of the publishers' desire to present Jewish interests and Jewish life to non-Jewish society. Thus, the history of the Jewish press manifests the linguistic and cultural pluralism that has characterized Jewish history and reflects the interdependencies between language and culture.[19] These interdependencies can be found throughout modern Jewish history, most significantly with the establishment of Hebrew as the Jewish national language within the Zionist context.[20]

Discovering the Jewish World: Connecting Jewish Places and Creating Jewish Spaces

The Jewish Diaspora consistently relied on transterritorial corporate and economic structures, family networks and transcultural communication, and the exchange of ideas and information. In premodern times, Jewish life had already been shaped by social mobility and migration, and this intensified dramatically from the nineteenth century onward.[21] The Jewish press reflected these developments directly and indirectly and grew, like the press in general, with social and technological innovations that accelerated the pace of modern life.[22]

Books and periodicals produced in important centers of Jewish and specifically Hebrew printing, such as Amsterdam, Berlin, Frankfurt am Main, Leipzig,

Kraków, Lemberg (L'viv), Vienna, and Warsaw, often targeted audiences beyond the local Jewish community, reinforcing the specific transterritorial and transnational patterns of Jewish communication. This wider targeting arose in part to expand the limited market for Jewish publications and thus address an economic problem. But there were other exigencies; the censorship of Hebrew and Yiddish printing, particularly in Tsarist Russia, is a case in point. Although a wide variety of general censorship regulations applied to the Jewish press and Jewish printing in the different European countries, these had been changing since the eighteenth century.[23] The specific regulations restricting Hebrew and Yiddish printing were partially eased in Western and Central Europe during the nineteenth century but were still in use in Tsarist Russia until the early twentieth century.[24] Editors of Jewish papers and journals in Russia, facing an illiberal atmosphere, attempted to get around these circumstances by using Jewish printing facilities outside the Russian border. For example, the Hebrew newspaper *HaMaggid* (1856–1903) was printed in Lyck, East Prussia, but targeted Russian Jews, as did the monthly *HaShahar* (1868–84), which was printed in Vienna.[25] Another example is the Orthodox journal *HaLevanon* (1863–86), the first Hebrew periodical printed in Palestine. *HaLevanon* principally addressed Orthodox Jews in the Old Yishuv (Palestine) and Eastern Europe. It relied on a transnational network of authors and supporters, which helps explain the extraordinary journey it took over the course of its existence, being published successively in a host of different cities including Jerusalem, Paris, London, and Mainz.[26]

The transterritorial and transnational dimension of Jewish history, as well as the role of personal and professional networks and of practices of communication and collaboration, influenced the production, distribution, and content of the Jewish press. The Damascus affair of 1840, generally regarded as a defining moment in the history of the Jewish press, exemplifies many aspects of these developments.[27] It highlights the acceleration of the dissemination of information then under way, the impact of the European public sphere, mainly constituted by the press, and the significance of Jewish networks[28] and Jewish diplomacy in the fight for the Jewish cause.[29] The affair involved the disappearance of an Italian monk and his servant and the allegation of ritual murder subsequently made against the Damascene Jews, over a dozen of whom were imprisoned and tortured. Against the backdrop of the "Eastern question,"[30] and with European diplomats involved, the affair was covered extensively in the European press. Reports ranged in their tone from sensationalistic fascination to judgmental certainty that the alleged "Jewish atrocities" had indeed taken place, prompting a unique Jewish reaction: Jewish leaders and communities challenged the allegations, along with the Europe-wide wave of anti-Jewish prejudice they unleashed, in numerous public statements published in the general and Jewish press. Urging their governments to intervene, leading French and British Jews launched a diplomatic mission to Alexandria to support the accused, who were finally acquitted and released if they had survived. Jonathan Frankel's groundbreaking study on the affair highlights the crucial role Jewish networks and public interventions by European Jews played.[31]

The Damascus affair is rightly perceived as a key event in modern Jewish history, when Jews went from engaging exclusively in *shtadlanut*, the traditional Jewish advocacy system,[32] to getting involved in modern Jewish diplomacy and new communicative patterns. The Jewish press coverage of the affair in general exemplifies how important this press was becoming as it brought a transnational Jewish public sphere to life. Although most of the Jewish periodicals reporting on the affair located themselves within a particular national setting and viewed themselves as mouthpieces to German, French, or English Jewry, they also reflected on their mutual bonds and embraced Jewish solidarity.[33]

Jewish responses to the affair were rooted not only in empathy for their persecuted brothers but also in anxiety about the potential impact these events could have on Jewish emancipation. The accusations, along with the prevalent implication that the alleged actions had been based on Jewish tradition, as well as the torture meted out to the accused Damascene Jews, contradicted reason and the Enlightenment ideas that Europe supposedly championed in contemporary Jews' perception. This broader context lent urgency to the affair, which further explains why political steps were taken—in particular, the diplomatic mission led by Aldophe Crémieux (1796–1880) and Moses Montefiore (1784–1885)[34] to Alexandria—and underscores its role in the emergence of a modern European Jewish public sphere. The affair highlighted the value of papers dealing with Jewish affairs as an instrument of Jewish agency; they provided information that was not necessarily accessible in the general press and eventually served as a forum for exchange within and among Jewish communities.

The Jewish Press and Responses to Modernity

In the course of the nineteenth century, Jewish periodicals were established all over Europe and North America, displaying an astonishing diversity in their programmatic approaches, contents, and formats. This multifaceted Jewish press reflected a broader historical process whereby modern societies came to be influenced by media based on technological innovations and by comparatively open public debate. The Jewish press gained particular importance as an instrument for representing and reflecting the full breadth of Jewish life and as a space for the negotiation of the very idea of modern Judaism. Jewish papers and journals constituted a public arena for discussing the challenges of modernity and for negotiating and defining the boundaries of various concepts of Judaism, whether understood as a religion, a nation, or a culture, as liberal, moderate, or Orthodox, Maskilic, Zionist, or socialist. Jewish periodicals often chose to advocate one of these positions to a broader public, thereby contributing to its formation and consolidation. Examining the German-Jewish press in the 1830s and 1840s thus reveals the extent to which it became an instrument of presentation and representation, of communication and negotiation of Jewishness and "Jewish space," as well as an indispensable Jewish space of its own.

Between the 1830s and the late 1840s, two partially intertwined developments impacted the history of German Jews: first, the emergence and dissemination of a Jewish public sphere,[35] and second, a tense intra-Jewish debate on the necessity, possibilities, and limitations of religious reform against the background of the emancipation process.[36] The debates on reform accelerated and intensified in the 1830s and 1840s and laid the foundations for the fragmentation of German Jewry, leading to the emergence of liberal, conservative, and neo-Orthodox movements. The German-Jewish press served as one of the main arenas for the debate around this process. Yet the parties were not equally represented from the beginning. Until the 1830s, the German-Jewish press almost exclusively promoted progressive ideas; publications in this vein included *Sulamith* (1806–48), which focused on Jewish education and religious reform, Ludwig Philippson's (1811–89) *Israelitisches Predigt- und Schulmagazin* (1834–36), and his influential *Allgemeine Zeitung des Judenthums* (1837–1922, continuing as a merger with the *C.V. Zeitung* until 1938), which covered almost all aspects of Jewish life. Further publications served the emergent field of modern Jewish scholarship[37] and represented the positions of the radical reform movement.[38]

In the mid-1840s, the German-Jewish press flourished,[39] largely prompted by the remarkable dynamic of the intra-Jewish debate on reform, which challenged extant ideas of Judaism and the foundations of Jewish life in general. In the course of these debates new publications began to challenge and counterbalance the mainly progressive voices in the German-Jewish public. Journals were founded giving a public voice to those who either promoted a moderate approach toward reform or opposed it on the basis of a professed commitment to upholding Jewish tradition. Two occasions were crucial in this development. The first was the so-called Geiger-Tiktin controversy, which spanned several years following the appointment of Abraham Geiger (1810–74) as the assistant rabbi of the Breslau community in 1838. Geiger's progressive ideas, promoted in his *Wissenschaftliche Zeitschrift für Jüdische Theologie* (1835–47) and elsewhere, were vehemently rejected by the Orthodox chief rabbi of Breslau, Salomon Tiktin (1791–1843), and his supporters. The ensuing discussions over Geiger's actual or alleged reform agenda gained attention in the broader German-Jewish public through continuous coverage in the German-Jewish press and numerous pamphlets published on behalf of both sides.[40] The intensity of the debate and the allegations involved drew an unfavorable picture of the Breslau community and affected the reputations of all involved. The course of the debate exemplified the uncertainties of publicity, demonstrating that public debates tend to develop a life and logic of their own, spinning out of the control of those who initiated it.[41] Moreover, the controversy brought significant attention to those critical of the reform agenda, particularly representatives of the Orthodox faction, within the broader public debate.

On the second occasion, the rabbinical conferences held between 1844 and 1846 in Braunschweig, Frankfurt am Main, and Breslau, respectively, the debate on religious reform reached a climax and as well resulted in an unprecedented level of opposition to reform. These events addressed the most pressing issues of the reform debate and were intended to help determine the extent to which the authority of Jewish tradition could and should be preserved and whether religious practices should adapt to the new era. The debates included the questions of circumcision and early burial, the role of Hebrew and the use of the organ in synagogue services, and the significance of the messianic idea.[42] The German-Jewish press frequently reported on the conferences and the disputes that arose around them.

Compared to Ludwig Philippson's *Allgemeine Zeitung des Judenthums* and Julius Fürst's (1805–73) *Der Orient* (1840–50/51), two of the most visible and most competitive German-Jewish weeklies of the 1840,[43] Zacharias Frankel's *Zeitschrift für die religiösen Interessen des Judenthums* (1844–46) added a notably different dimension to the public debate on religious reform and the efforts to redefine Judaism in the face of modernity. Prior to the opening of the Braunschweig conference, Frankel published an elaborate and ambivalent piece on the upcoming event, in which he reflected on the current state of German Jewry and the challenges of the reform debate. While he supported the idea of a meeting in principle, he challenged the claim that the conference resembled a "synod" that could issue binding decisions, as he felt it did not have enough legitimacy to exercise such authority.[44] Further, wishing to prevent reform from becoming radicalized, Frankel called for conference attendees to avoid implying anything that could endanger the unity of Judaism. For Frankel, the foundation of Judaism was divine revelation, a premise that could be neither questioned nor altered.[45] Although Frankel had previously announced he would attend the Braunschweig conference, he did not in fact do so; he also refused to sign the protest note issued by traditional (i.e., Orthodox) rabbis.[46] In another article published at the end of 1844, reviewing the proceedings of the conference, he again took a moderate position toward religious reform. He thought reform needed to be based on revelation and critical scholarship, yet he appreciated the efforts to establish a common forum for carrying out the ongoing debate.[47] At the second conference in Frankfurt am Main, Frankel tried and failed to unite the advocates of moderate reform, subsequently leaving the conference under protest due to disagreement about the use of Hebrew in the synagogue service. Thereafter, Frankel sought to establish a competing conference of Jewish theologians (*Theologenversammlung*), promoting it intensely in his journal, though he did not immediately meet with success.[48] Frankel's periodical was not a neutral place that stood above the controversies between reformers and the Orthodox but rather formed a strategic position in the debate in itself.[49]

Although Frankel's *Zeitschrift*, discontinued in 1846, proved to have only limited success, it had constituted a pioneering public voice for a moderate approach to reform and had provided Frankel with a forum for presenting his concept of moderate reform that eventually laid the foundation for Conservative Judaism.[50] The rabbinical conferences had the further significant effect of mobilizing the German-Jewish Orthodoxy and prompting it to develop means of public expression. The protest note referred to above, signed by 77 initially and eventually 116 rabbis, marked a first step toward bringing broader opposition to the reform agenda to light. It contributed substantially to Orthodox Judaism becoming an organized and publicly visible movement. Jakob Ettlinger (1798–1871), who had led the protest, became the editor of the first Orthodox journal, the weekly *Der Treue Zionswächter* (1845–54),[51] which clearly opposed those who promoted Enlightenment, critical scholarship, and the reform of Judaism.[52]

These examples indicate how the Jewish press, like Jewish printing in general, became a crucial space for carrying out the debates on the reform of Judaism, helping to reinforce the boundaries between its various factions. The press also fulfilled this function beyond German Jewry, extending the debates to Jewish communities in Europe and North America. Whereas the broader political and social context of these debates differed as their protagonists changed over time, they continued to focus on how Judaism and Jewish life were to be defined.[53] The press, as part of a broader printing culture, constituted a public forum for negotiating various responses to these issues, significantly contributing to the formation of modern Judaism and framing the boundaries of competing denominations and political factions.

Conclusion

In this brief excursion into the history of the Jewish press and its spatial dimensions, I have sought principally to highlight the historical and historiographical value of print media in general and the Jewish press in particular for our understanding of Jewish history. They are indispensable to a view of history that acknowledges the importance of publicity, public spaces, and public interaction in the modern age. As I have demonstrated, the Jewish press served the particular needs of the Jewish minority from the local to the transnational level in relation to the specific characteristics of the Jewish Diaspora. Secondly, I have sought to illuminate the significance of the press as a practice of communication—an instrument of presentation and representation—which not only became a space for discussion, debate, and negotiation but was part of a larger sphere of public interaction. In addition, approaching the history of Jewish press via its spatial dimensions acknowledges that communication is always related to space, by "communicating the space," as "communication in space," and in the "constitution of space via communication."[54]

The Jewish press reflected on Jewish places and spaces and, in its reporting on particular locations of Jewish life, contributed to the "discovery" of the many cultural and geographical faces of the Jewish Diaspora and a Jewish world broadened by migration. Furthermore, the Jewish press itself became a significant Jewish space, a forum for debate and the exchange of ideas and a site of the formation and negotiation of modern Judaism and its boundaries, whether in relation to the Gentile world or to intra-Jewish debates. The above examples show that Jewish representatives, editors, and authors felt the need to act and used the press to mobilize the Jewish public for their causes. Although the Jewish press was not the only tool of public interaction at this time, its significance derived from the *urgency* of specific issues and can be seen as an expression of *agency*. Given the social and political particularities of the Jewish minority's situation and the transterritorial bonds of the Jewish Diaspora, the Jewish press was crucial for distributing information and knowledge and as a public space for exchanging ideas and points of view, which could range from intellectual discussions to fierce religious and political controversies.

Kerstin von der Krone is a research fellow at the German Historical Institute in Washington, DC. Her fields of research include Jewish history in Central Europe in the modern era, the history of Jewish thought, the transformation of knowledge and knowledge production in Jewish education, and modern Jewish scholarship. She is the author of *Wissenschaft in Öffentlichkeit: Die Wissenschaft des Judentums und ihre Zeitschriften* (2012).

Notes

I would like to thank Mathias Berek and the editors of this volume for their valuable and insightful comments on early drafts of the chapter.

1. Mann, *Space and Place*, 5.
2. Ibid.
3. Roemer, "German Jewish Reading Cultures."
4. Sorkin, *Transformation;* Lässig, *Jüdische Wege.*
5. As Koselleck pointed out, the experience of time and space changed significantly at the threshold of modernity. Koselleck, *Zeitschichten* or, in English, *Practice*. This was related to the transformation of knowledge and knowledge production, including the paths and instruments of its distribution. See Burke, *A Social History;* Weingart, "Short History."
6. On the history of the printing press, see Eisenstein, *Printing Press.*
7. In relation to Jewish or German-Jewish history, Presner emphasized the role of "movement, migration, exile, exchange, encounter and contamination." Presner, "Remapping German-Jewish Studies," here 298.
8. These three dimensions resemble the relationship between communication and space as "communicating space," "communication in space," the "constitution of space via communication." Geppert, Jensen, and Weinhold, "Verräumlichung," here 28.
9. Pach, "Short-Lived Blossoming"; Fuks and Fuks-Mansfeld, *Hebrew Typography.*
10. Fuks and Fuks-Mansfeld see no call for a Spanish paper in the Jewish community of Amsterdam, as Sephardic Jews usually read Dutch in addition to Spanish or Ladino. Fuks and Fuks-Mansfeld, *Hebrew Typography,* 2:343. Pach argues that a Dutch paper printed in Spanish had to target a Jewish audience, as Jews were the only ones capable of reading Spanish in the Netherlands. Pach, "Short-Lived Blossoming," 33–34.
11. The *Kurant* was printed by Uri Faybesh Halevi (1625–1715) and later by David de Castro Tartas, the printer of the *Gazeta*. On Tartas and his work as a printer, see Fuks and Fuks-Mansfeld, *Hebrew Typography,* 2:339–82.

12. Pach, "Short-Lived Blossoming," 34. This argument is supported by Berger's conclusion that Amsterdam printers did not add much to the corpus of Yiddish literature but embraced a "division of labor: importing texts and exporting books." Berger, "Yiddish Book Production," here 203.
13. Gutwein, "Traditional and Modern Communication."
14. On *HaMeassef*, see Kennecke, "Die erste moderne Zeitschrift der Juden in Deutschland"; Pelli, *Sha'ar la-haskalah*. On the history of the *Haskalah* in general, see Feiner, *Jewish Enlightenment*.
15. Pelli, *Haskalah and Beyond*, 181–229. On the Hebrew press and choice of language, see Soffer, "Why Hebrew?"
16. Gilboa, *Leksikon ha-itonut*.
17. With regard to Jewish literacy in Eastern Europe, Stampfer emphasizes that all figures are to be interpreted carefully, as having the ability to read did not necessarily mean one could write. See Stampfer, *Families*, 145–274; on literacy, see 190–210.
18. The first Yiddish newspaper in Tsarist Russia was *Kol Mevaser* (The Herald, 1862–72), founded as a supplement to the Hebrew weekly *Ha-Melits* (The Advocate, 1860–1904). Its editor Alexander Zederbaum (1816–93) obtained a license to print a newspaper in German with Hebrew characters, which he used for *Kol Mevaser*. Fishman, *Rise of Modern Yiddish Culture*, 22. On the censorship of the Yiddish press, see 21–25. See also Cohen, "Yiddish Press."
19. The multilingualism of the Jewish press mirrored a crucial characteristic of the Jewish Diaspora. Harshav has highlighted the significance of multilingualism among Eastern European Jews and their descendants from the late nineteenth century onward and emphasized the role of migration and therefore mobility as a significant factor in its spread. Harshav, "Multilingualism," here 27.
20. On Hebrew and Zionism, see Harshav, *Language in Time of Revolution*.
21. Menache, "Communication in the Jewish Diaspora"; Thulin, "Jewish Networks."
22. On technological innovations that drove the emergence of print media, see Briggs and Burke, *A Social History of the Media*, 91–112. On the role of the postal system in revolutionizing communication patterns in early modern times, see Behringer, "Communications Revolutions."
23. On the history of censorship in the German context, see Wilke, "Censorship and Freedom of the Press." With regard to licensing and censorship regulations affecting the early German-Jewish press, see Schwarz, "'Ew. Exellenz wage ich … unterthänig vorzulegen.'"
24. On censorship of the Hebrew press, see Lederhendler, *Road to Modern Jewish Politics*, 95–100; for Yiddish, see Fishman, *Rise of Modern Yiddish Culture*, 21–25.
25. On *HaShahar*, see Gilboa, "Ha-itonut ha-yehudit," 51. On the historical contexts of the emergence of the Hebrew press in Eastern Europe, see Bartal, *Jews of Eastern Europe*, 107–11.
26. See Beer-Marx, "'Halevanon.'"
27. The affair and its aftermath, while they may not have triggered the emergence of the Jewish press as such, certainly amplified and shaped its development. A closer discussion can be found in Mevorah, "Ikvoteha shel 'alilat."
28. Thulin, "Jewish Networks."
29. Frankel, *Damascus Affair*; Krone, "Die Berichterstattung."
30. The "Eastern question" denotes political and diplomatic problems that arose following the disintegration of the Ottoman Empire from the early nineteenth century onward and the concomitant aspirations some of its territories had to become autonomous; these issues

drew the attention of European powers, which then began to extend their influence into the region.

31. Frankel, *Damascus Affair,* 233–56.

32. *Shtadlanut* refers to the traditional Jewish advocacy system. The *shtadlan* served as the official or unofficial representative of a Jewish community or a union of communities to the Gentile authorities. On the political concept of *shtadlanut,* see Klieman, "*Shtadlanut* as Statecraft by the Stateless."

33. Frankel, *Damascus Affair,* 311–28.

34. On Montefiore's role in the affair, see Parfitt, "'The Year of the Pride of Israel'"; Green, *Moses Montefiore,* 133–56.

35. Gutwein, "Traditional and Modern Communication"; Sorkin, *Transformation,* 79–104; Lässig, *Jüdische Wege,* 442–562.

36. In Prussia and most of the German states, conditions-based models of emancipation demanded that the Jews make far-reaching changes to their religious practice, customs, and social structure. See Katz, *Out of the Ghetto,* 161–175; Sorkin, *Transformation,* 86–90.

37. The emergence of modern Jewish scholarship—also known as *Wissenschaft des Judentums*—was closely intertwined with the debate on emancipation and reform. The first Jewish scholarly journal, the *Zeitschrift für die Wissenschaft des Judenthums,* appeared in 1822. By 1939, there were almost thirty journals and yearbooks being published that promoted a critical approach toward Jewish history and culture. See Krone, *Wissenschaft in Öffentlichkeit,* 56–67.

38. An example is *Der Israelit des 19. Jahrhunderts* (1840–48), edited by Mendel Hess and Samuel Holdheim. On the history of this journal, see Bunyan, "The Emancipation Debate."

39. Toury, "Das Phänomen," here 6d.

40. For a closer examination of the events, see Gotzmann, "Der Geiger-Tiktin-Streit."

41. Ibid.

42. Meyer, *German-Jewish History,* 163–67.

43. Philippson and Fürst each claimed intellectual leadership for their papers, framing them as neutral in the intra-Jewish debates. Nevertheless, Philippson, with his *Allgemeine Zeitung des Judenthums,* was a proponent of progressive reform, whereas Fürst opened up his journal to a broader range of voices.

44. Zacharias Frankel, "Ueber die projectirte Rabbinerversammlung," *Zeitschrift für die religiösen Interessen des Judentums (ZRIJ)* 1 (1844): 89–106.

45. Ibid., 101.

46. Brämer, *Rabbiner Zacharias Frankel,* 229; Lowenstein, "1840s," 264–65 and appendix, 276–85.

47. Zacharias Frankel, "Die Rabbinerversammlung zu Braunschweig," *ZRIJ* 1 (1844): 289–308.

48. Zacharias Frankel, "Aufruf zu einer Versammlung jüdischer Theologen," *ZRIJ* 2 (1846): 201–204; Frankel, "Ich erlaube mir hiermit den Wunsch auszusprechen … ," *ZRIJ* 3 (1847): 241; Frankel, "Einiges über die projectirte Theologen-Versammlung," ibid., 339–41; "Nachricht über die projectirte Theologen-Versammlung," ibid., 387.

49. Brämer, "Auf der Suche," 220–21.

50. Schorsch, "Zacharias Frankel."

51. From 1846 onward, *Der Treue Zionswächter* also appeared in Hebrew as *Shomer Zion ha-Ne'eman.*

52. "Die Redaktion an das Publikum," *Der Treue Zionswächter. Organ zur Wahrung der Interesse des orthodoxen Judenthums* 1 (1845): 1–3. On the history of the Orthodox press, see Bleich, "Emergence."
53. On the significance of the press for the politics of Russian Jewry, see Lederhendler, *Road to Modern Jewish Politics.*
54. Geppert, Jensen, and Weinhold, "Verräumlichung," 28.

Bibliography

Bartal, Israel. *The Jews of Eastern Europe, 1772–1881.* Philadelphia: University of Pennsylvania Press, 2005.

Beer-Marx, Roni. "'Halevanon': Re-interpretation of an Orthodox Organ." In *Tarptautinės Mokslinės Konferencijos "Vilniaus Žydü Intelektualinis Gyvenimas iki Antrojo Pasaulinio Karo" medžiaga—Proceedings of the International Scientific Conference "Jewish Intellectual Life in Pre-War Vilna,"* edited by Larisa Lempertienė, 108–12. Vilnius: Mokslo Aidai, 2005.

Behringer, Wolfgang. "Communications Revolutions: A Historiographical Concept." *German History* 24 (2006): 333–75.

Berger, Shlomo. "Yiddish Book Production in Amsterdam between 1650–1800: Local and International Aspects." In *The Dutch Intersection: The Jews and the Netherlands in Modern History,* edited by Yosef Kaplan, 203–12. Leiden: Brill, 2008.

Bleich. Judith. "The Emergence of an Orthodox Press in Nineteenth-Century Germany." *Jewish Social Studies* 42 (1980): 323–44.

Brämer, Andreas. "Auf der Suche nach einer neuen jüdischen Theologie—Die Zeitschrift für die religiösen Interessen des Judenthums (1844–1846)." *Menora* 12 (2001): 209–28.

———. *Rabbiner Zacharias Frankel. Wissenschaft des Judentums und konservative Reform im 19. Jahrhundert.* Hildesheim: Georg Olms Verlag, 2000.

Briggs, Asia, and Peter Burke. *A Social History of the Media: From Gutenberg to the Internet.* Cambridge: Polity Press, 2009.

Bunyan, Anita. "The Emancipation Debate in *Der Israelit des Neunzehnten Jahrhunderts* 1839–1845." In *The German-Jewish Dilemma: From the Enlightenment to the Shoah,* edited by Edward Timms and Andrea Hammel, 63–75. Lewiston, NY: Edwin Mellen Press, 1999.

Burke, Peter. *A Social History of Knowledge: From Gutenberg to Diderot.* Cambridge: Blackwell, 2000.

Cohen, Nathan. "The Yiddish Press and Yiddish Literature: A Fertile but Complex Relationship." *Modern Judaism* 28 (2008): 149–72.

Eisenstein, Elisabeth. *The Printing Press as an Agent of Change.* 2 vols. Cambridge: Cambridge University Press, 1979.

Feiner, Shmuel. *The Jewish Enlightenment.* Philadelphia: University of Pennsylvania Press, 2004.

Fishman, David E. *The Rise of Modern Yiddish Culture.* Pittsburgh: University of Pittsburgh Press, 2005.

Frankel, Jonathan. *The Damascus Affair: 'Ritual Murder,' Politics, and the Jews in 1840.* Cambridge: Cambridge University Press, 1997.

Fuks, Lajb, and Renate G. Fuks-Mansfeld. *Hebrew Typography in the Northern Netherlands: 1585–1815: Historical Evaluation and Descriptive Bibliography.* 2 vols. Leiden: Brill, 1987.

Geppert, Alexander T., Uffa Jensen, and Jörn Weinhold. "Verräumlichung. Kommunikative Praktiken in historischer Perspektive, 1840–1930." In *Ortsgespräche: Raum und Kommunikation im 19. und 20. Jahrhundert,* edited by Geppert, Jensen, and Weinhold, 15–49. Bielefeld: Transcript, 2005.

Gilboa, Menuha. "Ha-itonut ha-yehudit be-Vienna be-mahzit hashnija shel meah ha-tsha-esreh [The Hebrew Press in Vienna in the Second Half of the Nineteenth Century]." *Qesher* 10 (1991): 49–54.

———. *Leksikon ha-itonut ha-ivrit ha-meah shmonah-esreh we-ha-tsha-esreh* [Lexicon of Hebrew Periodicals in the Eighteenth and Nineteenth Century]. Jerusalem: Bialik Institute, 1992.

Gotzmann, Andreas. "Der Geiger-Tiktin-Streit: Trennungskrise und Publizität." In *In Breslau zu Hause? Juden in einer mitteleuropäischen Metropole der Neuzeit,* edited by Manfred Hettling et al., 81–98. Hamburg: Dölling und Galitz, 2003.

Green, Abigail. *Moses Montefiore: Jewish Liberator, Imperial Hero.* Cambridge, MA: Belknap Press of Harvard University, 2010.

Gutwein, Daniel. "Traditional and Modern Communication: The Jewish Context." In *Communication in the Jewish Diaspora: The Pre-Modern World,* edited by Sophia Menache, 409–46. Leiden: Brill, 1996.

Harshav, Benjamin. *Language in Time of Revolution.* Stanford: Stanford University Press, 1999.

———. "Multilingualism." In *The Polyphony of Jewish Culture,* 23–40. Stanford, CA: Stanford University Press, 2007.

Katz, Jacob. *Out of the Ghetto: The Social Background of Jewish Emancipation, 1770–1870.* New York: Schocken Books, 1978.

Kennecke, Andreas. "Die erste moderne Zeitschrift der Juden in Deutschland." *Das Achtzehnte Jahrhundert* 23 (1999): 176–99.

Klieman, Aaron. "*Shtadlanut* as Statecraft by the Stateless." *Israel Journal of Foreign Affairs* 2, no. 3 (2008): 99–113.

Koselleck, Reinhard. *Zeitschichten: Studien zur Historik.* Frankfurt am Main: Suhrkamp, 2000. Translated by Todd Samuel Presner et al. as *The Practice of Conceptual History: Timing History, Spacing Concepts* (Stanford, CA: Stanford University Press, 2002).

Krone, Kerstin von der. "Die Berichterstattung zur Damaskus-Affäre in der deutsch-jüdischen Presse." In *Jewish Images in the Media,* edited by Martin Liepach, 153–76. Vienna: Austrian Academy of the Sciences Press, 2006.

———. *Wissenschaft in Öffentlichkeit. Die Wissenschaft des Judentums und ihre Zeitschriften.* Berlin: De Gruyter, 2010.

Lässig, Simone. *Jüdische Wege ins Bürgertum: Kulturelles Kapital und sozialer Aufstieg im 19. Jahrhundert.* Göttingen: Vandenhoeck & Ruprecht, 2004.

Lederhendler, Eli. *The Road to Modern Jewish Politics: Political Tradition and Political Reconstruction in the Jewish Community of Tsarist Russia.* New York: Oxford University Press, 1995.

Lowenstein, Steven M. "The 1840s and the Creation of the German-Jewish Religious Reform Movement." In *Revolution and Evolution 1848 in German-Jewish History,* edited by Werner Mosse, Arnold Paucker, and Reinhard Rürup, 255–97. Tübingen: Mohr Siebeck, 1981.

Mann, Barbara. *Space and Place in Jewish Studies.* New Brunswick, NJ: Rutgers University Press, 2012.

Menache, Sophia. "Communication in the Jewish Diaspora: A Survey." In *Communication in the Jewish Diaspora,* edited by Sophia Menache, 15–57. Leiden: Brill, 1996.

Mevorah, Baruch. "Ikvoteha shel 'alilat Damesek behitpathuta shel ha'itonut hayehudit bashanim 1840–1846" [The Traces of the Damascus Affair in the Development of the Jewish Press in the Years 1840–1846]. *Zion* 23–24 (1958): 46–65.

Meyer, Michael A. *German-Jewish History in Modern Times.* Vol. 2, *Emancipation and Acculturation: 1780–1871.* New York: Columbia University Press, 1997.

Pach, Hilde. "The Short-Lived Blossoming of the Yiddish Press in the Netherlands." In *Leshonot, sifruyot, omanuyot* [*Languages, Literatures, Arts*], edited by Tamar Alexander-Frizer et al., 25–33. Jerusalem: World Union of Jewish Studies, 2007.

Parfitt, Tudor. "'The Year of the Pride of Israel': Montefiore and the Damascus Blood Libel of 1840." In *The Century of Moses Montefiore,* edited by Sonja L. Lipman and Vivian D. Lipman, 131–48. Oxford: Oxford University Press, 1985.

Pelli, Moshe. *Haskalah and Beyond: The Reception of the Hebrew Enlightenment and the Emergence of Haskalah Judaism.* Lanham, MD: University Press of America, 2010.

———. *Sha'ar la-haskalah: mafteach mu'ar le-Ha-Me'asef, ketav-ha-'et ha-'Ivri ha-rishon* [*The Gate to Haskalah: An Annotated Index to Hame'asef, the First Hebrew Journal*]. Jerusalem: The Hebrew University Magnes Press, 2000.

Presner, Todd. "Remapping German-Jewish Studies: Benjamin, Cartography, Modernity." *German Quarterly* 82 (2009): 293–315.

Roemer, Nils. "German Jewish Reading Cultures, 1815–1933." *Aschkenas* 18–19 (2008): 9–23.

Schorsch, Ismar. "Zacharias Frankel and the European Origins of Conservative Judaism." In *From Text to Context: The Turn to History in Modern Judaism,* 255–65. Hanover, NH: Brandeis University Press, 1994.

Schwarz, Johannes V. "'Ew. Exellenz wage ich … unterthänig vouzulegen.' Zur Konzessionierung und Zensur deutsch-jüdischer Periodika in den Königreichen Preußen und Sachsen bis 1850." In *Zwischen Selbstbehauptung und Verfolgung : deutsch-jüdische Zeitungen und Zeitschriften von der Aufklärung bis zum Nationalsozialismus,* edited by Michael Nagel, 101–38. Hildesheim: Olms, 2002.

Soffer, Oren. "Why Hebrew? A Comparative Analysis of Language Choice in the Early Hebrew Press." *Media History* 15 (2009): 253–69.

Sorkin, David. *The Transformation of German Jewry, 1780–1840.* New York: Oxford University Press, 1987.

Stampfer, Shaul. *Families, Rabbis and Education: Traditional Jewish Society in Nineteenth-Century Eastern Europe.* Oxford: Littmann Library of Jewish Civilization, 2010.

Thulin, Mirjam. "Jewish Networks." *European History Online (EGO).* Mainz: Institute of European History, 2010. Accessed 30 August 2014. http://www.ieg-ego.eu/thulinm-2010-en; URN: urn:nbn:de:0159-20100921358.

Toury, Jacob. "Das Phänomen der jüdischen Presse in Deutschland [Kaviim le-hitpatchut ha-itonut ha-yehudit be-Germania]." *Qesher* 5, special issue, *Jüdische Zeitungen und Journalisten in Deutschland* [*Itonut ve-Itonaim yehudim be-Germania*] (1989): 4d–13d / 4–11.

Weingart, Peter. "A Short History of Knowledge Formation." In *The Oxford Handbook of Interdisciplinarity,* edited by Robert Frodeman, Julie Thompson Klein, and Carl Mitcham, 3–14. Oxford: Oxford University Press, 2010.

Wilke, Jürgen. "Censorship and Freedom of the Press." *European History Online* (EGO). Mainz: Institute of European History, 2010. Accessed March 7, 2015, http://www.ieg-ego.eu/wilkej-2013a-en; URN: urn:nbn:de:0159-2013050204.

8

Out of the Ghetto,
Into the Middle Class

*Changing Perspectives on Jewish Spaces
in Nineteenth-Century Germany—
The Case of Synagogues and Jewish Burial Grounds*

Andreas Gotzmann

For understandable reasons, historians tend to characterize historical processes from their own standpoint, taking the future as the aim and indeed as a yardstick. Our vision of the developments German-Jewish society and culture experienced during the era of emancipation is one of thorough, rapid, and profound changes, a groundbreaking transformation that was barely restrained by the slow and strenuous process of legal and social emancipation. In the early modern period, Jews had survived in economic niches, pushed to the fringes of society, and often despised by their Christian neighbors. Their legal standing defined them explicitly as foreigners who were tolerated under certain preconditions for a specified period of time. By the end of the nineteenth century, German Jews had moved from the precarious position of living an endangered and often impoverished life into the emerging middle class, enjoying relative security. Not only had their legal and economic situation changed dramatically; their cultural outlook and self-perception differed so much from a few generations previously that these Jews retrospectively almost appear to be a new people. This historical perception actually meets the self-perception of Jewish contemporaries at the end of the era of emancipation. They had reinvented themselves and defined new cultural models that looked toward the middle class and the idea of a liberal society. This said, the question remains as to whether this process was such a revolutionary one after all, a swift cutting of all ties and a setting sail to new shores. Beyond all hopes and visions of the Jews of the time, can we really view this process as a radical break with the past, or were these changes cautious steps that sought to maintain, as far as possible, the previous status quo?

This chapter attempts to answer this very question with regard to the changes that characterized Jewish concepts of space during emancipation. Is there a connection between the early modern, rather bleak community synagogue in the Frankfurt ghetto and the new place of worship constructed after the former *Judengasse* had been renamed Börnestrasse? Was it just a step from the modest baroque synagogue in a backyard of the Heidereuthergasse in Berlin to the towering building of the Neue Synagoge at Oranienburger Strasse? And did the idea of belonging, that is, of being part of an essentially Jewish space, change at all, or as much and as swiftly as it appears? To understand these developments, we need to take a closer look at the cultural concepts espoused by German Jews during the early modern period. This proves to be quite a challenge: While many decades of research have yielded a wealth of information about the modern era, the early modern period has been rather neglected. Although interest in Jewish history from the sixteenth to the eighteenth centuries has grown during the last two decades, most approaches have remained restricted to particular localities and specific developments. Consequently, most research on Jews and space to date deals with issues characteristic of the modern period, with particular emphasis on the twentieth century.[1] Only the architectural history of the modern synagogue constitutes a clear exception to this trend.

This chapter picks up where this research has left off; it analyzes changes specific concepts of Jewish space underwent and the new ones that emerged. Our focus remains on the Jewish perspective considering that Jewish society, as a rather small minority, depended on external legal, economic, and social parameters as well as general cultural perceptions. Perspectives on Jewish spaces, therefore, were usually defined by non-Jews as well as Jews. Jewish cultural models of essentially Jewish places were often of a symbolic nature and rephrased pre-existing non-Jewish perspectives. Ideas around the Jewish character of central aspects of Jewish life nevertheless remained central to the discourse, as both Jewish and non-Jewish society in the early modern period remained committed to the idea of segregation and clearly distinct social and cultural spheres.[2] One compelling example of the simultaneous overlap and distinctiveness of differing cultural perspectives is the ghetto as an essentially Jewish space. The Jewish ghetto was a rather exceptional form of settlement in the Holy Roman Empire, largely restricted to areas like the Rhine-Main region and some cities, such as Prague, while most Jews elsewhere lived among the general population. Although Jews viewed the restrictions on space and mobility accompanying confinement to the ghetto as a form of imprisonment, the ghetto nevertheless had some qualities that made it a Jewish space.[3] The seclusion and the unusual occurrence of a Jewish majority within this restricted area granted the Jewish community an undisturbed space and allowed it to live out its own cultural ideal of autonomy. Thus, the imprisonment within the ghetto acquired qualities of an ideal Jewish space with the potential to approach the cultural idea of Jews having a place of their "own."

Yet on closer inspection, this "independent" place remained very much intertwined with the outside world in many respects. One example relates to the city of Frankfurt am Main, where Jews were understood as confined to a provided place. The legal arrangement was that of permanent leasehold; the city owned the houses, while the Jewish residents had to pay rent. Yet the houses were built and renovated by the inhabitants, who perceived themselves to be the owners, as they could mortgage, bequeath, and sell them as if they were their own property. This perception on the Jewish side was even accepted by Christian courts, as long as the implicit interchanging of the concepts of ownership and property was not spelled out.

This example is all the more telling as it highlights two central factors that defined Jewish life. First, there was the attribution of a quasi-natural foreignness to Jews, expressed in their legal disqualification from acquiring property. Second was the perception, held by Christians and Jews alike, that authority was tied to territory. But even the closed-off ghetto revealed an unexpected degree of ambiguity. The synagogue, for instance, was perceived by both sides as a Jewish space, as univocally as any location could be, because religious places were both legally and popularly regarded as pertaining only to one specific religion. However, the establishment or renovation of synagogues and the supervision of the rituals that took place within them was in the hands of the authorities, which sometimes meant the Christian Church. In Frankfurt, the general understanding was that—within certain limits—each religious group was to regulate its own religious rituals, yet the Protestantisches Konsistorium did intervene when the community wanted to renovate the synagogue; the city used the synagogue quite regularly for official announcements and for legal proceedings such as the swearing-in of Jewish litigants. Mirroring the early modern idea of an intrinsic relationship between empowerment and territory, the question of who could exercise authority within a specific space stood at the center of the early modern idea of a Jewish space.

Another feature that made a space essentially Jewish within early modern Jews' understanding was its safety. Early modern society was alarmingly violent measured by today's standards, but foreigners—all the people perceived as not "belonging," with Jews being the paradigmatic ones as the only non-Christian group—were especially endangered. As a result, Jews did not define locations known to be dangerous as Jewish spaces. This is most visible in the Jewish perception of the area beyond the boundaries of the city or the village. Just as peasants belonged to the rural landscape, Jews clearly belonged to the city. This general idea was likewise held by Jews themselves, even characterizing the thinking of the rural Jewish population. While the differences between the urban and rural environments were perceived in the early modern period as oppositions that repeated the antagonism of culture versus nature, Christian city-dwellers, and Jews even more so, believed that potential danger rose in proportion to one's distance from the fortified city walls. The green land beyond the roadside, and especially the woods, were fraught with danger, so Jews would usually go to some length to avoid such spaces. But as most Jewish men had to earn their livelihood as traveling salesmen, the dangerous road actually could acquire some spatiotemporal Jewish characteristics; examples are central trade routes and markets and fairs regularly attended by Jews, as documented by Jewish calendars listing fairs and markets relevant to Jews. These spaces thus undoubtedly acquired an abstract idea of temporary "Jewishness." Nonetheless, these routes and markets were never truly conceived of as Jewish, even in early modern times. It is telling that no single Jewish map from that period has been found presenting an independent Jewish perspective on the local landscape, pointing out Jewish communities as well as routes and places of Jewish interest. The reasons for this involved cultural perceptions of safety, as security always remained a question of authority over the particular space. Places that entailed high risks to Jews could never become truly Jewish.

A Home in Foreign Lands: The Synagogue and the Burial Ground in the Early Modern Era

The synagogue was certainly a prominent topographic center of Jewish life, above all since general society accepted the Jewish claim to it as a holy place. Other spaces also come to mind as particularly Jewish spaces, perhaps most readily Jewish cemeteries.[4] The Jewish perspective on these spaces differed somewhat from the Christian one, especially in situations of growing conflict. In accordance with Jewish religious law, Jews regarded their cemeteries as permanently Jewish spaces—until the "world came to an end"—because the graves belonged to the deceased, who could not dispose of their property. Christians, however, often failed to honor Jewish ownership of the ground. Not only were Jewish cemeteries dispossessed and dissolved after the expulsion of Jewish communities; the rights to usage of the cemeteries were often not solely in Jewish hands. A general passageway leading through the Jewish cemetery of Frankfurt am Main and public orders to build a gunpowder mill or ramparts on the burial plots document this fact. Still, cemeteries' religious character made them Jewish spaces for both communities, a notion that was supported by widespread superstition and the actual danger of being attacked by one of the "firstborn animals," the *bekhorim*, kept on the grounds.[5] The Jewish population also associated the cemetery with being an important Jewish place, even more than the synagogue. As Jews, in stark contrast to the outside world, perceived themselves not as foreigners or part of vagrant social groups but as legitimate "residents in foreign lands," their graveyards testified to their long-term presence. Social status and continuous presence dating back centuries were of special significance to these communities, as were family networks as the centerpiece of early modern Jewish concepts of community. These factors made the cemetery especially important as a lasting document of social status and as a means of Jews asserting their legitimacy within the Christian-dominated world.[6]

Similar patterns of social hierarchy inscribed in space or constituted by spatial movement also pertain to the synagogue.[7] Both synagogues and cemeteries symbolically and actually rendered Jews—who lived dispersed among the nations—one holy community; these sacred spaces representing the concept of Jewish autonomy united them. Both spaces directly reflected social patterns and "provided" something quite simple that was denied to Jews: real property as a token of belonging. The importance of ownership in marking social status is evident in the Jewish practice of selling the limited number of synagogue seats to full members of the community, excluding most of the poor as well as foreigners, domestics, and minors. Although these individual seats—the *stender*—were understood to be "real estate" (*immobilia*), they could be moved within the synagogue. This made them representations of social and cultural status, as one could lose a prominent position if one's social standing was impaired or even be prohibited from using one's seat at all.[8] Unambiguous "Jewishness" was central to both cemeteries and synagogues, particularly since Jews and non-Jews alike accorded them an undisputed sacredness and the government recognized and secured Jewish authority over them. These qualities made them the "best" Jewish places, while competing claims on other places made them ambiguous and less Jewish.

After the Advent of Emancipation

Much has been written about the emergence of modern synagogues in Germany. Central features of the synagogue as a Jewish space in the early modern period were that it was secluded from the dangerous Christian world and sheltered from it and its gaze. Up through the first decades of the nineteenth century, these characteristics remained intact, with many new synagogue buildings being built in backyards that acted as barriers against the outside world. The courtyards surrounding these buildings created a relatively sheltered Jewish public space for social gatherings and religious rituals, while their interiors were even more concealed. In the mid-nineteenth century, new aspiring groups began to promote increased visibility of synagogues, a development that non-Jewish authorities sometimes encouraged or even demanded.[9] In early modern times, synagogue design had generally adhered to given architectural forms, while clearly avoiding styles characteristic of Christian churches. Most synagogue façades were understated, barely indicating the building's purpose. Only synagogues in secluded areas such as the Frankfurt ghetto or in backyards could risk greater Jewish visibility, and even in such places, design tended fundamentally to be cautious. Yet in the mid-nineteenth century, most German states and provinces were reluctant to accede to Jewish communities' growing desire to have structures whose representative qualities matched those of other religious buildings. The Jewish community of Dresden, for instance, was denied central plots several times and finally had to accept a less prominent space somewhat outside the city center and remote from Christian places of worship. The same tendency was evident many years later in Munich. Nonetheless, both the Dresden and Munich synagogues would later become outstanding symbols of the general recognition of Jewish presence in these cities.[10]

These tendencies notwithstanding, visibility became central to modern synagogue architecture, evolving over the course of the nineteenth century into a public sign of Jewish empowerment and willingness to integrate into the wider society. While Christians—even royals—had occasionally attended Jewish religious services in the early modern period, synagogues had remained very secluded from the outside world. In stark contrast, the modern synagogue announced itself as an open space, visible, with large gates opening toward the city, even though Christian attendees at services continued to be rare. The dedication ceremonies for synagogues built during the nineteenth century clarify the expectations that both Jewish communities and the authorities had of them. Unlike in premodern times, these dedications were staged as public events with strict rules. Fixed orders, social hierarchies, and pronounced references to the authoritarian state, along with repeated affirmations of patriotism, characterize these solemn occasions. They served to integrate the Jewish community and its representatives symbolically into the hierarchies of the state and the local area. Christians were expected to take an active part in these events, especially as public representatives. In Frankfurt am Main, the names of such representatives were even added to the charter sealed in the synagogue's foundation stone as a lasting, yet invisible, memory.[11] Almost all reports and programs of such dedication ceremonies state the presence of Christian representatives, generally by name, title, and office. Sometimes even royals were present or had a congratulatory note read aloud. Both Jewish and general newspapers gave similar reports on these events, though the Jewish papers provided more details.[12] This tellingly reveals the extent to which these events became embedded in the general cultural fabric, as they tied Jews and non-Jews together in a paradigmatic way. The middle-class ideology, with specific ideas about respectability and social status, became characteristic of these ceremonies. Public interest in these dedication ceremonies and in the new buildings was so great that some communities announced visiting hours or even sold tickets, as most people would only be able to watch as bystanders, outside the building.[13]

While the dimensions of these dedication ceremonies and their openness to the public would have been unimaginable a few decades before, one aspect is especially striking: many newspaper reports listed members of the Christian clergy in attendance with their denominations, as Protestant or Roman Catholic. While most reports do not indicate whether all the invited clergy accepted the invitation, few reports explicitly criticized Roman Catholic priests for not attending when their Protestant counterparts had taken part. Thus, unanimous support and participation in dedication ceremonies by everyone, especially by the relevant official representatives, had come to symbolize that Jews and Christians could and did live harmoniously together. This very matter and Jews' readiness to integrate were also mentioned repeatedly in speeches at these events. Therefore, anything that undermined such sentiments, such as pointing out the absence of Catholic representatives, directly jeopardized the festivities' intention—at least on the Jewish side—to create a tolerant, pluralist society. This made it easier to ignore the fact that many Catholic priests quite probably would not have attended similar events at Lutheran, Waldenser, or Huguenot churches, either.[14]

The inauguration addresses and the sermons at these events were clearly expected to follow this general line of thought, which they usually did throughout the nineteenth century. Only rarely do we find mention of anything that troubled the symbolic ritual and its evocation of all-embracing harmony, which was merely wishful thinking at first and remained fragile even in later years. When Rabbi Victor Meyer Rawicz seized the opportunity in 1881 at the dedication of the synagogue of Ettenheim to denounce the rising anti-Semitic movement for jeopardizing mutual tolerance, several press reports criticized him for having desecrated the solemn occasion.[15] The public failed to understand that his politicized action actually accorded with the purpose of the event, to secure a tolerant social climate. This said, these ceremonies were not intended to be explicit battlegrounds for social recognition and equality. They were limited to a general public symbolism and to affirming something that was still in question.

These somewhat muted ceremonies, marking the remodeling and reconstruction of synagogues as central spaces of Judaism, were only part of the accelerating redefinition of Judaism. In the Austrian Empire, and especially in Hungary, the slightest "modernization" was understood as a defeat of traditional Judaism. Yet the overwhelming majority of German Jews elsewhere found the far-reaching remodeling of the sacred space acceptable. This remodeling—which also redefined its use, was referred to as "aesthetic reforms." These shifts become all the more interesting as the forms of architectural design that emerged in this period directly reflected changes in the concepts surrounding the religious community as well as the general idea of religion. Religion became a means for the moral improvement and self-examination of the individual.[16] The space of the modern synagogue underlined the importance of the religious sermon as indispensable for re-educating the Jewish population in accordance with the idea of Jewish "improvement."[17] The subdivision of the interior transformed what had hitherto been a rather individualistic ritual into the service of a community choreographed by religious specialists. Surprisingly, these important features were not much of an issue for most contemporaries. The conflicts that took place around forms of synagogue service at that time generally pertained to the perpetuation of the traditional liturgy and customs. The new buildings often went hand in hand with changes in the ritual and thus emerged as a major factor in the formation of different Jewish denominations. More than anything else, the introduction of the organ, theretofore a Christian instrument, into synagogues became the ultimate demarcation line between the Jewish Orthodoxy and the Reform movement.[18]

The degree to which religious space generated conflict relating to Jewish identity, such as the contemporary concept of "assimilation," transformed many new synagogues from spaces of togetherness into symbols of new affiliations. The severe religious conflicts that broke out could often only be eased by separating the dissenting groups; the spatial shift distinguished legitimate community members from dissenters. Internal quarrels usually culminated when the old communal synagogue became the center of the desire for reform; as long as its status was not questioned, a compromise could usually be found, whereas the demolition or remodeling of this central time-honored place of identification would rip communities apart. At the same time, secular authorities and Jewish community boards tried to introduce the concept of a centralized Jewish community that defined one specific synagogue as the only center of communal worship—usually the one owned by the community. Other places of prayer and religious learning were shut down, which aggravated the conflicts around the religious orientation of the central place. Most Orthodox Jews were not opposed to new central synagogues, but they rejected changes in liturgy, arguing that the relaxation of the strict separation of the sexes and, most of all, the introduction of the organ would strip them of their synagogue and their place in Jewish society. The question of ownership was central to many of the controversies around such refurbishments; one example is the replacement of individual seats, which could be shifted to denote changes in individual congregants' status, by fixed rows of pews. Even the Reformers made sure that this important "real property" was converted to adequate individual places in the new building.[19] Not only did Orthodox Jews lose their long-standing seating, but they also found themselves having to contribute to the cost of building a synagogue they would not use.

Frankfurt am Main serves as a telling example of this process of redefining the synagogue as an exclusive, communal place. The case presents all the extremes of the developments outlined here, alongside the typical compromises that became characteristic of many German communities. The community's synagogue had once literally stood at the center of the ghetto. Its demolition therefore symbolized the end of the period of forced seclusion even more than in other cities, although the ghetto walls had already fallen in 1796. The small and run-down community synagogue had consisted of the men's synagogue and an adjacent building for the women, plus another small prayer house at the side of the complex. When the liberal governing board of the Jewish community decided to rebuild it, the Orthodox Amschel Meyer Rothschild agreed in 1843 to give a generous endowment toward the cost of construction. After the board members disregarded parts of the agreement and hired the moderate Reform rabbi Leopold Stein against the protest of the community rabbi Salomon Salman Trier, Rothschild withdrew his financial support, and the plan collapsed. After many efforts to secure the financial means to continue, the board members had the old complex demolished, resulting in lasting fragmentation of the community. The Orthodox members now required another place because the new main synagogue, completed in 1860, was marked by liturgical reforms. The community board agreed to provide a renovated building in a yard that had previously housed the Jewish Reform school Philanthropin and one of the first Reform services. Other

Orthodox members continued to use the existing private places nearby, such as the Löw Elias Reiss'sche Synagogue.[20]

As more and more Orthodox members threatened to leave the Jewish community for an entirely independent, Orthodox counterpart, the additional compromise was to hire an Orthodox rabbi and provide them with appropriate institutions. This entailed establishing a new representative place of worship, which was opened in 1882. Other parts of the community's Orthodox sections had already established a private association after Stein's inauguration and opened another synagogue with the aid of the Rothschild family,[21] subsequently seceding from the general community in 1876 after legislation allowed their transformation into a fully independent community. The community's central place of worship, in existence since 1711, had fragmented into at least three main synagogues alongside several "lesser" places of worship, all ministering to different religious strands of an increasingly complex Jewish landscape. The grand new building had neither united the community nor replaced other places of worship.

The Old and the New

The inauguration ceremony of the first community synagogue under the auspices of Leopold Stein took place in the familiar way. Many sermons and speeches given on such occasions took the new synagogues not merely as replacements of older structures but as symbols of the reconstruction of Jewish society, distancing the old from the new. The call for Jewish integration into the majority society frequently referred to the abysmal conditions in which Jews had been forced to live until recently, thereby putting the blame foremost on the Christian side.[22] Many sermons emphasized the former forced isolation of the Jews—an isolation symbolized by the old synagogue as well as the ghetto; in this trope, the new synagogue was a contrasting symbol of a new beginning. All of this alluded to a specific concept of modern Jewish historical thought: the redefinition of the Jewish past as a "victimized history" of suffering and sacrifice.[23] This historicization of space, contrasting the old with the new, held a particular meaning for the Jews. Following this concept, Stein's inaugural sermon reminded congregants of the dark days of the ghetto and praised the new era embodied by the new building.[24]

Yet others bemoaned the loss of the old place. Lazarus Geiger, the son of one of the community's rabbis and nephew of Abraham Geiger, published a lengthy poem—"Tercets on the Fall of the Synagogue"—dedicated to the old community synagogue. It evokes an eerie picture of the last service in the place of his childhood memories.[25] He describes men praying, all dressed up as if for a Yom Kippur service in white shrouds, in *kittel* or *sargenes*. But these men are not actual community members celebrating a final farewell; they symbolize the eternal Frankfurt community consisting of the souls of the deceased who had come from the nearby cemetery for a last prayer of the dead. These are the ancestors who had so often interceded with God for the living on the Day of Atonement and in times of danger. Geiger, a staunch defender of Orthodox Judaism, combined his evocation of traditional ideas about the merits of the ancestors and their lasting concern for their families and the community with a critique of changes in liturgy and piety in general. He intended for the solemn, pompous poem to remind readers that the new building—and the new kind of Judaism that would be established there—departed from the centuries-old tradition for which the building had stood. At the end of the poem, the gathered spirits shout the religious creed that there is only one god: "And as they cried, the aged and young, the walls resounded moaning with a droning hum." The old synagogue becomes the embodiment of the timeless Jewish community that was to be understood as an extension of the past and not, as Stein would have had it, as a departure from it.

However, most members of the contemporary communities were less inclined to such spooky romanticism. Upon the demolition in 1887 of the remaining houses of the old Jewish quarter, the majority supported Leopold Stein's call for these relics of a horrible past to quickly disappear, which he had voiced at the synagogue's dedication: "We do not need this street any more that is still an aggravating neighbor to our house of God. Its existence constitutes a gloomy reminiscence of the dark night of the past; its extinction would elicit pleasant pictures of the ever brighter days of the present age."[26] This sentiment was shared even by most Orthodox, and only the old home of the Rothschild family, which was left to stand as a symbol, preserved the memory of the ghetto. Having secured the dilapidated structure and even had the whole house moved some meters to allow the street to be widened, the family set up a small museum testifying to the humble yet decent beginnings of their ancestors and the community.[27]

Alongside the trend toward a musealization of the Jewish past that took off during the final third of the nineteenth century, Lazarus Geiger's sentiment was not exceptional. However, he had picked the wrong place for his overstretched vision of a timeless Jewish community embodied by an intrinsically Jewish space; in this period, the synagogue, as an icon of Judaism, represented the aspired-to social position of the emancipated Jews and their hopes for the future. It stood for visibility and openness, yet—at least in large communities—it failed to serve as a unifying center for the whole community.

It was the cemetery that retained many more of its traditional characteristics and continued to symbolize the bond with the past. The graveyard had always stood at the center of the Jewish perception of belonging, as it had traditionally been one of the few actual spaces in which Jewish presence was documented at all, as the public space would not allow for Jewish places of memory. The gravestones recorded an overarching tradition, as well as the local one. This symbol of belonging to a specific place had always stood in harsh contrast to the Christian notion of Jewish "foreignness." The cemetery gained relevance because Christians and Jews alike considered time-honored practice, alongside long-time residency, to constitute legal claims and to act as vital factors in signaling social status. Jewish cemeteries also inscribed social relationships directly into the landscape. The size of the gravestones and the relationships evident among them gave witness to the status of various individuals and the important family networks that occupied central parts of the burial ground. Countless inscriptions attested to deceased individuals' ancestry, scholarship, benevolence, and piety, thus giving testimony of the achievements of the departed for generations to come. In comparison to other spaces such as the synagogue, where social status could be established via closeness and distance, the cemetery had been the only place Jewish society knew of that exhibited such a degree of permanence in relation to individuals' status. It was also the only truly exclusive Jewish place, as only Jews were buried there. The members of the community were united by the common space sacred to a timeless holy community, rather as in Geiger's poem.

Being buried with one's ancestors remained of lasting importance to German Jews into the modern period. Liberal community boards therefore sometimes resorted to early-modern procedures and utilized cemeteries to pressure the few Orthodox dissenters not willing to pay taxes to a Reform community. Few of these community members were prepared to sever bonds with the Reform community and thus forgo a respectable burial alongside their departed loved ones. The proponents of segregationism, foremost among them Rabbi Samson Raphael Hirsch, struggled to convince Orthodox believers that it was preferable and even required by religious law to split from the general community, even if this meant being buried apart from it. Even in the central segregationist community of Frankfurt am Main, the emotional bond of the community to its burial place remained so strong that it hampered the movement's success. Although the community opened an independent cemetery, many members of the Rothschild family, some of whom were ardent supporters of the segregationist cause, were unwilling to turn their backs on what they considered to be their "appropriate" final place. Their graves are not to be found in the adjacent cemetery of the segregationists, who claimed to be the true heirs of the Frankfurt tradition, but on the plot of the general Jewish community.

Popular Jewish ghetto literature at that time picked up on the issue of cultural and social change with its pitfalls and its opportunities, yet it hardly paid any regard to the Jewish cemetery as a cultural icon.[28] Instead, growing Jewish interest in the history of the local community prompted most of the attention given to this place. In a more abstract way, the cemetery served as a legitimization of Jewish presence that complemented the political statement made by the new synagogues. The old cemeteries again bore witness to the non-Jewish world as visible proof that Jews had been a part of the German nation from its beginnings. Just as Edwin Oppler considered the early modern synagogue of Worms to be representative of a common national style in his reflections on synagogue architecture, the small Jewish community of Worms became the paradigmatic age-old Jewish settlement.[29] Picking up on the new phenomenon of tourism, the city of Worms started to advertise its Jewish past with the synagogue, the ritual bath, and the Jewish cemetery, the Heiliger Sand, which dated back to the Middle Ages.[30] A poem by Ludwig Schulmann that refers to the medieval pogroms in Hildesheim reflects this historicization of the modern idea of belonging to one's *Heimat* that was symbolized by the graveyard. The narrator laments that Jews have been condemned to "walk astray from place to place, a stranger just to strangers," and then expresses a desire to reclaim the space defined as "home": "but I will win it back for me, one home, one hearth; … to rest my troubled head on the beloved ground … with the forefathers in the burial vault; but you [= the Jewish youth] will live and breathe homeland's delicious air."[31]

In the years that followed, associations for Jewish history repeatedly focused on the medieval period as the age in which the German nation had been forged, thus attesting to a past shared by Jews and Christians. Toward the end of the nineteenth century, publications about graveyards and epitaphs became a central topic for Jewish scholars, bringing together religious tradition with their new politicized historical interest. In 1855 the preacher of the community of Worms, Ludwig Lewysohn, published a volume on some of the inscriptions in the old cemetery. To secure financial support for this project, the renowned Orthodox community rabbi Jakob Koppel Bamberger had previously assured potential donors that a donation would honor not only the deceased but the donor as well, as the publication would "provide to Judaism the acknowledgment that it does not stand behind other religious denominations that attempt their utmost for the conservation of *their* monuments."[32] Traditional religiosity became identified not only with *Wissenschaft*—the new tradition of historical research—but also with a mutual interest in a shared past.[33]

In the first Jewish travel guide to Germany, published in 1884, Worms received one of the longest entries as "the oldest Jewish community in Germany. According to repeated reports the first Jewish inhabitants had been the offspring of Germans [*Vagionen*] who served in the army of Emperor Titus and Jewish women who had been kidnapped." The claim to original "Germanness" could not have been stated any more clearly. The account for Worms, like those for some other communities, mentions famous rabbis and their specific merits but also notes tourist attractions: "Of interest are: Rashi's seat, Rashi's wall, the old cemetery."[34] The need to provide proof of a long-standing history, which Heidingsfelder, the author of the guide, dated back to antiquity, prompted him to use these visible places as indicators of belonging and to add the Jewish story to the German national narrative. The sometimes uneasy coexistence of traditional aspects of Judaism with the new idea of a shared German nationality was symptomatic of the difficulties contemporaries had in redefining "their" place within their localities and the wider context.

The New against the Old

In the context of these struggles to define and establish Jewish space and place, the Jewish graveyard retained its characteristic exclusivity and seclusion throughout the nineteenth century; it remained an effectively secluded place related primarily to the family and the community. By contrast, publicity became a decisive feature of the synagogue. The ways in which nineteenth-century synagogues emerged into the public space bring to the fore an apparent desire on the part of German Jews to make up for their uncertain status. The location, style, and grandeur of these new synagogues mirrored the Jewish communities' aspirations to form a central and simultaneously harmonious element of society.[35] While the few debates about the political statement of architectural styles that took place at that time reiterated idiosyncratic Jewish claims of belonging and integration, contemporary newspaper reports reveal the aspects of synagogue architecture that were important to most Jews. They frequently mentioned the contribution the synagogues made to their overall settings and expressed the acknowledgment Jewish communities hoped their buildings would embody. These reports also stereotypically pointed to the magnificence of these buildings, which occasionally bordered on outright ostentatiousness. Many contemporaries criticized the enormous cost of the buildings, maintaining that their pomp contradicted the "nature of the Jewish religion." But even these critics—among them those Orthodox who denounced these "temples" as empty shells trading authentic Jewishness for worldly splendor—would not have less grand synagogues for themselves provided they had sufficient means; the segregationist community of Frankfurt am Main is a case in point.[36] Visibility and recognition rather than cautious reserve had become central characteristics of the synagogue.

All sides in the debates continued to agree that synagogues had to differ from churches, a view all the more urgent because the Romanic and Gothic styles now frequently adopted for synagogues remained intrinsically linked to church architecture. While some argued that these styles were foremost national ones, Jews and Christians alike continued to have difficulty with a cruciform ground plot for synagogues.[37] Further, the general look of some Jewish buildings with stained-glass and rose windows, which were characteristic of medieval cathedrals, provoked caustic remarks. Jewish ideas of religious spaces were primarily shaped by the same aspects as the synagogue inauguration ceremonies we have discussed above, that is, they aimed to advertise a will to integrate or a completed process of integration. Yet this central message was contradicted by the singularity of these buildings, more so than by their interchangeability, because to be Jewish also meant to differ. It is telling that newspapers stereotypically described these buildings as alien or outlandish in their intricacy and sumptuousness. Moreover, all the possible styles of synagogues were eventually perceived as overabundant, lavish, and elaborate in design, and, again, as foreign. Even early buildings that presented a rather unpretentious façade could have a dramatically different interior. The interior design of Gottfried Semper's famous Dresden synagogue, for example, involved colorful decoration in the neo-Islamic style combined with up-to-the-minute architectural techniques. Many later synagogues that opted for more established forms of religious architecture nevertheless featured splendid interior decoration as well. The foreignness of these buildings became one of their central features, seemingly counteracting the idea that they were made to fit in. Many adjustments were made to the Jewish religious service so that it could meet the general idea of bourgeois religiosity, wherein it was particularly linked to the new liberal Protestant form of service as a paradigm of modernity. All the same, Jewish religious ritual remained unfamiliar to the general public, so most found it quite natural that these places of worship should reflect this. A general and in-tractable-seeming problem arose from the contradiction thus embodied between the appeal for assimilation and the retention of exceptional visibility.

Against all arguments that the Romanic or Gothic style best symbolized German Jews' long-standing affiliation with the German nation, buildings in neo-Islamic, Moorish, and similarly mixed styles became a distinct trend during the second half of the century.[38] Adopting the general perception of Jews as the people of a foreign nation, these buildings used the widespread Orientalism of the period to give the idea of an "Oriental" or "Semitic" Jewish origin a positive connotation. Arguments were advanced linking these styles to the Temple of Jeru-salem or alluding to a supposedly harmonious existence of Christians, Jews, and Muslims in medieval Muslim Spain. Yet these were most probably rather irrele-vant compared to the basic intention of these buildings: to stand out. Certainly, most contemporary Christians would have had little interest in such strained de-ductions. The beauty of this artificial foreignness essentially appealed to a wide-spread enthusiasm and transmitted the message that the foreignness of Jews and Judaism was a harmless and fascinating contribution to national culture.[39]

In contrast to this drive for visibility, Jews and Jewish religious practices in the form known for centuries disappeared from the public space. While all of these synagogue buildings made appeals for attention, the religious service within them became the paradigm of a middle-class ceremony; it built upon older tradi-tions, introducing bourgeois concepts about strict discipline regarding time, the body, and demeanor. The immediate intention was to turn what had been a very individualized Jewish religious ritual into a regulated service of a community that was "in its place," whose members raised their voices only when they were sup-posed to and, most interestingly of all, did not publicly exhibit religious practices now deemed too blatantly Jewish. People frowned upon those who walked to the synagogue in their slippers, wrapped in a prayer shawl, wearing phylacteries, and entering the place of worship in stocking feet.[40] Jewish weddings, a public act that had previously always been performed outside the synagogue, were moved inside the sacred space; there are few accounts from this period of religious ritu-als in open spaces, such as the monthly prayers for the new moon. Although an abstract air of foreignness had become a central aspect of Jewish religion, actual displays of religious behaviors that diverged from the middle-class norm seem to have been avoided, as they presumably ran counter to Jews' efforts to affiliate with the rising middle class.[41] The risk of ridicule or even attack implicitly fueled old memories of violence. While a few strictly Orthodox believers clung to such rituals, traditional Jewish religious practice that testified to difference became less visible and thereby truly foreign.

This certainly does not mean that Jews and Judaism disappeared entirely from public space. Many Jewish shops were closed on holidays, and during the

High Holidays the streets around synagogues buzzed with people hurrying to prayer. Men usually had to wear a top hat and frock coat, upscale attire that indicated the solemnity of the occasion, because the "ragged Jew" caricature of the early nineteenth century gave way to an upper-middle-class lifestyle as the distinctive mark of the Jewish worshiper.[42] The old synagogue garb that all community members had worn at public and religious events was quickly put aside. Further, even the new ostentatious and demonstrative synagogue buildings barely bore any Jewish symbols on their exteriors; such emblems were restricted to the interior, while the façades usually featured, at most, abstract Jewish architectural elements, like the two columns of the Jerusalem Temple. Only the tablets of the covenant became a common icon of the synagogue, sometimes complemented by a Hebrew inscription or the seven-branched candelabrum.[43] The Star of David was rarely used then; it was a less attractive symbol at least in southern Germany, where it was used to advertise beer halls. The interior décor certainly celebrated Jewishness to a greater extent, yet strong Jewish symbolism came into trend only toward the turn of the twentieth century.

Conclusion

The quality of the spaces that were the most Jewish—the synagogue and the cemetery—underwent clear changes in nineteenth-century Germany as both places became directly associated with the central political project of integrating the Jewish population into the state and nation. For synagogues, seclusion gave way to visibility and openness as key hallmarks of their construction and architecture. The buildings' spatial language intentionally promoted formerly negative characteristics such as "foreignness" to underline the distinctiveness of Jewish religious tradition. The tension between integration and distinction, however, was mitigated by the new synagogues' added function as tools for transforming the Jewish population into an integral part of the new middle class, whose virtues, such as reservedness, were not compatible with outlandish appearances.

At the same time, the concept of symbolizing a centralized Jewish community with a single synagogue struggled to succeed. In early modern times, the community synagogue would have held an uncontested status even in larger communities. But in the nineteenth century, the quest for modernization of the Jewish religious tradition multiplied these ritual spaces and transformed them into places of specific affiliation. It robbed the communal centers of their uniqueness and created a growing landscape of different kinds of Judaism represented by a diversity of places of worship. Both the synagogue and the burial ground also became part of a politicized historical discourse that accentuated the positive changes in modern times, distancing the present from the past.

Even so, it became important to safeguard a distinct Jewishness, leading to the persistence of cultural traditions around community and belonging related to space and encapsulated in the timeless burial ground. Although the Jewish

cemetery was certainly affected by the rapid emergence of new middle-class sentiments and aesthetics, older Jewish concepts of belonging remained relevant; this
space continued to be characterized by permanence and seclusion and was able to
lend cohesion within a community by conveying values and status. Nevertheless,
Jewish burial grounds also often remained subject to outside authority over the
space; many a Jewish community defended its cemetery vehemently against the
demands of modern urban planning.

The essential question of whether Jews were really empowered over the
course of the nineteenth century by these changes in space is difficult to answer.
The time between the widespread pogroms of 1819 and 1848 and the rise of
political anti-Semitism in the 1870s has been understood as a period of relative
peace, although it has not been the subject of much attention from scholars of
anti-Semitism. Reluctance among German Jews to display symbols or behaviors conceived of as pronouncedly Jewish in their public spaces could indicate
that old fears were still ingrained in Jewish memory. Alternately, it could suggest
that they were engaged in balancing between assimilation and aspirations and
attempts to form a visible part of German public life and space.[44]

Andreas Gotzmann is full professor and chair for Jewish studies and religious
studies at Erfurt University. His research interests include German-Jewish history, Jewish religious and cultural history from early modern to modern time,
and the theory and method of cultural studies. Among his authored monographs
is *Jüdische Autonomie in der Frühen Neuzeit. Recht und Gemeinschaft im deutschen
Judentum* (2008). He has also edited or coedited several volumes, such as *Kaiser
und Reich in der jüdischen Lokalgeschichte* with S. Wendehorst and S. Ehrenpreis
(2013).

Notes

1. Some examples are Kümper et al., *Makom*; Jasper and Schoeps, *Deutsch-jüdische Passagen*;
   Schlör, *Das Ich der Stadt*.
2. Gotzmann, "Im Zentrum der Selbstverortung?"; Gotzmann, "In der Gemeinde—auf der
   Straße."
3. This development accelerated during the nineteenth century. One contemporary observer, for example, remarked, "Der Judengasse—die, ach, vordem für uns gewesen eine
   beengende Gasse, ein zweites Egypten." See "Hebräische Votivtafel," 245.
4. Gotzmann, *Jüdische Autonomie.*
5. Gotzmann, "Im Zentrum der Selbstverortung?"
6. Gotzmann, *Jüdische Autonomie,* 745–86.
7. Ibid., 133, 152, 284, 327, 332, 383.
8. Ibid., 245–46.
9. Sylvia Necker's contribution to this volume (chap. 9) discusses this development in detail.
10. Hammer-Schenk, *Synagogen,* 128–30; Fritsch, "Die neue Synagoge in München."

11. Twiehaus, *Synagogen*, V, 190–92. "Hebräische Votivtafel," 241–45; "Rundschau—'al Mi-schmarti …': Frankfurt am Main (Juli)," *Der israelitische Volkslehrer 7* (1855): 270.
12. Twiehaus, *Synagogen*, V, 13, 15, 17, 19, 23–26, 28, 30, 31, 36, 42, 44, 47–48, 50–51, 57, 70–71, 76, 78–79, 95–96, 99, 102, 104, 105, 111, 120, 124, 133–34, 139, 143, 145–53, 157, 165, 168–71, 176, 179, 181, 184, 191, 194, 201, 204, 207, 216, 229, 235–36, 238, 242, 246, 248–49, 255–56, 260, 262, 265, 268, 270, 272–73, 285, 290, 293, 296, 319, 320, 322, 324–25.
13. Twiehaus, *Synagogen*, V, 72.
14. Twiehaus, *Synagogen*, V, 12 [according to *Ortenauer Tageblatt und Lahrer Wochenblatt* 52 (1 Mar. 1868), 209]. Blaschke, *Katholizismus und Antisemitismus*, 133–35.
15. Baumann, *Zerstörte Nachbarschaften*, 70; Twiehaus, *Synagogen*, V, 80, 83, 202–4, 216, 274, 298, 306, 309, 311, 313, 315–16; 271; Rabbi Adolf Schwarz of Karlsruhe [according to *Badische Landesbote* 613 (31 December 1892), 2–3] made similar remarks.
16. Lässig, *Jüdische Wege*, 254–90.
17. Ibid., 290–325; Twiehaus, *Synagogen*, V, 297.
18. Gotzmann, *Jüdisches Recht*; Meyer, *Response to Modernity*, 184, 489; see the many index entries about the organ that indicate local conflicts.
19. Selig, "Die Synagogenbauten," 44–46, 81–84.
20. Mainz, "Aus der Geschichte der Gemeinde," 2–4. After the relocation of the "Reiss'sche shul," the new building held the *aron hakodesh* of the old community synagogue that had been preserved, thus linking the Orthodox part of the community back to the "true" tradition by means of historical relics.
21. Gotzmann, *Jüdisches Recht*, 303–16.
22. Twiehaus, *Synagogen*, V, 63, 66, 213.
23. Gotzmann, "Ambiguous Visions of the Past"; Gotzmann, "Historiography as Cultural Identity"; Gotzmann, *Eigenheit und Einheit*, 114–211; Twiehaus, *Synagogen*, V, 106, 213, 286, 326, 330.
24. Stein, *Bedeutung*; Twiehaus, *Synagogen*, V, 87–88, 90.
25. Geiger, *Terzinen*, 10–12.
26. Stein, *Bedeutung*, 135.
27. Some romanticism around the ghetto as a singularly Jewish place persisted throughout the period. See Kobler, *Jüdische Geschichte*, 81–82; "Die Frankfurter Judengasse," *Jeschurun* 4 (1883): 371–72; there are also references to the new non-Jewish, usually poor occupants and to prostitution.
28. Glasenapp, *Aus der Judengasse*.
29. Oppler, "Synagogen und jüdische Begräbnisplätze." An even earlier piece by Albrecht Rosengarten of 1840 argued against Orientalism: *Architektonische Stylarten*, 2nd ed. (Braunschweig, 1857). Hammer-Schenk, *Synagogen*, 202, 104; Hammer-Schenk, "Edwin Opplers Theorie"; Weissbach, "Buildings Fraught with Meaning"; Twiehaus, *Synagogen*, V, 39, 52.
30. Roemer, "The City of Worms"; Roemer, *German Cities*.
31. Schulmann, "Michael," 316.
32. Lewysohn, *Nafshot tsadikim*, 4–5.
33. Wolf, *Die jüdischen Friedhöfe*, follows a different format, yet combines the older literary genre with a historical account of the Vienna burial society and its history. A more tra-ditional approach is evident in Landshuth, *"Toldot Anshe Schem."* Toward the end of the century, more such publications and editions appeared; examples are Salfeld, *Der alte israelitische Friedhof*, and Horovitz, *Die Inschriften*.

34. "Rashi" stands for Rabbi Shlomo Yitzchaki, a leading medieval French commentator who, as tradition claims, had taught in Worms for some time: Heidingsfelder, *Allgemeines Lexicon*, 160–61.
35. Künzl, *Islamische Stilelemente*, 109–125; Hammer-Schenk, *Synagogen*, 58–74, 165–96, 251–58; Twiehaus, *Synagogen*, V, 65–66, 164, 276, 290, 322.
36. Twiehaus, *Synagogen*, V, 233.
37. Künzl, *Islamische Stilelemente*, 175–85; Hammer-Schenk, *Synagogen*, 199–203; Twiehaus, *Synagogen*, V, 48, 59, 79, 96, 99, 123, 136–39, 180, 232, 244, 247, 252–53, 261, 334–35; Schorsch, "The Myth of Sephardic Supremacy."
38. Hammer-Schenk, *Synagogen*, 251–310, 351–52; Künzl, *Islamische Stilelemente*, 186–424; Meyer, *Response*, 183.
39. Hammer-Schenk, *Synagogen*, 96–98, 288–90; Coenen Snyder, *Building a Public Judaism*, 25–86.
40. *"Diwre Shalom weEmet,"* 5–6; Lässig, *Jüdische Wege*, 262–70, 364–72; Gotzmann, "Der Geiger-Tiktin Streit"; see the criticism of Abraham Geiger's public appearance in Breslau: Selig, *Synagogen*, 48–55; the regulation of 1826 for the community of Munich (ibid. 54 Cap. IX, § 1) even prohibited groups from standing together on the street in front of the synagogue. See Breuer, *Jüdische Orthodoxie*, 283–84; not writing in school on Shabbat was already understood as a *kiddush hashem*, a sanctification of God.
41. Gotzmann, "Reconsidering Judaism."
42. Haibl, "Im Widerschein."
43. Hammer-Schenk, *Synagogen*, 109.
44. Twiehaus, *Synagogen*, V, 82, 85.

Bibliography

Baumann, Ulrich. *Zerstörte Nachbarschaften: Christen und Juden in badischen Landgemeinden (1862–1940)*. Hamburg: Dölling & Galitz, 2000.

Blaschke, Olaf. *Katholizismus und Antisemitismus im deutschen Kaiserreich*. Göttingen: Vandenhoeck & Ruprecht, 1997.

Breuer, Mordechai. *Jüdische Orthodoxie im Deutschen Reich (1871–1918): Die Sozialgeschichte einer Minderheit*. Frankfurt am Main: Athenäum, 1986.

Coenen Snyder, Saskia. *Building a Public Judaism: Synagogues and Jewish Identity in Nineteenth-Century Europe*. Cambridge, MA: Harvard University Press, 2013.

"Diwre Shalom weEmet". Worte des Friedens und der Wahrheit. Ansprache des israelitischen Gemeindevorstandes zu Offenbach am Main an seine Gemeindemitglieder über dessen Anordnungen und Einrichtungen seit 1821 bis auf die Gegenwart. Offenbach, 1843.

Fritsch, K. E. O. "Die neue Synagoge in München." In *Synagogen und jüdische Friedhöfe in München*, edited by Wolfram Selig and Gabriele Dischinger, 131–48. Munich: Aries, 1988.

Geiger, L. *Terzinen beim Fall der Synagoge zu Frankfurt am Main*. Frankfurt am Main: Auffarth, 1854.

Glasenapp, Gabriele von. *Aus der Judengasse: Zur Entstehung und Ausprägung deutschsprachiger Ghettoliteratur im 19. Jahrhundert*. Tübingen: Niemeyer, 1996.

Gotzmann, Andreas. "Ambiguous Visions of the Past: The Perception of History in Nineteenth Century German Jewry." *European Journal of Jewish Studies* 1, no. 2 (2007): 365–94.

————. "Der Geiger-Tiktin Streit: Trennungskrise und Publizität." In *In Breslau zuhause? Juden in einer mitteleuropäischen Metropole der Neuzeit*, edited by Manfred Hettling, Andreas Reinke, and Norbert Conrads, 81–98. Hamburg: Dölling & Galitz, 2003.

————. *Eigenheit und Einheit: Modernisierungsdiskurse des deutschen Judentums der Emanzipationszeit.* Leiden: Brill, 2002.

————. "Historiography as Cultural Identity: Towards a Jewish History Beyond National History." In *Modern Judaism and Historical Consciousness: Identities—Encounters—Perspectives*, edited by Andreas Gotzmann and Christian Wiese, 494–528. Boston: Brill, 2007.

————. "Im Zentrum der Selbstverortung? Das Ghetto als jüdischer Raum." In *Frühneuzeitliche Ghettos in Europa im Vergleich*, edited by Fritz Backhaus, Gisela Engel, Gundula Grebner, and Robert Liberles, 333–67. Berlin: Trafo, 2012.

————. "In der Gemeinde—auf der Straße: Zur Raumwahrnehmung der jüdischen Bevölkerung in der Frühen Neuzeit." *Jüdische Topographien.* Berlin, forthcoming.

————. *Jüdische Autonomie in der Frühen Neuzeit: Recht und Gemeinschaft im deutschen Judentum.* Göttingen: Wallstein, 2008.

————. *Jüdisches Recht im kulturellen Prozeß: Die Wahrnehmung der Halacha im Deutschland des 19. Jahrhunderts.* Tübingen: Mohr Siebeck, 1997.

————. "Reconsidering Judaism as Religion: The Religious Emancipation Period." *Jewish Studies Quarterly* 7 (2000): 352–66.

Haibl, Michaela. "Im Widerschein der Wirklichkeit: Die Verbürgerlichung und Akkulturation der deutschen Juden in den illustrierten Zeitschriften zwischen 1850–1900." In *Juden—Bürger—Deutsche: Zur Geschichte von Vielfalt und Differenz (1800–1933)*, edited by Andreas Gotzmann, Rainer Liedtke, and Till van Rahden, 217–40. Tübingen: Siebeck, 2001.

Hammer-Schenk, Harold. "Edwin Opplers Theorie des Synagogenbaus: Emanzipationsversuche durch Architektur." *Hannoversche Geschichtsblätter*, n.s., 33 (1979): 99–117.

————. *Synagogen in Deutschland: Geschichte einer Baugattung im 19. und 20. Jahrhundert (1780–1933).* Hamburger Beiträge zur Geschichte der deutschen Juden 8. Hamburg: Hans Christians, 1981.

"Hebräische Votivtafel, verfasst vom Herausgeber und eingelegt in den Grundstein der Hauptsynagoge zu Frankfurt am Main. 'Anokhi LaShem, Anokhi Ashira.'" *Der israelitische Volkslehrer* 7 (1853): 241–45.

Heidingsfelder, B. *Allgemeines Lexicon sämmtlicher jüdischer Gemeinden Deutschlands: Nebst statistischen und historischen Angaben, sowie öffentlichen Mittheilungen der jüdischen Hotels, öffentlichen und Privat-Restaurants zum Gebrauche für Behörden, Gemeindevorstände, Reisende, Gewerbetreibende etc.* Frankfurt am Main: Kauffmann, 1884.

Horovitz, Marcus. *Die Inschriften des alten Friedhofs der israelitischen Gemeinde zu Frankfurt am Main.* Frankfurt am Main: Kauffmann, 1901.

Jasper, Willi, and Julius H. Schoeps, eds. *Deutsch-jüdische Passagen: Europäische Stadtlandschaften von Berlin bis Prag.* Hamburg: Hoffmann und Campe, 1996.

Kobler, Franz. *Jüdische Geschichte aus in Briefen aus Ost und West: Das Zeitalter der Emanzipation.* Vienna: Saturn, 1938.

Kümper, Michael, Barbara Rösch, Ulrike Schneider, and Helen Thein, eds. *Makom: Orte und Räume im Judentum: Real—Abstrakt—Imaginär.* Hildesheim: Olms, 2007.

Künzl, Hannelore. *Islamische Stilelemente im Synagogenbau des 19. und frühen 20. Jahrhunderts.* Frankfurt am Main: Lang, 1984.

Landshuth, L. *"Toldot Anshe Schem wePeulatam beedat Berlin."* Berlin: Poppelauer, 1884.

Lässig, Simone. *Jüdische Wege ins Bürgertum: Kulturelles Kapital und sozialer Aufstieg im 19. Jahrhundert.* Göttingen: Vandenhoeck & Ruprecht, 2004.

Lewysohn, L. *Nafshot tsadiḳim: Sechzig Epitaphien von Grabsteinen des israelitischen Friedhofes zu Worms; regressiv bis zum Jahre 905 übl. Zeitr. Nebst biographischen Skizzen und einem Anhang; mit einer Abbildung der äußern und innern Ansicht der Raschi-Kapelle zu Worms.* Frankfurt am Main: Baer, 1855.

Mainz, Michael M. "Aus der Geschichte der Gemeinde: Erinnerungen an die Judengasse." *Gemeindeblatt der Israelitischen Gemeinde Frankfurt am Main* 6 (1925): 2–4.

Meyer, Michael A. *Response to Modernity: A History of the Reform Movement in Judaism.* New York: Oxford University Press, 1988.

Oppler, Edwin. "Synagogen und jüdische Begräbnisplätze." In *Deutsches Bauhandbuch II, Baukunde des Architekten II,* 270–85. Berlin: Deutsche Bauzeitung GmbH, 1884. First ed., 1882.

Roemer, Nils. "The City of Worms in Modern Jewish Traveling Cultures of Remembrance." *Jewish Social Studies* 11, no. 3 (2005): 67–91.

———. *German Cities—Jewish Memories: The Story of Worms.* Hanover: Brandeis University Press, 2010.

Salfeld, Siegmund. *Der alte israelitische Friedhof in Mainz und die hebräischen Inschriften des Mainzer Museums.* Berlin: Simion, 1898.

Schlör, Jocahim. *Das Ich der Stadt: Debatten über Judentum und Urbanität.* Göttingen: Vandenhoeck & Ruprecht, 2005.

Schorsch, Ismar. "The Myth of Sephardic Supremacy." *Leo Baeck Institute Yearbook* 34 (1989): 47–66.

Schulmann, L. "Michael." *Der israelitische Volkslehrer* 12 (1856): 315–22.

Selig, Wolfram. "Die Synagogenbauten der Neuzeit." In *Synagogen und jüdische Friedhöfe in München,* edited by Selig, 35–120. Munich: Aries, 1988.

Stein, L. *Die Bedeutung des Synagogen-Weihefestes: Zwei Predigten gehalten bei der Einweihung der neuen Hauptsynagoge zu Frankfurt am Main.* Separate print, 23 and 24 March 1860.

Twiehaus, Christiane. *Synagogen im Großherzogtum Baden (1806-1918).* Heidelberg: Winter, 2012.

Weissbach, L. Shai. "Buildings Fraught with Meaning: An Introduction to a Special Issue on Synagogue Architecture in Context." *Jewish History* 25 (2011): 1–11.

Wolf, Gerson. *Die jüdischen Friedhöfe und die "Chewra Kadischa" (fromme Bruderschaft) in Wien.* Vienna: Hölder, 1879.

9

Spatial Variations and Locations

Synagogues at the Intersection of Architecture, Town, and Imagination

Sylvia Necker

Since the nineteenth century, synagogues in Germany have increasingly moved from backyards and alleyways into the urban landscape, their topographical shift toward the town proper mirroring the history of Jewish communities. Wherever the synagogue was located, the community and its infrastructure—such as kosher shops, clubs, and other community institutions—were typically close by. Thus, a synagogue's "place," its concrete structure and location, could be set by the topography, yet over the centuries, urban places, whether synagogues or other architectural structures, have also undergone constant and far-reaching changes. In Hamburg's Neustadt, for example, a Jewish center has been evolving since the seventeenth century. Freedom of movement was first guaranteed in 1860, when Hamburg's new constitution granted residents citizenship rights, regardless of their confession. Many Jews took this opportunity to move from the dilapidated Neustadt to a new area of the city. For the first time, a majority of the Jewish community shifted to a new location, the Grindel Quarter, which by the 1920s had developed into a "Jewish space" within the city's landscape. The new location brought a new Jewish space into being.[1]

Unlike locations and places, which are clearly definable within the topography, "spaces" remain vague and highly varied in how they are identified and perceived. Spaces are delineated and experienced in different ways and are under constant renegotiation and production.[2] The stasis of "place" stands in contrast to the performative malleability of spaces. In other words, "a Jewish place is determined by its situation, a Jewish space through its performativity,"[3] or through ascription and perception. In this contribution, I will investigate historical German synagogues both as places *and* spaces. While the synagogue as a place was characterized by its visibility or invisibility, or whether it was oriented toward or away from the town—I limit myself here to urban examples—as a Jewish space, it was often charged with various longings. For members of the Jewish community, the synagogue could be, among other things, a place of childhood and perhaps also a sanctuary as a closed, protected space; for non-Jews, it was a place that remained secretive and, thus, even more charged with projections. After the pogroms of November 1938, synagogues became a "memory place" (*Erinnerungsort*) for both Jews and non-Jews.[4] Although synagogues remained inscribed on the urban topography, they had vanished and were buried or "forgotten" in many cities up to the 1980s; they first had to be rediscovered, labeled, and "remembered."[5] A volume called *Erinnerungsorte*,[6] in fact, despite emphasizing a German-Jewish cultural tradition, included both Auschwitz and the Protestant vicarage as German "memory places," but not the synagogue. This was because the synagogue remained a Jewish space rather than becoming a German one, and one that German Jews had increasingly distanced themselves from in their successful campaign to achieve a middle-class, emancipated Jewish existence. After 1945, German Jews—at least those who were not strictly Orthodox—also tended to regard the synagogue as a memory *space,* used up to 1989 more to commemorate the November pogroms and less for religious observance.

The Synagogue as a Place and as a Space

This chapter describes and analyzes the development of synagogues in Germany from the late eighteenth to the early twenty-first century based on various topological criteria. First, it covers the architectural form—certainly the most common historical perspective applied to synagogues. There are few publications in German on this topic: *Synagogen in Deutschland* (1981),[7] a systematically arranged collection by Harold Hammer-Schenk; *Die Architektur der Synagogen* (1988),[8] an essay collection for the exhibition in the German Architecture Museum in Frankfurt; and the publications of the Bet Tfila–Research Unit for Jewish Architecture in Europe.[9] However, synagogues cannot be understood in isolation. Their architecture is always shaped by the interplay between the functional requirements of Jewish communities, these communities' desire to represent a particular self-image, and, not least, the expectations the non-Jewish majority has of the Jewish minority. This was as true of the synagogues of the past centuries as it is for newly constructed ones.

Consequently, analysis of synagogues must extend beyond the perspective of art and architectural history to incorporate the location of synagogues and their function for urban society. The topos and the topography join form as criteria for analysis,[10] inquiring into "Jewish urban landscapes." Other academic investigations focus predominantly on metropolises. For example, studies in the European context concentrate on London, Paris, and Berlin, and American studies primarily on New York. The striking frequency of studies exploring the intersection of Jewishness and urban areas[11] derives from the primarily urban setting of the process of Jewish emancipation. Moreover, today the stereotype of the "urban Jew" persists, as Joachim Schlör points out.[12] The present study looks at big cities like Berlin, Frankfurt am Main, Hamburg, and Munich alongside midsize cities such as Karlsruhe and smaller towns such as Dessau.

A further perspective involves looking into the meaning-making process.[13] Have synagogues bestowed identity on Jewish communities? Do synagogues thus represent a form of built community? The "spatial turn" of the past three decades prompted research into Jewish places and spaces,[14] including synagogues, though mostly focused on religious traditions.[15] These Jewish spaces are characterized by the interplay of construction and imagination, as shown by Barbara E. Mann in her study of Tel Aviv.[16] This contribution investigates the mutual relatedness of architecture and aesthetics, on the one hand, and the uses and ascriptions of buildings, as well as the impact these have on identity formation, on the other.[17] Mann describes the interrelationship of space and identity formation in relation to the ideas of sociologist Henri Lefebvre:

> The process is both mutual and dialectical: society produces space, and these very spaces produce the society. In a sense, we may find here a relation like that between identity and representation; just as identity does not exist a priori to language, but is always produced and understood through language, space is never a given or static entity, but always the product of its usage.[18]

Although Mann interprets space here as a social product, it is debatable whether the synagogue can be viewed as a Jewish "identity space," as Eva Lezzi suggested regarding the Sabbath celebrations in the middle-class living rooms of the nineteenth century.[19] Lezzi argued that beside the Jewish family, the bourgeois house itself became a symbol of emancipation and therefore an identity space. The question for us is, rather, what relationship obtained for synagogues of the last 150 years between situation, continuation, and representation and also, to an extent, between lasting, visible non-integration and thus the (building) culture of the urban periphery? Since about the year 2000, a new trend in Jewish spaces is apparent. For one thing, since 1989, a rediscovery of Jewish spaces has been under way as Jewish communities have been re-established in cities such as Berlin and Krakow; this has, meanwhile, prompted tourism of these spaces.[20] The effect of this tourism can be seen in the new permanent exhibition in the Jewish Museum in Warsaw, where there is a meticulously crafted reconstruction of the synagogue of Gwozdziec. Western European museums typically eschew such replicas, yet this one embodies a wish on the part of curators and politicians to assert the position of Jews and make them visible in a nation-state where only a few Jewish places, such as synagogues, survived on account of destruction by the National Socialists. These reconstructions and "invented Jewish traditions"—the title of a November 2013 symposium in Hamburg[21]—stand counter to these losses. Attempts to create new Jewish topographies in virtual realms can also be counted among these (re)inventions.[22] For example, in the the computer game "Second Life," Jewish use suggests that people "build" synagogues and Jewish cafés first, thus turning immediately to traditional Jewish places.[23]

From the Back Streets to the Main Streets: Signaling Nineteenth-Century Aspirations for Emancipation

Up to the nineteenth century synagogues were a protective space for Jewish communities, isolated from non-Jewish society.[24] Under constant threat of animosity and anti-Judaic and anti-Semitic attacks, with scant legal protection,[25] synagogues were a place of refuge. Accordingly, their architecture was guarded and largely unrecognizable in the cityscape, even for synagogues in important communities. This changed over time. The construction of a synagogue on Kronenstrasse in Karlsruhe, Baden, between 1798 and 1810 marked an end to synagogues built in hidden places. Non-Jewish architect Friedrich Weinbrenner (1766–1826)[26] may not have moved the synagogue itself—this he placed back from the main street—but he designed an imposing entryway onto Kronenstrasse as a visible symbol. The classicist structure also created a connection to the central buildings in the marketplace, which Weinbrenner, as Baden's chief town planner, designed in the same style at almost the same time: the Protestant state church (1807–16) and the town hall (1805–6). Admittedly, the synagogue was on one of the eastern spokes of the city's fan-shaped layout and thus lay toward the periphery of the urban landscape; stylistically, however, Weinbrenner's design linked the "Jewish" space with the urban space.

Another nineteenth-century development saw synagogues aligned with other buildings so that their façades were visible from the streets. The synagogues in Kassel and Dresden represent this trend. Their master builders borrowed from contemporary religious design and became specialists in synagogue building. Albert Rosengarten (1810–93), one of the first Jewish architects in Germany, designed the synagogue in Kassel, which was erected between 1833 and 1839 on Lower Königstrasse in the town center. Rosengarten chose the basilica of late antiquity as a model, recognizable in the round Roman arches and other expected features.[27] Not only does the synagogue align with other buildings, but a mere two-meter iron fence separates the structure from the urban space; the building was surrounded by so much open space that one can almost speak of a freestanding synagogue, or at least one that is very visible. Even though it could barely be discerned whether the building was a church or a synagogue from the outside, it demonstrated the Jewish desire for legal equality and emancipation, which only became attainable in 1848. Weinbrenner's example in Karlsruhe made the stylistic unity of Jewish and urban space apparent; Rosengarten's building in Kassel cleared the way for synagogues' inclusion in town centers; and Gottfried Semper's synagogue in Dresden on the Elbe in 1840 even placed a synagogue in the city's skyline and exemplified an influential contemporary architect concerning himself with the task of synagogue-building.[28] It had previously been unusual for church steeples, castle towers, and synagogue domes to receive equal treatment. Semper established the dome as a recognizable marker for synagogues in Germany—a marker that flourished until the beginning of the Weimar Republic.

Searching for a Style in the Nineteenth Century

As Jewish communities differentiated into Reform, Conservative, and Orthodox varieties, synagogue interiors took different forms. Reform and sometimes also Conservative communities began to include an organ in the synagogue space. Regarding outward representation, the question arose as to whether a synagogue should display a particularly Jewish style that would mark it—for all and from a distance—as a Jewish place. Until World War I, the dome fulfilled this function. Architects could make synagogues more visible in the urban landscape by alluding to Moorish architecture with rich ornamentation and horseshoe arches and by making exteriors particularly colorful, clearly distinguishing them from Christian religious architecture. The synagogue in Berlin's Oranienburger Strasse built by Eduard Knoblauch (1801–65) speaks this language, although both visibility and invisibility come into play: a striking façade reveals the location of the synagogue, yet the building itself is hidden from the outside. The main synagogue of the Frankfurt community in the *Judengasse,* consecrated in 1860, provides another example of urban middle-class aspirations in synagogue design by the mere size of the building, with its one thousand seats, as well as an impressive façade. Around thirty years later, the organization Neue Dammtorsynagoge, established in 1894, built a prayer-house with Moorish elements designed by architects Schlepps and Rzekonski. However, its façade was hidden on the back side of the late nineteenth-century *Gründerzeit* buildings of the Grindel Quarter.

Synagogues in a German Style?

While Rosengarten's synagogue in Kassel had already suggested a freestanding building, that of Edwin Oppler (1831–80) in Hanover was the first to stand completely clear in its plot.[29] Oppler, the son of a Jewish merchant and a member of the neo-Gothic Hanover school of architecture, designed many prestigious villas in the Hanover region for the aspiring middle class. Oppler was largely responsible for a "German style" for synagogues becoming established—a style that found its expression in the synagogue erected in Hanover between 1864 and 1870. It boasted a combination of Gothic and Romanesque elements that symbolized Jews' belonging in mainstream society and their ability to assimilate. Conversely, the building was to be identity forming for the community. In Oppler's words, "The German Jew in the German state must also build in the German style!"[30]

The emancipation movement also fought for this form of political, societal, and cultural involvement. To be sure, opponents of the movement criticized such synagogues as "foreign"—referring to the Moorish influences of their façades; they also objected to the presence of synagogues in the urban landscape. Although the aforementioned structures displayed a desire for clearly visible representation, Jewish communities' everyday experience was often very different. Aiming to get along with the local authorities, they demanded as little public space as possible for the construction of new synagogues. The limits of integration were, thus, quickly reached.[31] The nineteenth century had certainly brought the "place" of the synagogue more from the periphery to the city center and from the back streets into full view. But even when the synagogues were located within their communities—say, in the "Jewish" quarter of German cities—it would be a long time before they were oriented toward the city itself, as the coexistence of different religiously and culturally coded spaces on equal terms was very rarely achieved in the Wilhelmine and Weimar periods.[32]

Synagogues as Places of Representation and Arrival in the Architecture?

What Edwin Oppler's structures in Hanover and Breslau had implied—the representation of Jewish communities as part of urban society and (building) culture via synagogues—some larger Jewish communities achieved around the turn of the twentieth century. In the early 1900s, synagogues in Essen, Berlin, Darmstadt, Dessau, and Hamburg arose whose exteriors made this claim toward the outside world. Between 1904 and 1906 Semmy Engel (1864–1948) erected a freestanding synagogue at Bornplatz in Hamburg's Grindel Quarter, where almost 40 percent of Hamburg's Jewish population resided. A Romanesque building of golden-brown brick and topped with a large dome, it was the largest synagogue in northern Germany at that time. As with Oppler's designs, the choice of style symbolized Jewish integration into Germany's Christian-oriented society. The interior space followed a classical layout: there was a gallery for women, while the central space of the synagogue was reserved for men. A closer examination of these "gendered spaces" is certainly worthwhile.[33] Generally speaking, the synagogue represented Jewish public space and, as a rule, was thus designated for men, while Jewish "women's spaces" were located in nonpublic areas such as the home.[34] With this came the spatial division inside the synagogue "where the presence of women [was] rather more tolerated than desired."[35] However, the Reform movement systematically changed this in the nineteenth century; for example, in Reform synagogues spatial barriers were removed, and the services were feminized through the introduction of choirs and organs.[36]

In the Weimar Republic, the construction of two synagogues caused a furor in the early 1930s: the unadorned structure in Plauen (completed in 1930), designed by Fritz Landauer (1883–1968), and the Hamburg temple on Oberstrasse (completed in 1931), designed by Felix Ascher (1883–1952) and Robert Friedmann (1888–1940). The former, built as a cuboid, recognizably adhered to the ideals of the Neues Bauen movement, with horizontal bands of windows and a Star of David as the only decoration on its façade. With the Oberstrasse synagogue, the Hamburg Reform community aimed for a plain, monumental structure, entirely rejecting the borrowing of historical styles. When the Oberstrasse synagogue was inaugurated just two years prior to the National Socialist takeover of power, there was no more talk of a German style. Instead, Rabbi Bruno Italiener issued the following printed statement to his congregation:

> When one observes the building as a whole … one gains the strong impression that the entire, so harmoniously divided structure in a way represents a living organism as a symbol of that spirit that are supposed to be manifested in its walls: Judaism not as something finalized, but as something living, which finds itself in an organic state of flux.[37]

The temple on Oberstrasse was the last synagogue construction completed before 1933.

Synagogues as a Space of Longing

We will now briefly digress to discuss two paintings to focus on synagogues as imagined places, or spaces of longing. As with architecture, paintings reveal the development of synagogues from being hidden and inconspicuous to becoming portals onto the urban space—that is, from being interior to exterior spaces. The first painting was by Carl Spitzweg (1808–85), who was primarily a landscape painter but also very pointedly characterized ideal figures and atmospheres of his time, as found in *The Poor Poet* and *The Cactus Lover*. Over the five years following his visit to the Old-New Synagogue in Prague, he created a painting titled *In the Synagogue*. The observer sees a dimly lit room where Jews are reading or praying while dressed in *tallitot*. The scene grants only a fleeting glance into the synagogue's interior; it is impossible to form a concrete, "realistic" impression of this "Jewish space"—it remains a fantasy, and it reinforces ideas, potent still today, of Jews as exotic and foreign. The painting places Spitzweg within the European artistic tradition of Orientalism,[38] in which representations of the Orient included Romantic European attributions such as foreignness, wild nature, or, in motifs such as the Turkish bath, sensuality from the viewpoint of travelers and sometimes colonialists, overlapping with an image of antiquity peculiar to Romanticism.

Max Beckmann's (1884–1950) painting *The Synagogue in Frankfurt am Main* (1919) shows us the synagogue from a different perspective, wherein interior space shifts to exterior space. His 1919 depiction of the Frankfurt Synagogue, which can be easily recognized by its green dome, shows it as an integral part of the city's landscape. Not the wooden fence opposite the synagogue, the crooked façades of the houses in the background, nor the street scene itself indicates that

Figure 9.1. Carl Spitzweg, *In the Synagogue* (1855–60). Used courtesy of Museum Georg Schäfer, Schweinfurt

the synagogue is perceived, or is intended to be perceived, as "foreign" to Beck-mann's colorful city: "The whole picture is dominated by its architecture, which defies the laws of gravity. Synagogue and houses tilt and lean toward one another in wedge formations."[39] The painting portrays the artist himself with friends, still in costume after a night at Karneval, as they walk past the synagogue's façade. According to art historian Dieter Bartetzko, Beckmann's painting may have been

Figure 9.2. Max Beckmann, *The Synagogue in Frankfurt am Main* (1919). © 2016 Artists Rights Society (ARS), New York / VG Bild-Kunst, Bonn

intended as a tribute to Jewish Frankfurt, as well as "a warning about the insecure position of Jews in Germany, which was once again growing increasingly apparent."[40] In this interpretation, does the crooked dome under the crescent moon not appear ex post to be menacing, and the synagogue itself become a surface for the projection of German-Jewish history in the twentieth century?

Destroyed Synagogues as a Caesura

This development toward Jewish matter-of-factness and presence in Germany, perhaps indicated by a widely visible synagogue dome as in Beckmann's composition, was abruptly halted by the "Night of Broken Glass" (*Kristallnacht*) in November 1938, during which the Nazis systematically destroyed hundreds of synagogues. In many towns and cities, the public dismantling of the buildings lasted for months. Literally before the eyes of the urban population, synagogues were gutted and removed from the urban space. This destruction laid the groundwork for a very particular perception of synagogues in Germany after 1945. In the Weimar Republic, the perception of synagogues had been characterized by architectural and structural presence; through their visible façades and partially exposed locations, synagogues had at least superficially become part of the normal inventory of urban structures. After 1945, they became perceived as *Erinnerungsorte,* their presence now only maintained in one's imagination or in illustrations. For many Jewish and non-Jewish Germans, the synagogue idea was connected not to the new building projects of Jewish communities from the 1950s onward, but with images from 1938 of burning synagogues, which became a visual staple of anniversaries and days of remembrance. Consequently, the destroyed and vanished place became a space of negotiation for Germany's culture of remembrance.

Withdrawn and Unobtrusive: Synagogues in Germany after 1945

For many survivors, a "life in the land of perpetrators" after the Shoah was unthinkable,[41] as was reconstructing community infrastructure and synagogues. The first prayer-houses were, therefore, often provisional in nature and housed in private or unused spaces, the makeshift synagogue in the displaced persons (DP) camp at Trutzhain being a case in point. Only at the beginning of the 1950s did the first communities begin to consider new construction. Ernst Guggenheimer (1880–1973) built the Stuttgart Community Synagogue in 1952 and Willy Nöckel built the Erfurt one a year later.[42] Nöckel incorporated a community center into his religious structure—an architectural solution common after 1945. The whole complex—the only newly built synagogue in the German Democratic Republic—aligns with the height of the eaves of the surrounding structures; the community center is set back slightly from the street and set off by a low fence. With its pitched roof, the building is inconspicuous enough to pass for a residential structure. As this example shows, synagogue architecture in the early postwar period followed the style of postwar modernism, which responded to shortages in building materials with improvised construction methods and styles.[43] At the same time, architectural historians have also read into this architectural withdrawal a desire for seclusion and a wish to attract a minimum of attention.[44] In the early 1960s, West Germany experienced a real wave of new synagogue construction; communities inaugurated new synagogues in Karlsruhe, Bonn, Hanover, and Hamburg, among others. Clearly inspired by the architectural language of postwar modernism, these new synagogues symbolized a new start for the communities, just as postwar modernism itself represented the "new city" in West Germany. A prime example of this was the newly constructed synagogue at the corner of Hohe Weide and Heymannstrasse in the residential suburb of Eimsbüttel in Hamburg, which was geographically removed from the "Jewish quarter" of the Weimar era, the Grindel Quarter. The modern, luminous white structure was built in 1960, designed by architects Karl Heinz Wongel and Klaus May.[45] The pentagonal synagogue, a community center, and living quarters for the community's staff are grouped around an atrium and appear to the outside world as an enclosed complex, separated from the street by a grass verge. As with the Erfurt synagogue, two possible interpretations come under consideration. From the outside, no portal into the complex is visible, only an entranceway that allows visitors access to an inner courtyard leading into the synagogue. This insularity can be interpreted as deliberate isolation from the city and as an architectural expression of the skepticism many Jewish communities felt toward their non-Jewish postwar surroundings. As a stylistic development, the free space around the synagogue can be read as characteristic of the architecture of postwar modernism and of contemporary urban planning. At the beginning of the 1990s the street was partially closed due to fears of possible attacks. The synagogue on Hohe Weide is far removed from the confident urban representation of the synagogue at Bornplatz over fifty years earlier.

Up to 1989 few synagogues were newly constructed, as membership in Jewish congregations in West Germany had stagnated or declined. The dedication of a new synagogue and community center in Darmstadt in 1988 was an exception. The postmodern structure by architect Alfred Jacoby, who went on to build other synagogues in Aachen, Kassel, and Chemnitz, was the first newly constructed synagogue in the postwar period to recall the traditional form of a central, domed building.[46] However, despite this connection and obvious coding as a Jewish place within the town, the synagogue remains set back from the street, hidden behind a white metal fence, robbed of its public impact.

Not all synagogues were destroyed in and following the November pogroms. Some remained intact but were repurposed. In the 1970s and 1980s, numerous small local history societies, principally, adopted these buildings. In many places such initiatives cooperated with authorities to protect the historical monuments,[47] securing the buildings and finding appropriate uses that would respect the synagogues' history, such as for museums, cultural centers, or memorials. In this way, formerly religious Jewish spaces were changed by non-Jews, through engagement with German-Jewish history and tradition, into secular places that acted as surrogate spaces for imagining Jewishness. An example of this is the Friends' Association of the Former Synagogue in Kippenhein in southern Baden, which uses the synagogue space as a venue for readings, public discussions, and concerts.[48]

This process of coming to terms with the loss of synagogues after 1938 is particularly evident in artistic projects. In 1983 Hamburg sculptor Margit Kahl created a floor mosaic on the site of the destroyed synagogue at Bornplatz, which visibly replicates the dimensions of the ceiling vault on the ground. After 1945, the Universität Hamburg had used this location as a parking lot. The mosaic, as an expression of dealing with structural loss, could hardly be more meaningful. In the 1980s, awareness of Jewish spaces and places was growing in Germany, shaped by numerous local initiatives whose research and historical work countered the history politics of the Kohl administration, leading to the erection of memorial stones, plaques, and places of remembrance. The Platz der Jüdischen Deportierten on Hamburg's Edmund-Siemens-Allee was dedicated in 1983 and contains a stone sculpture by Ulrich Rückreim. A further example is the memorial site "Neuer Börneplatz" opened three years later near the Jewish Museum in Frankfurt am Main, containing a collection of memorials by Nikolaus Hirsch, Wolfgang Lorsch, and Andrea Wandel. These and other initiatives enabled the previously vanished and invisible places to be made visible.[49] An attempt to cast this loss like a photographic negative image onto the urban space was made by

Jochen Gerz and Esther Shalev-Gerz in 1986. Over the course of a year the two artists gradually sank a column of lead into the ground as a "monument against fascism." The public was invited to inscribe the column with messages of remembrance and admonition, which nonetheless disappeared as the column descended underground. A visible "reverse reconstruction" was created by artist Horst Hoheisel in Kassel by sinking the negative form of a fountain into the ground. This was a memorial to the Aschrott Fountain, which had been donated by a Jewish businessman and was destroyed by the Nazis.[50]

Between Reinvention and Representation: New Synagogues in Germany after 1989

Synagogue building in West Germany stagnated in the 1980s, but since the 1990s Jewish communities have built several community centers and synagogues. Increasing congregation membership brought about by the so-called quota refugees (*Kontingentflüchtlinge*) from Eastern Europe made these construction projects necessary. From a present-day perspective the new synagogues—whose representative style I will illustrate with examples in Duisburg, Dresden, and Munich—mark a caesura in the history of Jewish communities in Germany, because their style connects them to the buildings from around 1900 and they could also be interpreted at least as an expression of Jewish communities' newfound confidence. At the same time, the new synagogues changed the view of these buildings from being a sort of photographic negative, either invisible or barely acknowledged before now, since so few Jewish communities rebuilt their old synagogues and because so few members of the Jewish communities, or none at all, had been spared at the locations of large synagogues. The wave of new construction attracted mostly positive media coverage, bringing synagogues back into view as current construction projects.

Architect Zvi Hecker's synagogue design in Duisburg is an expressionistic complex in exposed concrete laid out within a park in a fan shape.[51] Hecker envisioned the synagogue as a book with five pages, seeking to remind visitors of the presence of the "spirit of a book" in Jewish tradition: "This book binds together the separated communities of the Diaspora through customs and traditions. The book contains the strength necessary for the Jewish nation to survive in violent, hostile times. Constantly updating, it incorporates all historical events."[52] Hecker was neither in search of a "German style," as Oppler had been in the nineteenth century, nor did he aim to form a connection with Judaism's geographical roots, as the Moorish-revivalist style of the synagogue on Berlin's Oranienburger Strasse had sought to do. Hecker also completely decoupled his synagogue from past traditions in religious architecture—whether Jewish or Christian—and found his own path to creating a contemporary synagogue. The history of the Diaspora and its constant migrations find expression in the mobile artifact of the book and are then translated into architectural form. The reference is not stylistic but historical-imaginary, providing a location for Jewish history and identity and thus constituting a self-willed Jewish space. In the case of the structure in Duisburg, completed between 1996 and 1999, one can truly say that it stands alone.

The Dresden synagogue and community center built by the office of Wandel Hoefer Lorch + Hirsch between 1997 and 2001, by contrast, alludes to earlier skyline-changing religious buildings. As viewed from the Elbe, the complex fits into the cityscape between the fortifications of the Brühlschen Terasse, the Frauenkirche in the background, and the bridgehead of the Carola-Brücke. In designing it, the architects dispensed with the dome as a visual marker but integrated the plot of Gottfried Semper's nineteenth-century synagogue into the plan, thus connecting it to the earlier tradition.[53] Even though this modern synagogue, not being as tall as early twentieth-century examples, does not have the same impact—the dome of Dessau's New Synagogue (completed in 1908) dominated the view of the skyline from the banks of the Mulde, for example[54]—it is clear that urban planners meant to send a signal by positioning it on the Elbe.

The new synagogue on Munich's Jacobsplatz, dedicated in 2007 as a center for the Jewish community, with a synagogue and museum, also ties in to a former synagogue design—specifically, the idea of a freestanding building as had been successfully realized in the late nineteenth and early twentieth centuries in Hanover and Hamburg.[55] Brought away from the line of the street and given its own space, it demonstrates the will to be seen. The connection between community structure and museum it entails, at the same time, creates an interface between public and nonpublic Jewish space in the center of the city. This space asserts the normalcy of a presence that could no longer be taken for granted after 1938.

As characteristic as the above-mentioned synagogue structures in Duisburg, Dresden, and Munich may be, Berlin became the most important Jewish center in Germany and a "hot spot" for German-Jewish memorial culture after 1989.[56] At least in the year 2015, there is no lack of marking Jewish places. Quite the contrary: a great many of the former shuls, synagogues, and community centers have been rebuilt and repurposed in one form or another. And yet for me it is the empty space—as Margit Kahl shows with her mosaic at Hamburg's former Bornplatz (now Joseph-Carlebach-Platz)—that seems to be the most timeless form of memorializing Jewish places.

Conclusion

At the moment, Berlin seems to be the most "Jewish" space in Germany. With restaurants, cafés, community facilities round about the quarter of the New Synagogue on Oranienburger Strasse, and the Jewish Museum as a new hot spot, a form of "popular Jewish space" has been developing since the 1990s that attracts Jews and non-Jews alike.[57] The dynamics of this development exemplify how many actors have become involved in the construction of Jewish spaces. No longer is it constructed by the Jewish community itself—whose members, in the nineteenth century, demonstrated their desire to remain inconspicuous and to protect themselves by placing their synagogues in the back streets, and who, in the twentieth century, demonstrated their will to be seen and acknowledged with their freestanding synagogues. Place and location are no longer decisive for the meaning of a space, but rather performance within it. Consequently, the internationally successful Jewish Museum in Berlin now appears to be making a greater contribution than synagogues to the discussion of what a Jewish place is rather than a Jewish space—at least to the discussions in Munich in the mid-2000s. And yet it is the *places* such as synagogues, museums, and cafés that launch the imagination and spatialization of urban Jewish *spaces*, however and by whomever they may be shaped.[58]

Sylvia Necker is a research associate and curator at the Obersalzberg Documentation Centre/Intitute for Contemporary History in Munich. Her main fields of research are modern German and Jewish history, the history of National Socialism, and the history of architecture and urban planning since 1900. Her book publications include written *Konstanty Gutschow (1902–1978). Modernes Denken und volksgemeinschaftliche Utopie eines Architekten* (2012) and *"Hitlerbauten" in Linz. Wohnsiedlungen zwischen Alltag und Geschichte. 1938 bis in die Gegenwart* (2012).

Notes

1. For the discussion about Jewish "place" and "space," see Kümper et al., eds., *Makom.*
2. Lefebvre, *Production of Space.* For the construction of the concept of "space" and the various research perspectives, see, e.g., Günzel and Kümmerling, *Raum,* 121–320; Rau, *Räume.*
3. Lipphardt and Brauch, "Gelebte Räume," 28.
4. Nora, *Les lieux de mémoire.*
5. Bartetzko, *Eine verschollene Architektur.*
6. François and Schulze, *Deutsche Erinnerungsorte.*
7. Hammer-Schenk, *Synagogen in Deutschland.*
8. Schwarz and Hammer-Schenk, *Die Architektur.*
9. E.g., Cohen-Mushlin and Thies, *Synagogenarchitektur in Deutschland.*
10. Lipphardt, Brauch, and Nocke, "Exploring Jewish Space." This applies as much to urban space as rural topography. A specifically Jewish example of topography is the *Judenweg.* See Rösch, "Judenwege in Franken."
11. Mann, *Space and Place,* 116ff.; Schlör, "Jews and the Big City." Source editions for urban Jewish culture are popular in the literature. See, e.g., Schlör, *Wenn ich dein vergesse, Jerusalem.*
12. Schlör, "Jews and the Big City"; Schlör, *Das Ich in der Stadt.*
13. Gruber, "Jewish Identity."
14. Fonrobert, "New Spatial Turn."
15. Lipphardt, Brauch, and Nocke, "Exploring Jewish Space," 1.
16. Mann, *A Place in History.*

17. Ernst and Lamprecht, "Jewish Spaces," 7.
18. Mann, *A Place in History*, 1.
19. Lezzi, "Ein jüdischer Ort?"
20. Gantner and Kovács, "The Constructed Jew"; Waligórska, "Jewish Heritage Production."
21. Schlör, "Tagungsbericht."
22. Voloj, "Virtual Jewish Topography."
23. Lipphardt and Brauch, *Gelebte Räume*, 29; Gruber, "Jewish Identity," 21–31.
24. See Andreas Gotzmann's contribution in this volume for a more detailed account of early-modern Jewish synagogues.
25. Jersch-Wenzel, "Rechtslage und Emanzipation."
26. A majority of Jewish communities in the nineteenth century commissioned non-Jewish architects who could produce references from other projects. On the one hand, the engagement of Friedrich Weinbrenner or Gottfried Semper bespoke a particularly representative architectural solution; on the other, the professions of "master-builder" and "architect" only became available to Jews in the course of the nineteenth century. Until the mid-1800s, training took place at colleges of art that long remained closed to Germany's Jews. See Nerdinger and Böhm, *Der Architekt*. See also the thoughts on "Jewish architects" in the nineteenth and twentieth centuries presented by Sylvia Necker and Ulrich Knufinke in their paper "Jüdische Architekten—Lebenswege in der ersten Hälfte des 20. Jahrhunderts" at the *Tag der jüdischen Architekten* in Berlin 10 October 2013, to be published in 2016.
27. Przystawik,"Kassel"; Hass, Link, and Wegner, *Synagogen in Kassel*.
28. Müller, "Dresden, Synagoge Brühlscher Garten."
28. Keßler and Knufinke, "Hannover—Neue Synagoge Bergstrasse."
29. Ibid.
30. Ibid., 215.
31. Schüler-Springorum, "Liberale Stadtkultur."
32. Carol Herselle Krinsky views the end of World War I as a watershed for France, Norway, the Netherlands, and Britain. Many communities in Central and Western Europe chose a uniquely expressive and town-facing synagogue form after 1918 in part to counter the tangibly increasing political and social anti-Semitism and assert their presence in the urban landscape. Krinsky, *Synagogues of Europe*, 96.
33. Lezzi, "Ein jüdischer Ort?"
34. Schüler-Springorum, *Geschlecht und Differenz*, 87–109; Kaplan, *Jüdisches Bürgertum*, 93–111.
34. Schüler-Springorum, *Geschlecht und Differenz*, 87–109; Kaplan, *Jüdisches Bürgertum*, 93–111.
35. Lässig, "Religiöse Modernisierung," 51.
36. Ibid., "Religiöse Modernisierung," 51–61.
37. Knufinke, "Hamburg—Synagoge (Tempel) Oberstraße."
38. MacKenzie, *Orientalism*; Bertz, *Les juifs*.
39. Bartetzko, *Eine verschollene Architektur*, 3.
40. Ibid., 6.
41. Schoeps, *Leben im Land der Täter*.
42. Reinsberg, "Erfurts Neue Synagoge"; Knufinke, "Zur Geschichte der Synagogen," 35.
43. Architect Benedikt Boucsein called such architecture from the immediate postwar period to the early 1950s "gray." See his *Graue Architektur*.
44. See, among others, Knufinke, "Zur Geschichte der Synagogen," 19–52.

45. Lorenz, "Jüdische Gemeinde (1945–1989)," 138.
46. Knufinke, "Zur Geschichte der Synagogen," 42. For reflections by the architect himself, see Jacoby, "Vom Bau der Synagoge."
47. Bongartz, "Synagogen als Gegenstand der Denkmalpflege"; Landesamt für Denkmalpflege im Auftrag des Kultusministeriums Rheinland-Pfalz, *Rheinland-Pfalz: Synagogen und Denkmalpflege.*
48. See Förderverein Ehemalige Synagoge Kippenheim e. V., accessed 6 February 2015. http://www.ehemalige-synagoge-kippenheim.de/.
49. Bussmann, ed. *Arbeit in Geschichte*; Endlich, "Denkmale und Mahnmale." Regarding Jewish places in Hamburg, see Necker, "Orte jüdischer Geschichte."
50. Endlich, "Denkmale und Mahnmale," 77–79; Young, *Formen des Erinnerns.*
51. Pehnt, "Jewish Cultural Center Duisburg."
52. Stiftung Baukultur Rheinland-Pfalz, *Gebauter Aufbruch*, 95–96.
53. Sachs, "New Synagogue Dresden," 124–31; Stiftung Baukultur Rheinland-Pfalz, *Gebauter Aufbruch*, 88–93.
54. Brülls, *Synagogen in Sachsen-Anhalt*, 198ff.
55. Sachs, "Jewish Center Jakobsplatz," 132–35; Stiftung Baukultur Rheinland-Pfalz, *Gebauter Aufbruch*, 128–31.
56. Neumärker, "Monumente und Stolpersteine."
57. Berlin attracted many Jews from the United States and Israel; see Sapir, "Berlin, Berlin!" Regarding Russian-speaking Jews in Berlin, see Gromova, *Generation "koscher light."* Travel guides, too, bear witness to Berlin's popularity as a "Jewish space." See Roth and Frajman, *Goldapple Guide.*
58. Pinto, "Jewish Spaces versus Jewish Places?"

Bibliography

Bartetzko, Dieter. *Eine verschollene Architektur. Über Synagogen in Deutschland.* Frankfurt am Main: Frankfurter Bund für Volksbildung, 1988.

Bertz, Inka. *Les juifs dans l'orientalisme* [cet ouvrage a été édité à l'occasion de l'exposition présentée au Musée d'Art et d'Histoire du Judaïsme, Paris, du 7 mars au 8 juillet 2012]. Paris: Musée d'art et d'histoire du judaïsme, 2012.

Bongartz, Norbert. "Synagogen als Gegenstand der Denkmalpflege." *Denkmalpflege in Baden-Württemberg* 32, no. 1 (2003): 121–24.

Boucsein, Benedikt. *Graue Architektur. Bauen in Westdeutschland der Nachkriegszeit.* Cologne: Walther König, 2010.

Brauch, Julia, Anna Lipphardt, and Alexandra Nocke, eds. *Jewish Topographies: Visions of Space, Traditions of Place.* Aldershot: Ashgate, 2008.

Brülls, Holger. *Synagogen in Sachsen-Anhalt.* Berlin: Verlag Bauwesen, 1998.

Bussmann, Georg, ed. *Arbeit in Geschichte, Geschichte in Arbeit* [published in connection with the exhibition of the same name at the Kunsthaus und Kunstverein Hamburg, 23 September–13 November 1988]. Berlin: NiSHEN, 1988.

Cohen-Mushlin, Aliza, and Harmen H. Thies. *Synagogenarchitektur in Deutschland. Dokumentation zur Ausstellung "… und ich wurde ihnen zu einem kleinen Heiligtum …"—Synagogen in Deutschland in Deutschland. Dokumentation zur Ausstellung "… und ich wurde ihnen zu einem kleinen Heiligtum …"—Synagogen in Deutschland.* Petersberg: Michael Imhof, 2008.

Endlich, Stefanie. "Denkmale und Mahnmale zur NS-Diktatur." In *Denkmäler demokratischer Umbrüche nach 1945*, edited by Hans-Joachim Veen and Volkhard Knigge, 61–88. Cologne: Böhlau, 2014.

Ernst, Petra, and Gerald Lamprecht, eds. *Jewish Spaces. Die Kategorie Raum im Kontext kultureller Identitäten*. Innsbruck: StudienVerlag, 2010.

———. "Jewish Spaces: Die Kategorie Raum im Kontext kultureller Identitäten. Einleitende Anmerkungen zum Thema." In Ernst and Lamprecht, *Jewish Spaces*, 7–12.

Fonrobert, Charlotte Elisheva. "The New Spatial Turn in Jewish Studies." *AJS Review* 33, no. 1 (2009): 155–64.

François, Etienne, and Hagen Schulze, eds. *Deutsche Erinnerungsorte*. 3 vols. Munich: Beck, 2001.

Gantner, Eszter B., and Mátyás Kovács. "The Constructed Jew: A Pragmatic Approach for Defining a Collective Central European Image of Jews." In Šiaučiunaitė-Verbickiene and Lempertienė, *Jewish Space in Central and Eastern Europe*, 211–23.

Gromova, Alina. *Generation "koscher light". Urbane Räume und Praxen junger russischsprachiger Juden in Berlin*. Bielefeld: transcript, 2013.

Gruber, Samuel D. "Jewish Identity und Modern Synagogue Architecture." In Sachs, van Voolen, and Gruber, *Jewish Identity in Contemporary Architecture*, 21–31.

Günzel, Stephan, and Franziska Kümmerling, eds. *Raum. Ein interdisziplinäres Handbuch*. Stuttgart: Metzlar, 2010.

Hammer-Schenk, Harold. *Synagogen in Deutschland. Geschichte einer Baugattung im 19. und 20. Jahrhundert*. 2 vols. Hamburg: Hans Christians, 1981.

Hass, Esther, Alexander Link, and Karl-Hermann Wegner. *Synagogen in Kassel* [Exhibition in the Stadtmuseum Kassel on the occasion of the inauguration of the new synagogue in 2000.] Schriften des Stadtmuseums Kassel 9. Kassel: Jonas, 2000.

Jacoby, Alfred. "Vom Bau der Synagoge. Reflexionen eines Architekten." In *"Soviel Aufbruch war nie …". Neue Synagogen und jüdische Gemeinden im Ruhrgebiet. Chancen für Integration und Dialog*, edited by Manfred Keller, 30–56. Berlin: Hentrich & Hentrich, 2011.

Jersch-Wenzel, Stefi. "Rechtslage und Emanzipation." In *Deutsch-jüdische Geschichte in der Neuzeit*, vol. 2, *Emanzipation und Akkulturation 1780–1871*, edited by Michael Brenner, Stefi Jersch-Wenzel, and Michael A. Meyer, 15–56. Munich: Beck, 1996.

Kaplan, Marion A. *Jüdisches Bürgertum. Frau, Familie und Identität im Kaiserreich*. Hamburg: Dölling und Galitz, 1997.

Keßler, Katrin, and Ulrich Knufinke. "Hannover—Neue Synagoge Bergstrasse." In Cohen-Mushlin and Thies, *Synagogenarchitektur in Deutschland*, 213–16.

Knufinke, Ulrich. "Hamburg—Synagoge (Tempel) Oberstraße." In Cohen-Mushlin and Thies, *Synagogenarchitektur in Deutschland*, 265–68.

———. "Zur Geschichte der Synagogen in Deutschland." In Stiftung Baukultur Rheinland-Pfalz, *Gebauter Aufbruch*, 19–52.

Krinsky, Carol Herselle. *Synagogues of Europe: Architecture, History, Meaning*. Cambridge: Dover, 1985.

Kümper, Michal, Barbara Rösch, Ulrike Schneider, and Helen Thein, eds. *Makom. Orte und Räume im Judentum: Real—Abstrakt—Imaginär. Essays*. Hildesheim: Olms, 2007.

Landesamt für Denkmalpflege im Auftrag des Kultusministeriums Rheinland-Pfalz, ed. *Rheinland-Pfalz: Synagogen und Denkmalpflege*. Mainz: Kultusministerium Rheinland-Pfalz, 1989.

Lässig, Simone. "Religiöse Modernisierung, Geschlechterdiskurs und kulturelle Verbürgerlichung. Das deutsche Judentum im 19. Jahrhundert." In *Deutsch-jüdische Geschichte als*

Geschlechtergeschichte. Studien zum 19. und 20. Jahrhundert, edited by Kristen Heinsohn and Stefanie Schüler-Springorum, 46–84. Göttingen: Wallstein, 2006.

Lefebvre, Henri. *The Production of Space*. Oxford: Blackwell, 1991.

Lezzi, Eva. "Ein jüdischer Ort? Die bürgerliche Wohnstube in der deutsch-jüdischen Literatur und Kultur des 19. Jahrunderts." In Ernst and Lamprecht, *Jewish Spaces*, 173–89.

Lipphardt, Anna, and Julia Brauch. "Gelebte Räume—Neue Perspektiven auf jüdische Topographien." In Ernst and Lamprecht, *Jewish Spaces*, 13–32.

Lipphardt, Anna, Julia Brauch, and Alexandra Nocke. "Exploring Jewish Space: An Approach." In Brauch, Lipphardt, and Nocke, *Jewish Topographies*, 1–23.

Lorenz, Ina. "Jüdische Gemeinde (1945–1989)." In *Das Jüdische Hamburg. Ein historisches Nachschlagewerk*, edited by Institut für die Geschichte der deutschen Juden and revised by Kristen Heinsohn, 135–38. Göttingen: Wallstein, 2006.

MacKenzie, John M. *Orientalism: History, Theory and the Arts*. Manchester: Manchester University Press, 1995.

Mann, Barbara E. *A Place in History: Modernism, Tel Aviv, and the Creation of Jewish Urban Space*. Stanford, CA: Stanford University Press, 2006.

———. *Space and Place in Jewish Studies*. New Brunswick, NJ: Rutgers University Press, 2012.

Müller, Hans Martin. "Dresden, Synagoge Brühlscher Garten." In Cohen-Mushlin and Thies, *Synagogenarchitektur in Deutschland*, 173–78.

Necker, Sylvia. "Orte jüdischer Geschichte und Gegenwart in Hamburg. Untersuchung zum topographischen Netzwerk von Institutionen, Museen und Privatinitiativen zur jüdischen Geschichte in der Freien und Hansestadt Hamburg. A report for the Institut für die Geschichte der deutschen Juden (IGdJ), commissioned by the Alfred Töpfer Stiftung F.V.S. Hamburg 2008/2009." Accessed 6 February 2015. http://epub.sub.uni-hamburg .de/epub/volltexte/2012/15564/pdf/juedischeortehamburg_neckerIGdJ_April2009_web .pdf.

Nerdinger, Winfried, and Hanna Böhm, eds. *Der Architekt—Geschichte und Gegenwart eines Berufsstandes* [marking the exhibition "Der Architekt—Geschichte und Gegenwart eines Berufsstandes" in der Architekturmuseum der TU München in der Pinakothek der Moderne, 27 September 2012 to 3 February 2013]. Munich: Prestel, 2012.

Neumärker, Uwe. "Monumente und Stolpersteine. Erinnerungslandschaft Berlin." In *Die Zukunft der Erinnerung*, edited by Wolfgang Benz and Barbara Distel, 206–13. Dachau: Dachauer Hefte, 2009.

Nora, Pierre. *Les lieux de mémoire*. Paris: Gallimard, 1986.

Pehnt, Wolfgang. "Jewish Cultural Center Duisburg." In Sachs, van Voolen, and Gruber, *Jewish Identity in Contemporary Architecture*, 116–23.

Pinto, Diana. "Jewish Spaces versus Jewish Places? On Jewish and Non-Jewish Interaction Today." In *Der Ort des Judentums in der Gegenwart 1989–2002*, edited by Hiltrud Wallenborn, Michael Kümper, Anna Lipphardt, Jens Neumann, et al., 15–25. Berlin: Bebra, 2004.

Przystawik, Mirko. "Kassel. Synagoge Untere Königsstraße." In Cohen-Mushlin and Thies, *Synagogenarchitektur in Deutschland*, 167–70.

Rau, Susanne. *Räume. Konzepte, Wahrnehmungen, Nutzungen*. Frankfurt am Main: Campus, 2013.

Reinsberg, Julius. "Erfurts Neue Synagoge." *moderneREGIONAL. Online-Magazin für Kulturlandschaften der Nachkriegsmoderne*, 2 February 2015.

Rösch, Barbara. "Judenwege in Franken: ein kulturgeschichtliches Phänomen." In *Die Juden in Franken*, edited by Michael Brenner and Daniela F. Eisenstein, 115–38. Munich: Oldenbourg, 2012.

Roth, Andrew, and Michael Frajman. *The Goldapple Guide to Jewish Berlin*. Berlin: Goldapple, 1998.

Sachs, Angeli. "Jewish Center Jakobsplatz." In Sachs, van Voolen, and Gruber, *Jewish Identity in Contemporary Architecture*, 132–35.

———. "New Synagogue Dresden." In Sachs, van Voolen, and Gruber, *Jewish Identity in Contemporary Architecture*, 124–31.

Sachs, Angeli, Edward van Voolen, and Samuel Gruber, eds. *Jewish Identity in Contemporary Architecture. Jüdische Identität in der zeitgenössischen Architektur*. New York: Prestel, 2004.

Sapir, Jaev. "Berlin, Berlin! Junge Israelis und die deutsche Hauptstadt. Kritische Auseinandersetzung eines Befangenen." *Aus Politik und Zeitgeschichte* 65, no. 6 (2015): 41–46.

Schlör, Joachim. *Das Ich in der Stadt. Debatten über Judentum und Urbanität 1822–1938*. Göttingen: Vandenhoeck & Ruprecht, 2005.

———. "Jews and the Big City: Explorations on an Urban State of Mind." In Brauch, Lipphardt, and Nocke, *Jewish Topographies*, 223–38.

———. "Tagungsbericht: Invented Jewish Traditions. Jüdisches Erbe in Europa zwischen Erinnerung und Inszenierung [17 Nov. 2013–20 Nov. 2013 in Hamburg]," accessed 8 February 2015, http://www.hsozkult.de/conferencereport/id/tagungsberichte-5146.

Schlör, Joachim. *Wenn ich dein vergesse, Jerusalem. Bilder jüdischen Stadtlebens*. Leipzig: Reclam, 1995.

Schoeps, Julius H., ed. *Leben im Land der Täter. Juden im Nachkriegsdeutschland (1945–1952)*. Berlin: Jüdische Verlagsanstalt Berlin, 2002.

Schüler-Springorum, Stefanie. *Geschlecht und Differenz*. Paderborn: Ferdinand Schöningh, 2014.

———. "Liberale Stadtkultur und die Grenzen der Integration." In *Liberalismus und Emanzipation. In- und Exklusionsprozesse im Kaiserreich und in der Weimarer Republik*, edited by Angelika Schaser and Stefanie Schüler-Springorum, 109–22. Stuttgart: Steiner, 2010.

Schwarz, Hans-Peter, and Harold Hammer-Schenk, eds. *Die Architektur der Synagoge*. [Exhibition in the Deutschen Architekturmuseum Frankfurt am Main 11 November 1988–12 February 1989.] Stuttgart: Klett-Cotta, 1988.

Šiaučiunaitė-Verbickiene, Jurgita, and Larisa Lempertienė, eds. *Jewish Space in Central and Eastern Europe: Day-to-Day History*. Newcastle: Cambridge Scholars Publishing, 2007.

Stiftung Baukultur Rheinland-Pfalz, ed. *Gebauter Aufbruch. Neue Synagogen in Deutschland*. Regensburg: Schnell und Steiner, 2010.

Voloj, Julian. "Virtual Jewish Topography: The Genesis of Jewish (Second) Life." In Brauch, Lipphardt, and Nocke, *Jewish Topographies*, 345–56.

Waligórska, Magdalena. "Jewish Heritage Production and Historical Jewish Spaces: A Case Study of Cracow and Berlin." In Šiaučiunaitė-Verbickiene and Lempertienė, *Jewish Space in Central and Eastern Europe*, 225–50.

Young, James E. *Formen des Erinnerns. Gedenkstätten des Holocaust*. Vienna: Passagen, 1997.

10

Jewish Philanthropy and the Formation of Modernity

Baron de Hirsch and His Vision of Jewish Spaces in European Societies

Björn Siegel

Vered Shemtov and Anna Lipphardt have shown that an approach to Jewish history that negates the significance of space fails to take account both of the influence of physical places and of metaphorical spaces of reference.[1] Following Shemtov's and Lipphardt's ideas around these notions of space and place, we can raise questions relating to the influence of location and reality in a local and physical understanding of place, as well as around performance and opportunity in a more elusive perception of space. Even though such a focus on the importance of space in Jewish studies calls into question the long-established connection between Judaism/Jewish history and time, it does not deny the importance of the temporal dimension but rather incorporates it into the general approach.[2] Thus, similar to the ideas of Barbara Mann, this study will demonstrate that alongside the two crucial terms of "place" and "space," time performs an equally important function, reminding us that place and space are produced entities, neither static nor preordained.[3] In this triangle of time, place, and space, this chapter will seek to explore the complexity of Jewish philanthropy in the long nineteenth century and examine the motives and strategies of the philanthropist Baron de Hirsch (1831–96) in a European context. It will discuss the interrelationship between Jewish and non-Jewish spaces and study the role of Jewish transnational philanthropy in the processes that created modern European societies.

An International Debate: Modern Philanthropy
as a New Evolving Space

Baron de Hirsch's vision of philanthropy was a response to contemporary discourses on wealth, poor relief, and the state. As a businessman and philanthropist, he clearly adhered to the religious and social values that determined his
actions throughout his life.[4] In an article for the *North American Review* entitled
"My Views on Philanthropy" of 1891, he expressed his belief that the "possession of great wealth lays a duty upon the possessor"[5] to carry out philanthropic
actions.

Hirsch's understanding of philanthropy was evidently linked to other previously published ideas on the subject. Two years earlier, Andrew Carnegie (1835–
1919), a Scottish-born US-steel and railroad magnate, had declared that the
"proper administration of wealth" had become one of the major challenges of
modern civilization and was the key factor in maintaining the "ties of brotherhood between rich and poor";[6] in so doing, he initiated a wide-ranging public
discussion.[7] While he defended the concentration of wealth in the hands of a
few as a result of talent and ability, he also acknowledged the problematic consequences of industrialization (loss of empathy for the other) and modernization
(loss of social homogeneity).[8] He reasoned, as Hirsch did, that rich people hold
the solution in their hands and thus advocated the idea that wealth had to be
administered by its possessors for public benefit during their lifetimes,[9] coining
the maxim that "the man who dies ... rich, dies disgraced."

William Ewart Gladstone (1809–98), the four-time liberal-leaning prime
minister of Great Britain,[10] criticized the strong aversion evident in Carnegie's
argument toward hereditary transmission of wealth, which Carnegie rejected as
one of the most incorrect ways to deal with surplus wealth.[11] That Gladstone responded to Carnegie's statement reflects the broad attention given to the topic on
both sides of the Atlantic. In addition, Gladstone also discussed the millionaire's
"right to his place in the world" and the accompanying duties of such a status.
In contrast to Carnegie's vision, Gladstone promoted the foundation of a general
aid organization modeled on the charitable Universal Beneficent Society in order to reduce moral and social inequalities and foster collective action on social
welfare.[12] Gladstone's notion of "irresponsible wealth," that is, financial resources
that were not used for the purposes of economic progress or social advancement,
acted as a particular stimulant to the debate.[13]

Joining the discussion, Henry Edward Manning (1808–92), the Catholic
archbishop of Westminster and head of the Catholic hierarchy in the United
Kingdom from 1865 to 1891, asserted that money could not be held responsible for societal failings but that its possessors bore responsibility, thus following
Carnegie in his idea of an inherent connection between wealth and social duty.[14]
He acknowledged the "law of Israel" as an important but outdated philanthropic
tool, offering other laws in its place:

We are free from the law of Israel, but we are not free from a more perfect, searching, constraining, and even peremptory law which is the law of liberty; that is, the law of charity, of generosity, of watchful consideration of the needs of others, of temperate content with our own lot, and self-denying efforts for the help of our neighbours.[15]

Hugh Price Hughes (1847–1902), another important participant in the discussion, a clergyman, religious reformer, and first superintendent of the West London Methodist Mission,[16] respected Carnegie's lifestyle and philanthropic attitude but also attacked him as one of the millionaires whom he described as "an anti-Christian phenomenon, a social monstrosity, and a grave political peril."[17] Hughes's view was that millionaires had no beneficent raison d'être and were not a necessary result of modern industrial enterprise. In contrast to their visions, Hughes supported the redistribution of wealth and saw ideas of philanthropic action as a possible beginning to a transition from a society guided by "social heathenism to [one based on] social Christianity."[18] The generally accepted view that Christian values could solve the fundamental problems of a modernizing, industrializing, and diverging world was also supported by James Gibbons (1834–1921), the Catholic archbishop of Baltimore, who encouraged people to take on civic responsibility.[19] Thus, various ideas of modern philanthropy were seen as being in the tradition of core Christian values—perhaps even a version of Christian socialism—and reflected the strong connection between religion and relief work.[20]

Jewish Philanthropy: A Space of Modernization

The arguments on the distribution of wealth, philanthropy, and poor relief presented by Carnegie, Gladstone, Manning, Hughes, and Gibbons did not exclusively dominate the contemporary debate. Herman Adler (1839–1911), chief rabbi of the United Hebrew Congregations of the British Empire, also responded to Carnegie's and Gladstone's ideas and drew attention to the old concept of Jewish welfare, if one may use this modern term, which Manning had already referenced. Adler summarized an article on the complexity of the traditional Jewish communal welfare system and stated, "Give to God the tithe which is His: for the sake of the poor, that they may live; for the sake of society, that it may endure; for your own sake, that the work of your hands may be blessed; for the sake of duty, honesty and honour."[21] Adler even regarded Carnegie and Gladstone as standing in the tradition of core Jewish values and having done an "excellent service in enforcing this olden lesson under a new name."[22]

Another indication of how actively Jews were participating in the contemporary discussion is Hirsch's contribution to the debate. Although he did not wish to enter into either theoretical or philosophical discussions on wealth and social duties, he was eager to set forth the practical method of his philanthropic ideas, which partly contrasted with Adler's understanding of philanthropy but resembled others.[23] He agreed with Carnegie's notion that the duty of wealthy members of society was to "busy themselves in organizing benefactions from which the masses of their fellows will derive lasting advantage, and thus dignify their own lives."[24] Further, Hirsch "contend[ed] most decidedly against the old system of alms-giving which only makes so many more beggars" and, like Carnegie, advocated a more modern approach.[25] Carnegie had stressed that the highest purpose of life could not be reached by a simple imitation of the life of Christ but rather by recognizing "the changed conditions of this age, and adopting modes of expressing [Christ's] spirit suitable to the changed conditions under which we live …, but labouring in a different manner."[26] In a similar tone, but without the strong Christian connotations, Hirsch also lobbied for modernization of philanthropic actions.

Like Carnegie and Gladstone, who clearly connected the modernization of poor relief with religious values, Hirsch recognized the importance of Jewish traditions of alms-giving, philanthropy, and poor relief to Jewish communal life.[27] He understood that Jewish welfare principles, based on the concept of *tzedaka* (righteousness), had secured Jewish communal solidarity and stabilized Jewish communities.[28] The complex of rabbinical teachings concerning Jewish unity, solidarity, and welfare, which also included *gemilut hasadim* (acts of loving-kindness), were perceived as a perfect scheme for contemporary philanthropists to adhere to the implementation of their charitable acts. This specific framework, in which the recipient could retain pride and honor, was taken up as a blueprint for further action.[29] In formulating his beliefs along these lines, Hirsch followed traditional Jewish values, which called for the care of the Jewish poor as a religious commandment in a world that had to deal with secularization and the creation of nation-states, and demonstrated that modern philanthropy was not a movement that had abandoned religious teachings.[30] Moreover, Hirsch's aims, which attempted to counterbalance anti-Semitic images of Jews as "ruthless and disloyal wanderers" and aimed to implant the idea of Jews as integral members of societies, also illustrate his ambitions beyond any sense of religious obligation. In propagating education, new occupational structures, and a reform of religious practices, he lobbied not only for a reform of religious ideas but also for a new understanding of Jewish philanthropy that could preserve Jewish unity and also support Jewish integration into modern society.

As a consequence of this balancing act between a strong reliance on core Jewish values and a modern vision of Jewish integration and emancipation, Hirsch detached his benevolent actions from the traditional Jewish system of aid and relief efforts. Even though he also broadly supported Jewish communal life, Hirsch was eager to establish Jewish philanthropic organizations outside the traditional Jewish communal sphere, in a manner similar to what Gladstone had proposed. Hirsch believed that philanthropy should be used not only to meet the necessity of aesthetic pleasures but also as a tool to relieve human suffering and "furnish humanity with much new and valuable material."[31] In addition, Hirsch, as an emancipated Western European Jew, also regarded his benevolent activities as having the potential to secure Jewish spaces in Europe and elevate thousands of

Jews to the status of citizens of their states and nations. Thus, Hirsch used his philanthropic institutions as tools for strengthening the place of Jews in European societies and states without negating Jewish loyalty to religious tenets.

Consequently, Hirsch's acceptance of the invitation extended to him by the editors of the *North American Review* to join the contemporary international and interreligious discussion on wealth, philanthropy, poor relief, and the state indicated, as Derek J. Penslar has observed, that Jews and Gentiles were confronted with similar problems and sought to solve them with analogous resources and ideas;[32] further, it demonstrated that Jewish voices and perspectives were seen as a valuable contribution to the discussion. In the Jewish context, this new role of participation in public debates as equal citizens made Jewish elites and especially philanthropists aware of the importance of the emergent national and transnational public spheres as spaces within which they might put forward and advocate their vision of modernity. Hirsch, who defined the core values of "his" modernity as centering on the ideas of equality, liberty, national integration, and Jewish emancipation, was eager to defend these values in times of anti-Semitism, Jewish pogroms in Eastern Europe, and increasing Jewish mass migration—phenomena that, albeit for different reasons, challenged Jewish and non-Jewish communities. For Hirsch, it became very clear that the European Jewish public and the European political elites, or the "European Aeropag,"[33] as Joseph von Wertheimer, a Viennese philanthropist and a supporter of Hirsch's ideas, put it, had to be won over. In this way, philanthropy additionally emerged, alongside its status as a tool for the promotion of Jewish solidarity and unity, as an instrument for national emancipation and integration of Jews into the modern European states.

Envisioning and Creating the New Jewish Transnational Space

As one major step in the very complex processes of modernization that were unfolding at this time, Hirsch, like other philanthropists, advanced the foundation of independent organizations and associations in Europe, which embodied and represented the new attitudes of Western European Jewry. The foundation in 1860 of the Alliance Israélite Universelle (AIU) by a circle of French Jews clearly encapsulated this new understanding of Western European Jewry and its vision of participation and integration. In its first publication, the AIU declared its aim as being to work for the emancipation and moral progress of Jews, defend Jews suffering because of their status as Jews, and encourage publications to the end of achieving the first two aims.[34] The Anglo-Jewish Association (AJA), founded in Britain in 1871, and the Israelitische Allianz zu Wien (IAzW), created in Vienna and officially recognized by the Austrian authorities in 1873, were based on similar concepts and called for Jewish solidarity and unity, the defense of Jewish rights, and the improvement of religious education.[35] The Hilfsverein der Deutschen Juden, founded in 1901, also followed similar ideas and cooperated broadly with the other three organizations on an international level.[36]

Hirsch supported the newly founded associations from their beginnings. He donated one million francs to the AIU and became an AIU committee member in 1876. From 1879, Hirsch annually donated 50,000 francs to the AIU, and in 1882 he began to balance the French organization's annual deficit. In the same year, he also donated one million francs for the refugees from the pogroms in Russia. It was at this time that Hirsch, as noted by O. S. Straus, former US ambassador in Turkey and friend of the philanthropist, began to perceive his highest purpose in the support and aid of Eastern European Jewry.[37] Along with his broad support for the AIU, the IAzW, and other associations, Hirsch focused on the "Jewish question" in Eastern Europe. In 1891, he announced the foundation of the Jewish Colonization Association (JCA), which represented a manifestation of his vision of Jewish associations' status and function in the Jewish sphere.[38] For Hirsch, the JCA, alongside other associations, became the cornerstone of a transnational Jewish network whose purpose was to facilitate transatlantic migration as well as integration of Eastern European Jews.

Thus, a Jewish relief and aid network emerged on the basis of cooperation between Paris, London, Vienna, and Berlin, as well as Brussels, Rotterdam, Turin, New York, and other centers of Jewish life.[39] Despite criticisms of the national focus of these philanthropic associations, the AIU, as Yaacov Iram points out, initiated a development that founded a "pattern of Jewish politics and philanthropy on the international level."[40] Eugen F. Miller even argues that a "global philanthropy" evolved.[41]

In the contemporary context, the Jewish associations emerged as new players on the scene, became "diplomatic intercessors and international relief agencies,"[42] and, in line with Hirsch's vision, joined the Jewish communities as representatives of Jewish interests and affairs. The reasons they did so included the emergence of a more universal and not specifically religious approach toward philanthropic actions, the perceived necessity of professionalizing philanthropy according to business strategies, and the general influence of bourgeois ideas and values. In addition, the changes taking place in Europe in the late nineteenth and early twentieth centuries called for more flexible and transnational structures in response to contemporary issues such as labor migration, human trafficking, prostitution, and related phenomena.[43] The growth of international trade and increasing numbers of migrants and émigrés, too, gave rise to economic interdependencies. In the face of these changes, as L. S. Weissbach shows in the French context, many non-Jewish and Jewish philanthropic organizations promoted values that appeared to them to fit the times: ideas of order, productivity, thriftiness, discipline, cleanliness, and—to a certain degree—also religious observance.[44] Further, they framed their support of Jewish demands for emancipation as not exclusively a Jewish issue but as a fight for human rights and thus for a core value of European civilization.[45]

Hirsch, who also conceived of his philanthropic actions as a "work for all humanity"[46] and even called "humanity [his] heir,"[47] supported this advancement of Jewish issues not only into a transnational and public sphere where theoretical ideas could be discussed, as happened in relation to the debate on philanthropy, but also into the national contexts where the debated matters could be put into practice. The establishment of a transnational Jewish philanthropic space was therefore based on Jewish communities or associations with strong national roots and followed contemporary ideas on secular engagement in philanthropy and charitable measures.

As mentioned above, the Jewish associations regarded their activities as supporting universal human rights. This attitude became especially clear during a conference held in Brussels in 1872, organized by major Jewish aid associations in the context of the riots in Romania (1871–73), and also during a conference in Berlin in 1878 whose aim was to counteract the persecution of Jews in the Eastern European context. On both occasions, the Jewish associations called upon the European nations to act in accordance with their own expressed ideals of equality, righteousness, and humanity and to impose them onto the newly emerging nations in Eastern Europe.[48] As early as 1870, the Viennese Jewish weekly *Die Neuzeit* had given voice to this idea, arguing that Europe could not be complicit in a failure to propagate core European values for all humankind, in this case with specific relation to Romanian Jews.[49] Three years later, the same newspaper called for the "natural human rights of our [Jewish] co-religionists,"[50] stressing again the link between Jewish and European rights. Such discourse clearly followed the strategies that had evolved after the Damascus and Mortara affairs and sought to make instances of Jewish persecution known to the European public as crimes against humanity.[51] The Jewish associations and elites clearly developed an understanding of the importance of persuading the European sphere that the Jews' cause was a righteous one.[52]

Emigration and Education: Hirsch's Tools for Modernization

Unlike traditional ideas of alms-giving, then, Hirsch's ideas of philanthropy and his support for major Jewish aid organizations were evidently linked to a constructive plan,[53] one whose realization relied not only on the formation of Jewish associations and the resettlement of Jews outside Europe but also on changes in European social and political settings. As a result, Hirsch and the Jewish associations advocated their ideas via the established or supporting media and pushed their agenda at international conferences. One such congress was held in Vienna in 1882 in response to the arrival of Jewish refugees in the Austrian-Hungarian border town of Brody after the pogroms in the Russian Empire in 1881 and 1882.[54] The events in Brody illustrate the necessity of transnational organizations and the cross-border cooperation of associations as "religious-humanitarian"[55] agencies, as the annual report of the IAzW described them. Yet they also point to the ambitions the agencies at the time had to effect change in social and political settings in Eastern Europe. In 1891, Hirsch's scheme was described as an attempt to colonize America with "the Jews for whom there is no place in Europe" in a comment published in the *New York Times*.[56] However, Hirsch's vision went beyond the notion of emigration. Charles Netter, the AIU's representative in Brody, wrote to the organization's head office in Paris, stating that there would be no better mission for "our rich" than to financially support their suffering co-religionists.[57] Emmanuel Felix Veneziani, Hirsch's personal envoy to the conference, who discussed the events in Brody and was also an envoy to Brody itself, supported this view and worked to ensure that the associations would respond in a concerted way. This included supporting emigration but also repatriation to and reintegration into the Russian Empire.[58] While the language of the reports and letters also reflected a strong fear of Eastern European Jews "flooding ... Brody"[59] and possibly becoming "troublemakers" in Western Europe as professional "beggars," or *luftmenschen,* the overall need for Western European Jewries to promote Jewish emancipation and integration in the various Eastern European societies emerged as a key theme.

Hirsch's and the associations' activities were clearly driven by the increasing fear of anti-Semitism in the East, but also in the West. In 1899, the twenty-fifth annual report of the IAzW called anti-Semitism an "international mental disease"[60] that had to be "cured." Adolf Jellinek, the first preacher of the Viennese Jewish community and a supporter of the IAzW, argued that core concepts of European civilization, such as liberty and equality, and the success of the fight against anti-Semitism could only be secured by promoting and expanding education and knowledge (*Bildung* and *Wissen*).[61] Hirsch, who became one of Jellinek's major hopes, had assented to Carnegie's idea that the highest purposes of philanthropy resided in the creation of amenities such as free libraries, green parks, and beautiful churches.[62] However, Hirsch primarily advocated moral and physical regeneration by means of crafts, trades, agriculture, and the idea of returning to the soil. In 1873, addressing the board of the AIU, Hirsch had formulated ideas similar to Jellinek's:

> During my repeated and extended visits to Turkey, I have been painfully impressed by the misery and ignorance in which the Jewish masses live in that empire ... progress had bypassed them, their poverty stems from lack of education, and only the education and training of the young generation can remedy this dismal situation.[63]

To combat this eastern Jewish situation, further initiatives were undertaken. For example, Hirsch's envoy Arnold White attempted to implement a highly selective, planned, and organized migration scheme for Russian Jews with the intention of avoiding mass movement, but it was unsuccessful.[64] Although Hirsch saw emigration as part of the solution in the East, he supported Jellinek's approach of promoting and expanding education and knowledge. In 1888, Jellinek and Hirsch commenced negotiations on the conditions for a new foundation in the Austro-Hungarian Empire. According to Joseph Bloch, an Austrian rabbi, this was to become a model for similar endeavors in the Russian and Ottoman Empire, or, in other words, be the gateway to educating Eastern European Jews.[65]

After three years of consultations, the Baron Hirsch Stiftung (BHS) was founded, its purpose being to establish schools, kindergartens, and educational training academies for agriculture and craft trades in the eastern crownlands of the Habsburg Empire, Galicia, and Bukovina.[66]

The foundation of Jewish schools was nothing new. The AIU had already established numerous schools across North Africa, the Balkans, and the Middle East, and other organizations and private individuals had founded educational and agricultural networks in various regions. The Rothschild family's activities in Palestine are one example of this.[67] By bringing education and knowledge to the East, the BHS, in a manner similar to other Jewish educational networks, sought to uplift whole societies and give Jewish communities the opportunity to participate in processes of modernization. Hirsch, who supported these ideas, sought to get them implemented in the two Austrian-Hungarian crownlands of Galicia and Bukovina, which were both perceived as strongly affected by poverty, "social and cultural backwardness," and "medieval religiosity." The first annual report of the BHS therefore clearly propagated embourgoisement and promoted the turn toward agriculture and craft trades already adopted by the AIU and IAzW.[68] The educational network of the BHS and IAzW, which Hirsch heavily supported, expanded over the next decade, with over fifty schools, kindergartens, evening educational sessions, and academies being opened. Their purpose was to transform young men and women regarded as weak, uneducated, and malnourished into capable, well-educated, hard-working, and multilingual members of society.[69] This "*Culturwerk*"[70] was celebrated as a major step toward the improvement and cultivation of the Eastern European Jewish masses and societies and was regarded as a part of Europe's modernization.

While the BHS predominantly focused on young males, Baroness Clara de Hirsch, whom some studies depict as the most influential mind behind Hirsch's philanthropic activities, also embraced philanthropy as a field of action.[71] She founded the Baronin Clara von Hirsch-Kaiser Jubiläums-Stiftung (1899) in the Austro-Hungarian context and continued her husband's philanthropic activities after his death. She followed ideas based on religion, acculturation, and integration and dedicated herself to the advancement of conditions for women.[72]

Conclusion

Baron de Hirsch's philanthropy was singled out by the periodical the *Lancet* as an example "well worthy of imitation by other magnates of the turf"[73] and praised by many contemporaries. In 1896, after Hirsch died, the *Lancet* wrote of him: "His sympathy for the poor of his own faith is well known, and his scheme of colonisation will prove a lasting memorial of his name among the Semitic race."[74] In the same year, the *New York Times* also celebrated Hirsch's philanthropy, deeming it as "unlimited as his wealth."[75] Arnold White also praised Hirsch as a man who "gave a great deal more than his money. He gave his time, attention, and intellect to the minute study of the problems he attacked for the benefit of his co-religionists and others."[76]

Hirsch, as one example of an "elite philanthropist," participated in and contributed to the contemporary national and transnational discourses on state, citizenship, and welfare.[77] These discussions, however abstract, reflected Jewish engagement with a specific mid-nineteenth-century philanthropic spirit that gave rise to the general growth of associations and foundations across Europe.[78] Consequently, the foundation of transnational organizations outside traditional religious frameworks was just one example of the increasing integration of the Jews into the European intellectual and social sphere.[79] It demonstrated how Jews across Europe participated in debates on modernization and became vital players in both local and transnational contexts. Jewish philanthropic associations and the push for education and knowledge were, therefore, not just parts of a subculture, as David Sorkin argues, but also, to quote Simone Lässig, a "vital laboratory for Jewish *Bürgerlichkeit*."[80] The educational missions and the foundation of schools, and indirectly Hirsch's support for emigration, were designed to give the Eastern European Jews (or Oriental Jews) social capital, which enabled them to participate in the modernization processes in which their countries were engaged. In other words, the function of schools extended beyond their immediate educational purposes, as they became key instruments for integrating poor and uneducated Jews into their societies, finalizing Jewish acculturation, and fostering ongoing Jewish emancipation.[81]

Despite the increased breadth and application of philanthropic activities that some Jews engaged in, it remained important within the Jewish sphere for Jewish philanthropic organizations to take care of the Jewish poor, in accordance with the religiously grounded law of Israel, as Nancy L. Green has shown.[82] Because emerging state welfare systems were still seen as insufficient and ineffective, Jewish emancipation and integration into the Western European nations did not lead to the dissolution of Jewish aid institutions across Europe in the nineteenth century but instead resulted in their increasing modernization. Moreover, supporting traditions were adjusted and incorporated into them. Fears of anti-Semitic attacks based on an image of Jews as a burden to society and as "un-rooted wanderers" remained high, so the protection of "vulnerable parts" of Jewish society continued to be a core Jewish interest.[83] The schools, kindergartens, and training academies the aid associations founded, as well as the associations themselves, became distinctly Jewish spaces even without demanding exclusivity. Jewish associations and organizations that were founded outside of the traditional realm of Jewish communities and even emerged as major representatives of Jewish interests sought to become respected parts of bourgeois societies while they also pushed for the existence of an "exclusively Jewish sphere" which "by its nature ... reinforced separateness."[84] Rainer Liedtke has argued that the "separate Jewish culture of welfare" should be understood as a product of the tensions between the Jewish aspiration for societal integration and the need for self-preservation.[85] Whether this paradoxical position emerged "from a self-imposed segregation or a lack of acceptance [from outside],"[86] as Susan L. Tananbaum discusses, it demonstrates the unique situation of modern Jewish philanthropy.

Hirsch's philanthropy, then, can be said to have been based on a new understanding of state and society but also on a continuingly relevant "transnationalist sensibility,"[87] as well as on a "collective responsibility" or "intracommunal solidarity."[88] The existential difficulty of Jewish self-definition, which derived "from the time that Mendelssohn first openly confronted the problem of being both Jewish and European,"[89] was probably not resolved by Hirsch's philanthropic actions and the foundation of Jewish associations, but these clearly created a new "principal mode of Jewish identification."[90]

In sum, Hirsch's ideas fused a Jewish and a bourgeois understanding of philanthropy, combining a sense of *noblesse oblige* as members of a privileged class, a religious imperative, an intent to construct a position of societal respectability, a new sense of national identity, and a concept of self-help.[91] While the *New York Times* reprinted a very aggressive article that accused Hirsch of having "wasted millions in trying to buy the favor of society,"[92] we might instead view his scheme of philanthropy as being his way to solve the "Jewish question" in Europe. He evidently regarded offering thousands of Jews education and knowledge—and thus a chance to join in national processes of modernization without relinquishing their specific religious identity—as providing them with a place and space as well as with an "admission ticket"—an *Entréebillett*—to European societies.

Björn Siegel is a researcher at the Institute for the History of the German Jews in Hamburg (IGdJ). He authored *Austrian Jewry Between East and West: The Austrian Jewish Alliance 1873–1938* (Campus, 2010) and coauthored *Kurt Fritz Rosenberg: Einer der nicht mehr dazugehört* (Wallstein, 2012). His research interests focus on Jewish migration, the evolution of Jewish philanthropy, and Jewish maritime studies. His current research project deals with the ship as a place in Jewish history and re-evaluates migrations to Palestine in the Mandate period.

Notes

1. Shemtov, "Between Perspectives of Space," 142; Brauch, Lipphardt, and Nocke, "Introduction: Exploring Jewish Space—An Approach," in Brauch, Lipphardt, and Nocke, *Jewish Topographies*, 1–23, here 1, 4.
2. For an overview of the dimensions of time, place and space, see Foucault, "Of Other Spaces," 22; Fonrobert and Shemtov, "Introduction," 4.
3. Mann, *Space and Place*, 17–19.
4. On Hirsch, see Thane, "Hirsch, Maurice de"; Grunwald, *Türkenhirsch*.
5. Hirsch, "My Views on Philanthropy," 1.
6. Carnegie, "Wealth," 653.

7. See Wren and Greenwood, "Business Leaders"; Harvey et al., "Andrew Carnegie," 433–35; Lambkin, "Emotional Function," 322–24.
8. Carnegie, "Wealth," 654–56.
9. Ibid., 657.
10. On Gladstone, see Shannon, *Gladstone: Heroic Minister*; Shannon, *Gladstone: God and Politics*; Windscheffel, "Politics, Religion and Text"; Faught, "An Imperial Prime Minister?"
11. Gladstone, "Mr. Carnegie's Gospel of Wealth," 682–84.
12. The Universal Beneficent Society was founded by John Villier Stuart (1831–99) and was designed to help the poor and needy. Gladstone, "Mr. Carnegie's Gospel of Wealth," 690–92.
13. Ibid., 679.
14. Selby, "Henry Edward Manning," 163–70.
15. Manning, "Irresponsible Wealth (No. I)," 876.
16. See Oldstone-Moore, *Hugh Price Hughes*, 238–40; Hughes, *Life of Hugh Price Hughes*, 481–507.
17. Hughes, "Irresponsible Wealth (No. III)," 891.
18. Ibid., 897.
19. On Gibbons, see Noll, "Bishop James Gibbons."
20. Manning, "Irresponsible Wealth (No. I)," 881, 884. See also Manning, "The Catholic Church and Modern Society," 103.
21. Adler, "Irresponsible Wealth (No. II)," 888.
22. Ibid.
23. Hirsch, "My Views on Philanthropy," 1.
24. Carnegie, "Wealth," 661.
25. Hirsch, "My Views on Philanthropy," 1. On criticisms of traditional alms-giving, see also Carnegie, "Wealth," 663.
26. Carnegie, "Wealth," 661.
27. On the continuing importance of *tzedakah*, see Mesch, Moore, and Ottoni-Wilhelm, "Does Jewish Philanthropy Differ," 81–83.
28. For a short introduction to modern philanthropy, see Penslar, *Shylock's Children*, 90–123.
29. Liedtke, *Jewish Welfare*, 13–15.
30. Abigail Green demonstrated the strong continuation in her study on Moses Montefiore; see Green, "Rethinking Sir Moses Montefiore," 658. See also Bartal, "Moses Montefiore."
31. Hirsch, "My Views on Philanthropy," 2.
32. Penslar, *Shylock's Children*, 92.
33. Wertheimer, *Zur Emanzipation unserer Glaubensgenossen*, 6.
34. Chouraqui, *Cent ans d'histoire*, 38.
35. For the principles of the different associations, see Anglo-Jewish Association, *Ninth Annual Report*, 9; Israelitische Allianz zu Wien, *Bericht und Ausweise*, 3, 8.
36. Brinkmann, *Migration und Transnationalität*, 80–85.
37. Strauss, *Under Four Administrations*, 95. For further detail, see Thane, "Hirsch, Maurice de," 3.
38. See Sokolow, *History of Zionism*, 253–54.
39. Israelitische Allianz zu Wien, *Vierzigster Jahresbericht*, 11. See also Alliance Israélite Universelle, *Bulletin*, 76.
40. Iram, "History," 584. For criticism of the nationalization of the Alliances, see *Die Allgemeine Israelitische Allianz*, 9–10.
41. Miller, "Philanthropy and Cosmopolitanism," 51.

42. Penslar, *Shylock's Children,* 196.
43. On human trafficking and prostitution, see Gartner, "Anglo-Jewry," 171–78.
44. Weissbach, "The Nature of Philanthropy," 201.
45. It has been claimed that the awareness of "other" Jews with a less secure legal and social status was part of the Jewish modernization process; see Brinkmann, Penslar, and Rechter, "Introduction: Jews and Modernity," 284.
46. Hirsch, "My Views on Philanthropy," 4.
47. Strauss, "Baron de Hirsch," 561.
48. Israelitische Allianz zu Wien, *Sechster Jahresbericht,* 5–7.
49. *Die Neuzeit,* 11 February 1870, 63.
50. *Die Neuzeit,* 27 June 1873, 262.
51. For more detail, see Brinkmann, *Migration und Transnationalität,* 38–39.
52. Brinkmann, Penslar, and Rechter, "Introduction: Jews and Modernity," 284.
53. Adler-Rudel, "Moritz Baron Hirsch," 30.
54. For more detail on Brody, see Kuzmany, *Brody,* 237–46.
55. Israelitische Allianz zu Wien, *Neunter Jahresbericht,* 3–4, 7.
56. *New York Times,* 12 September 1891. It had originally appeared in the London *Times.*
57. Goldenstein, *Brody und die russisch-jüdische Emigration,* 12.
58. On repatriation measures, see Israelitische Allianz zu Wien, *Protokolle,* 1–8. For a general discussion of the economic implications of Jewish mass migration and the influence of non-Jewish actors, see Brinkmann, *Migration und Transnationalität,* 68–79. See also Brinkmann, "'Mit Ballin unterwegs,'" 75–96; Alroey, "'And I Remained Alone,'" 39–72. On the emigrants' hope and disillusionment, see Green, "To Give and to Receive," 208–11.
59. Israelitische Allianz zu Wien, *Zehnter Jahresbericht,* 87–105.
60. Israelitische Allianz zu Wien, *Fünfundzwanzigster Jahresbericht,* 14.
61. Jellinek, *Rede,* 6–12.
62. Carnegie, "Wealth," 663; Hirsch, "My Views on Philanthropy," 1–2.
63. Leven, *Cinquante ans d'histoire,* 23–24.
64. Perlmann, "Arnold White"; Johnson, "'A Veritable Janus,'" 42–43; 63; Sokolow, *History of Zionism,* 254–55.
65. Österreichische Wochenschrift, 1 January 1887, 4.
66. See Grunwald, "Note," 227–36; Siegel, *Österreichisches Judentum,* 97–119.
67. On schools and Palestine, see Raichel and Tadmor-Shimony, "Jewish Philanthropy."
68. Baron Hirsch Stiftung, *Bericht des Curatoriums,* 9–11.
69. Israelitische Allianz zu Wien, *Siebzehnter Jahresbericht,* xii.
70. Israelitische Allianz zu Wien, *Achtzehnter Jahresbericht,* ix.
71. On gender and philanthropy, see Tananbaum, "Philanthropy and Identity," 946–48. On Clara von Hirsch, see Alderman, "Hirsch, Clara de," 1–2.
72. For a general discussion on differences between male and female philanthropists and gender roles, see Tananbaum, "Democratizing British-Jewish Philanthropy," 64; Goldberg, "Sacrifices," 44; Modena, "Jewish Women," 25.
73. "Baron Hirsch's Gift to Medical Charities," *Lancet,* 7 April 1894, 879.
74. "The Late Baron Hirsch," *Lancet,* 25 April 1896, 1152.
75. *New York Times,* 22 April 1896.
76. *English Illustrated Magazine,* no. 153, June 1986, 1–4, here 3.
77. For the concept of the elite philanthropists, see Sperber, "Philanthropy and Social Control," 88–90.

78. For Germany, see Penslar, *Shylock's Children,* 92.
79. For a broader discussion, see Karlinsky, "Jewish Philanthropy," 160–62, 164.
80. Lässig, "How German Jewry Turned Bourgeois," 62–66; Lässig, *Jüdische Wege,* 442–504. On the notion of "subculture," see Sorkin, *Transformation.*
81. Lässig, "How German Jewry Turned Bourgeois," 62–66; Penslar, *Shylock's Children,* 107–23.
82. Green, "To Give and to Receive," 198–99.
83. Penslar, *Shylock's Children,* 103–4.
84. Kidd, "Review: *Jewish Welfare,*" 536.
85. Liedtke, *Jewish Welfare,* 242.
86. Tananbaum, "Philanthropy and Identity," 952.
87. Penslar, "An Unlikely Internationalism," 309.
88. Green, "To Give and to Receive," 197.
89. Meyer, *Origins of the Modern Jew,* 182.
90. Kaplan, "*Unter uns,*" 60.
91. Weissbach, "The Nature of Philanthropy," 192–93; Tananbaum, "Philanthropy and Identity," 939.
92. *New York Times,* 3 May 1896. It had originally appeared in the *London Speaker.*

Bibliography

Adler, Hermann. "Irresponsible Wealth (No. II)." *Nineteenth Century* 28, no. 166 (1890): 886–89.

Adler-Rudel, Shalom. "Moritz Baron Hirsch: Profile of a Great Philanthropist." *Leo Baeck Institute Yearbook* 8 (1963): 29–69.

Alderman, Geoffrey. "Hirsch, Clara de, Baroness de Hirsch (1833–1899)." *Oxford Dictionary of National Biography.* Cambridge: Oxford University Press, 2004. Accessed 17 June 2016. www.oxforddnb.com/view/article/56692.

Alliance Israélite Universelle. *Bulletin d'Alliance Israélite Universelle 1. Semestre (1. January 1873).* Paris: Siege de la Société, 1873.

Alroey, Gur. "'And I Remained Alone in a Vast Land': Women in the Jewish Migration from Eastern Europe." *Jewish Social Studies* 12, no. 3 (2006): 39–72.

Anglo-Jewish Association. *The Ninth Annual Report of the Anglo-Jewish Association in Cooperation with the Alliance Israélite Universelle.* London: Offices of the Anglo-Jewish Association, 1880.

Baron Hirsch Stiftung. *Bericht des Curatoriums der Baron Hirsch-Stiftung zur Beförderung des Volksschulunterrichtes im Königreiche Galizien und Lodomerien mit dem Grossherzogthume Krakau und im Herzogthume Bukowina für das erste Verwaltungsjahr 1891.* Vienna: M. Waizner, 1892.

Bartal, Israel. "Moses Montefiore: Nationalist before His Time, or Belated Shtadlan?" *Studies in Zionism: Politics, Society Culture* 11, no. 2 (1990): 111–25.

Brauch, Julia, Anna Lipphardt, and Alexandra Nocke, eds. *Jewish Topographies: Visions of Space, Traditions of Place.* Aldershot: Ashgate, 2008.

Brinkmann, Tobias. *Migration und Transnationalität.* Paderborn: Ferdinand Schöningh, 2012.

———. "'Mit Ballin unterwegs': Jüdische Migranten aus Osteuropa im Transit durch Deutschland vor dem Ersten Weltkrieg." *Aschkenas* 17, no. 1 (2007): 75–96.

Brinkmann, Tobias, Derek Penslar, and David Rechter. "Introduction: Jews and Modernity — Beyond the Nation." *Journal of Modern Jewish Studies* 7, no. 3 (2008): 283–86.

Carnegie, Andrew. "Wealth." *North American Review* 148, no. 391 (1889): 653–65.

Chouraqui, André. *Cent ans d'histoire – L'Alliance Israélite Universelle et la renaissance juive contemporaine 1860–1960*. Paris: Presse Universitaires de France, 1965.

Die Allgemeine Israelitische Allianz: Bericht des Central-Comités über die ersten fünfundzwanzig Jahre 1860–1885. Berlin: Commissionsverlag J. Kaufmann, 1885.

Faught, C. Brad. "An Imperial Prime Minister? W. E. Gladstone and India 1880–1885." *Journal of the Historical Society* 6, no. 4 (2006): 555–78.

Fonrobert, Charlotte E., and Vered Shemtov. "Introduction: Jewish Conceptions and Practices of Space." *Jewish Social Studies* 11, no. 3 (2005): 1–8.

Foucault, Michel. "Of Other Spaces: Utopias and Heterotopias." *Diacritics* 16, no. 1 (1986): 22–27.

Gartner, Lloyd P. "Anglo-Jewry and the Jewish International Traffic in Prostitution 1885–1914." *AJS Review* 7/8 (1982/83): 129–78.

Gladstone, William E. "Mr. Carnegie's Gospel of Wealth: A Review and a Recommendation." *Nineteenth Century* 28, no. 165 (1890): 677–93.

Goldberg, Idana. "'Sacrifices upon the Altar of Charity': The Masculinization of Jewish Philanthropy on Mid-Nineteenth Century America." *Nashim* 20 (2010): 34–56.

Goldenstein, Leo. *Brody und die russisch-jüdische Emigration*. Frankfurt am Main: Commissionsverlag J. Kauffmann, 1882.

Green, Abigail. "Rethinking Sir Moses Montefiore: Religion, Nationhood, and International Philanthropy in the Nineteenth Century." *American Historical Review* 110, no. 3 (2005): 631–58.

Green, Nancy L. "To Give and to Receive: Philanthropy and Collective Responsibility among Jews in Paris 1880–1914." In *The Uses of Charity: The Poor on Relief in the Nineteenth-Century Metropolis*, edited by Peter Mandler, 197–226. Philadelphia: University of Pennsylvania Press, 1990.

Grunwald, Kurt. "A Note on the Baron Hirsch Stiftung Vienna 1888–1914." *Leo Baeck Institute Yearbook* 17, no. 1 (1972): 227–36.

———. *Türkenhirsch: A Study of Baron Maurice de Hirsch, Entrepreneur and Philanthropist*. Jerusalem: Israel Program for Scientific Translations, 1966.

Harvey, Charles, Mairi Maclean, Jillian Gordon, and Eleanor Shaw. "Andrew Carnegie and the Foundation of Contemporary Entrepreneurial Philanthropy." *Business History* 53, no. 3 (2011): 425–50.

Hirsch, Maurice de. "My Views on Philanthropy." *North American Review* 153, no. 416 (1891): 1–4.

Hughes, Dorothea Price. *The Life of Hugh Price Hughes*. London: Hodder and Stoughton, 1904.

Hughes, Hugh Price. "Irresponsible Wealth (No. III)." *Nineteenth Century* 28, no. 166 (1890): 890–900.

Iram, Yaacov. "The History of Franco-Jewish Educational Philanthropy in North Africa and the Levant." *Paedagogica Historia* 28, no. 3 (1992): 580–88.

Israelitische Allianz zu Wien. *Achtzehnter Jahresbericht der Israelitischen Allianz zu Wien erstattet an der achtzehnten ordentlichen General-Versammlung am 4. Juni 1891*. Vienna: M. Waizner, 1891.

———. *Bericht und Ausweise der Israelitischen Allianz zu Wien vorbereitet für die erste ordentliche General-Versammlung am 7. Juni 1874*. Vienna: Verlag der Israelitischen Allianz zu Wien, 1874.

———. *Fünfundzwanzigster Jahresbericht der Israelitischen Allianz zu Wien erstattet in der fünfundzwanzigsten ordentlichen General-Versammlung am 8. Mai 1898: Mit einem Rückblick auf die fünfundzwanzigjährige Wirksamkeit der israel. Allianz 1873–1898.* Vienna: M. Waizner, 1898.

———. *Neunter Jahresbericht der Israelitischen Allianz zu Wien erstattet in der neunten ordentlichen General-Versammlung am 8. Juni 1882.* Vienna: M. Waizner, 1882.

———. *Protokolle der internationalen Conferenz in Wien am 2., 3. und 4. August 1882 zu Gunsten der russisch-jüdischen Flüchtlinge.* Vienna: Verlag "Styrermühl," 1882.

———. *Sechster Jahresbericht der Israelitischen Allianz zu Wien erstattet in der sechsten ordentlichen General-Versammlung am 25. Mai 1879.* Vienna: M. Waizner, 1879.

———. *Siebzehnter Jahresbericht der Israelitischen Allianz zu Wien erstattet in der siebzehnten ordentlichen General-Versammlung am 9. Juni 1890.* Vienna: M. Waizner, 1890.

———. *Vierzigster Jahresbericht der Israelitischen Allianz zu Wien erstattet an die XL. ordentliche Generalversammlung am 16. April 1913 nebst einem Rückblick auf die vierzigjährige Wirksamkeit des Vereins 1872–1912.* Vienna: Verlag der Israelitischen Allianz zu Wien, 1913.

———. *Zehnter Jahresbericht der Israelitischen Allianz zu Wien erstattet in der zehnten ordentlichen General-Versammlung am 24. Mai 1883.* Vienna: M. Waizner, 1883.

Jellinek, Adolf. *Rede zur Förderung der israel. Allianzen—Samstag, den 19. Juli 1879—im israelitischen Bethause der inneren Stadt Wien gehalten.* Vienna: Jacob Schlossberg, 1879.

Johnson, Sam. "'A Veritable Janus at the Gates of Jewry': British Jews and Mr. Arnold White." *Patterns of Prejudices* 47, no. 1 (2013): 41–68.

Kaplan, Marion A. "*Unter uns*: Jews Socialising with Other Jews in Imperial Germany." *Leo Baeck Institute Yearbook* 48, no. 1 (2003): 41–65.

Karlinsky, Nahum. "Jewish Philanthropy and Jewish Credit Cooperatives in Eastern Europe and Palestine up to 1939: A Transnational Phenomenon?" *Journal of Israeli History: Politics, Society, Culture* 27, no. 22 (2008): 149–70.

Kidd, Alan J. "Review: *Jewish Welfare in Hamburg and Manchester* by Rainer Liedtke." *Victorian Studies* 43, no. 3 (2001): 535–36.

Kuzmany, Börries. *Brody: Eine galizische Grenzstadt im langen 19. Jahrhundert.* Vienna: Böhlau Verlag, 2011.

Lambkin, Brian. "The Emotional Function of the Migrant's Birthplace in Transnational Belonging: Thomas Mellon (1813–1908) and Andrew Carnegie (1835–1918)." *Journal of Intercultural Studies* 29, no. 3 (2008): 315–29.

Lässig, Simone. "How German Jewry Turned Bourgeois: Religion, Culture, and Social Mobility in the Age of Emancipation." *Bulletin of the German Historical Institute* 37 (2005): 59–73.

———. *Jüdische Wege ins Bürgertum: Kulturelles Kapital und sozialer Aufstieg im 19. Jahrhundert.* Göttingen: Vandenhoeck & Ruprecht, 2004.

Leven, Narcisse. *Cinquante ans d'histoire. L'Alliance Israélite Universelle (1860–1910).* Vol. 2. Paris: Libraire Felix Alcan, 1920.

Liedtke, Rainer. *Jewish Welfare in Hamburg and Manchester c.1850–1914.* Oxford: Clarendon Press, 1998.

Mann, Barbara E. *Space and Place in Jewish Studies.* New Brunswick, NJ: Rutgers University Press, 2012.

Manning, Henry E. "The Catholic Church and Modern Society." *North American Review* 130, no. 279 (1880): 101–15.

———. "Irresponsible Wealth (No. I)." *Nineteenth Century* 28, no. 166 (1890): 876–85.

Mesch, Debra, Zach Moore, and Mark Ottoni-Wilhelm. "Does Jewish Philanthropy Differ by Sex and Type of Giving?" *Nashim* 20 (2010): 80–96.

Meyer, Michael A. *The Origins of the Modern Jew: Jewish Identity and European Culture in Germany, 1749–1824.* Detroit: Wayne State University Press, 1967.

Miller, Eugene F. "Philanthropy and Cosmopolitanism." *Good Society* 15, no. 1 (2006): 51–60.

Modena, Luisa Levi D'Ancona. "Jewish Women in Non-Jewish Philanthropy in Italy (1870–1938)." *Nashim* 20 (2010): 9–33.

Noll, Mark A. "Bishop James Gibbons, the Bible, and Protestant America." *U.S. Catholic Historian* 31, no. 3 (2013): 77–104.

Oldstone-Moore, Christopher R. *Hugh Price Hughes: Founder of a New Methodism; Conscience of a New Nonconformity, 1847–1902.* Cardiff: University of Wales Press, 1999.

Penslar, Derek J. *Shylock's Children: Economics and Jewish Identity in Modern Europe.* Berkeley: University of California Press, 2001.

———. "An Unlikely Internationalism: Jews at War in Modern Western Europe." *Journal of Modern Jewish Studies* 7, no. 3 (2008): 309–23.

Perlmann, Moshe. "Arnold White: Baron Hirsch's Emissary." *Proceedings of the American Academy for Jewish Research* 46/47 (1928–29/1978–79): 473–89.

Raichel, Nirit, and Tali Tadmor-Shimony. "Jewish Philanthropy, Zionist Culture, and the Civilizing Mission of Hebrew Education." *Modern Judaism* 34, no. 1 (2013): 60–85.

Rozenblum, Serge-Allain. *Le baron de Hirsch: Un financier au service de l'humanité.* Paris: Punctum, 2006.

Selby, D. E. "Henry Edward Manning and the Catholic Middle Class: A Curricular Study." *Paedagogica Historica* 10, no. 1 (1970): 149–70.

Shannon, Richard T. *Gladstone: God and Politics.* London: Hambledon Continuum, 2007.

———. *Gladstone: Heroic Minister, 1865–1898.* London: Allen Lane, 1999.

Shemtov, Vered. "Between Perspectives of Space: A Reading in Yehuda Amichai's 'Jewish Travel' and 'Israeli Travel.'" *Jewish Social Studies* 11, no. 3 (2005): 141–61.

Siegel, Björn. "Das 'Es werde Licht' ist gesprochen; … : Die Bildungsmissionen der Israelitischen Allianz zu Wien, der Baron Hirsch-Stiftung und der Alliance Israélite Universelle im Vergleich, 1860–1914." *Transversal* 12, nos. 1–2 (2011): 83–112.

———. *Österreichisches Judentum zwischen Ost und West: Die Israelistische Allianz zu Wien 1873–1938.* Frankfurt: Campus, 2010.

Sokolow, Nahum. *History of Zionism: 1600–1918.* London: Longmans, Green, 1919.

Sorkin, David J. *The Transformation of German Jewry, 1780–1840.* New York: Oxford University Press, 1987.

Sperber, Haim. "Philanthropy and Social Control in the Anglo-Jewish Community during the Mid-Nineteenth Century (1850–1880)." *Journal of Modern Jewish Studies* 11, no. 1 (2012): 85–101.

Strauss, Oscar S. "Baron de Hirsch." *Forum,* July 1896, 558–65.

———. *Under Four Administrations: From Cleveland to Taft.* Boston: River Side Press, 1922.

Tananbaum, Susan L. "Democratizing British-Jewish Philanthropy: The Union of Jewish Women (1902–1930)." *Nashim* 20 (2010): 57–79.

———. "Philanthropy and Identity: Gender and Ethnicity in London." *Journal of Social History* 30, no. 4 (1997): 937–61.

Thane, Pat. "Hirsch, Maurice de [formerly Moritz von Hirsch]." *Oxford Dictionary of National Biography.* Cambridge: Oxford University Press, 2004. Accessed 17 June 2016. http://www.oxforddnb.com/view/article/49024.

Weissbach, Lee Shai. "The Nature of Philanthropy in Nineteenth-Century France and the *mentalité* of the Jewish Elite." *Jewish History* 8, nos. 1/2 (1994): 191–204.

Wertheimer, Joseph von. *Zur Emanzipation unserer Glaubensgenossen.* Vienna: M. Waizner, 1882.

Windscheffel, Ruth C. "Politics, Religion and Text: W. E. Gladstone and Spiritualism." *Journal of Victorian Culture* 11, no. 1 (2006): 1–29.

Wren, Daniel A., and Ronald G. Greenwood. "Business Leaders: A Historical Sketch of Andrew Carnegie." *Journal of Leadership & Organizational Studies* 5, no. 4 (1999): 106–13.

11

Reconstructing Jewishness, Deconstructing the Past

Reading Berlin's Scheunenviertel
over the Course of the Twentieth Century

Anne-Christin Saß

Berlin's Scheunenviertel (literally "barn quarter"), located northwest of Alexander-platz between Rosa-Luxemburg-Platz and Rosenthaler Platz, is widely regarded as the city's historic Jewish quarter. From the earliest days of the twentieth century, this poverty-stricken part of the Spandauer Vorstadt had been increasingly considered a "Jewish" district, dominated by Jewish immigrants from Eastern Europe. For many contemporaries, Jews and non-Jews alike, the visible presence of Orthodox migrants on its streets made it *the* Jewish space of Berlin.[1] Consequently, the neighborhood represented a highly symbolic locality in the composition of Jewish Berlin and the modern metropolis.

During the 1990s, the Scheunenviertel also played a prominent role in emerging discussions about the experiences and realities of Jewish life in the postwar era and the place of Jewish culture in reunified Germany.[2] As a city district with a distinct history, the Scheunenviertel was central to Berlin's image as a multicultural and cosmopolitan city. The rediscovered Scheunenviertel was not restricted to its historical limits but had been extended to include other parts of the central Berlin district of Mitte, encompassing the area from Hackescher Markt to the Neue Synagoge at Oranienburger Strasse, which had been one of the major centers of Jewish community life in Berlin before the Shoah. Extensive redevelopments accompanied rapid gentrification; the quarter's imagined authenticity gave it a particular appeal. Suddenly the Scheunenviertel was, in publicist André Meier's words, "always and everywhere."[3]

This chapter retraces the ways in which the Scheunenviertel was constructed, experienced, and perceived as a Jewish space in the course of the twentieth century. What made this urban district "Jewish," and what did "Jewish" mean for individual actors, Jews and non-Jews, citizens and migrants, Germans and Eastern Europeans, in different periods? Which role does Jewishness play in the constant rewriting, deconstruction, and reconstruction the Scheunenviertel has undergone, and how is the relationship between the real and imagined urban space structured?[4] This chapter argues that a main characteristic of the district, as perceived throughout the twentieth century, was a continuous inconsistency between real and imagined urban space. Moreover, the various presentations from the first third of the twentieth century had considerable influence upon the picturesque image attached to the Scheunenviertel in reunified Berlin.

Taking the long view of the area, this chapter seeks to join two strands of research: first, the study of urban districts as historical Jewish spaces; and second, the study of former Jewish spaces as places of remembrance. Although there is no strict dividing line between scholarly works dealing with questions of memory culture and the analysis of historical Jewish spaces, historical research, in particular, tends to focus on practices of commemorative culture or the reconstruction of urban spaces viewed from a historical perspective.[5] The research on Berlin's Scheunenviertel is no exception.[6] Simultaneously analyzing historical and cultural views of the neighborhood—specifically those concerned with cultures of memory—seems promising for identifying common factors, differences, and interrelations between the various images of Jewishness and related self-representations of German society in great detail and also for gaining new insights into the relationship between real and imagined urban spaces. On a more general level, the following analysis attempts to evaluate the complex relationships between space, urban experiences, social power relations, and the human need for identification and belonging.

From a Slum District to a "Jewish Ghetto": The Making of a Jewish Quarter, 1900–1918

Around 1900, the Scheunenviertel was one of the *Elendsquartiere* (quarters of misery) of the expanding capital of the German Empire. The twenty-seven grain barns that gave the area its name had been razed in the first third of the nineteenth century, making way for apartment buildings with two, three, or sometimes four floors on small plots of land. Poor housing conditions and cheap building materials soon rendered the quarter one of Berlin's poorest residential areas; from the 1880s, it became synonymous with social decline and misery. Although Jews had been living there since the mid-eighteenth century, it was not explicitly perceived as Jewish.[7] This changed only with growing Jewish emigration from Eastern Europe from about 1865.

At that time, Berlin was a central transit point for Jewish migrants from Eastern Europe en route to the principal overseas ports of Antwerp, Bremerhaven, Hamburg, and Rotterdam. The vast majority only passed the transit camp in Ruhleben, then outside Berlin, and were not allowed to stop over in Berlin itself. The few thousand migrants who obtained residency status often chose to live in the Scheunenviertel; its low rents, proximity to the city center, and Jewish community institutions such as the Old and New Synagogues made it attractive to them.

An early reference to the Scheunenviertel as a place with a distinct Eastern European Jewish minority can be found in a petition the Bezirksverein Victoria submitted to the city council in 1892. Calling for the speedy implementation of a road construction project, the association's members depicted the continuous immigration of Romanians, Galicians, and Poles to Berlin as a possible health risk to citizens.[8] The petition followed the popular stereotype of Eastern European Jews as dirty and backward and attributed Jewishness to the area solely on the strength of increasing numbers of Jewish inhabitants.

By contrast, Hans Ostwald's ethnographic portrayal *Abend im Scheunenviertel*, published in the first volume of his series *Großstadtdokumente*, can be understood as a first attempt to explore the area's assumed "Jewishness."[9] As a journalist, Ostwald deduced its ghetto-like character from its physical structure alone, which resembled the layout of a Jewish ghetto in a provincial town. In Ostwald's account, the district seemed cut off from the noise of the rest of the city, and the narrow, old, gray streets were crowded with people from underprivileged social classes, petty criminals, and prostitutes. The notable presence of Jewish migrants and the existence of distinctly Jewish beer halls and lodgings gave it a unique atmosphere:

> Old, disheveled women. Bearded men with gaunt faces—or fat shaven heads. Four, five young girls are standing in front of a cellar entrance.... Here and there one can see a young Jewish man in the latest English fashion with suit, garish tie, and top hat. Almost every one of them is surrounded by an air of Galicia, Poland, and Russia.[10]

The interior design of the "kosher distilleries" did not differ from typical Berlin drinking establishments. Again, in Ostwald's narrative, the people transformed the place into a Jewish space. The migrants, although diverse in age, gender, appearance, and their relationship to Jewish traditions, had one thing in common: their Eastern European origin. This background, primarily visible and audible through their clothing, the Yiddish language they used, and an assumed difference in their national character, set them apart from the non-Jewish inhabitants. It is striking that Ostwald's observations on the Jewish tavern scenes resembled those on non-Jewish hostelries. Both accounts describe quarrels, flirtations, and other everyday occurrences. Evidently, Ostwald described migrants and non-migrants, Jews and Gentiles with the same interest and humor, thereby presenting them as essential parts of the modern metropolis.

Subsequent literary depictions of the Scheunenviertel as a Jewish space were linked to Ostwald's observations but set different emphases. The unknown author Ernst Seiffert, for example, focused on migrants' economic activities, describing in detail the monopolization of the used-clothing trade by Russian Jews there. He highlighted the "enormous diligence and toughness" of the migrants who built a livelihood out of nothing and also supplied their suffering brethren in Russia.[11] In Seiffert's view, the influx of Russian Jews gave the quarter a special Jewish note but did not transform it into a Jewish ghetto. Instead, German, Russian, English, and French people lived "all mixed together" in the imperial German capital.

Contrastingly, German-Jewish author Hans Ermy entitled his Scheunenviertel portrayal *The Ghetto of Berlin*.[12] Like Ostwald and Seiffert, he constructed an opposition between the quarter and big-city Berlin. While the other two authors contrasted the hectic pace of the metropolis with the less intense bustle of small-town life, Ermy stressed the difference between change and continuity: whereas the physiognomy of the big city changed daily, Scheunenviertel streets looked the same year after year. By exaggerating and omitting the real numbers of Jews and Gentiles there, Ermy drew a static picture of traditional Jewish everyday life. For him, this "ghetto culture" was traditional, unenlightened, and backward. Thus, Ermy represented the majority view middle-class Jews held of the stereotyped *Ostjude*, a figure that was antithetical to their self-perception as modern, highly acculturated German Jews.[13] However, many Scheunenviertel migrants did not fit the image of the Orthodox shtetl Jew. Among them were class-conscious workers who had left Russia after the revolution of 1905 as well as secular Jews from the industrial centers in Congress Poland.

In contrast to Ermy, who at least described the migrants with a kind of ethnological curiosity, political scientist and Deutschkonservative Partei member Adolf Grabowsky observed the presence of Eastern European Jews in Berlin with a mixture of fear and repulsion. His account *Ghettowanderung* combined literary urban description with strong negative stereotyping, inextricably intertwining them.[14] Walking down Kaiser-Wilhelm-Strasse, Grabowsky felt more and more deeply enclosed in the "eternal East with its vast plains and dark soil, lazy mammoth rivers and intense dances." The dark lanes and alleys corresponded to and symbolized the alleged low cultural standard of Eastern European Jews: "These figures! Is this Berlin? Wearers of kaftans and stout women ... ragged children with expectant eyes, sneaky men who, timidly and in ecstasy, sway sparkling stones in their hands."[15]

Grabowsky's piece exemplifies how the visible presence of Eastern European Jews in Berlin "reinforced the Germanness of German Jews and strengthened the stereotype of the *Ostjude*."[16] The exaggerated description of the Berlin ghetto and the presentation of the Eastern Jew as dirty, uncultured, and uneducated were not far removed from anti-Semitic writings depicting Eastern European Jews as a serious danger to society.[17] As early as 1906, the German national daily newspaper

Deutsche Zeitung had discovered Grenadierstrasse as a place where poverty, vice, and crime went hand in hand.[18]

Galician-born writer Salomon Dembitzer took a strongly contrasting view in his essay collection *Aus engen Gassen,* published in 1915.[19] Like the article in *Deutsche Zeitung,* he turned his attention solely to Grenadierstrasse, depicting kosher restaurants and hotels alternating in the streets with Jewish-owned grocery stores, butchers, and junk shops. Bakeries sold Galician specialities, and a bookshop provided the residents with prayer books, fiction, and the latest trashy novels. This infrastructure included everything obligatory in a Jewish shtetl and turned the street into a Jewish space. Many of the Jewish residents worked all day in the better-off parts of Berlin and thus had frequent contact with big-city life; Grenadierstrasse, by contrast, appeared set apart from the metropolis and "formed a small town of its own, with its sufferings, joys and hopes, with its own language, clothing, habits and customs and is in no way related to big booming Berlin."[20] For Dembitzer, who grew up in the village of Lancut near Krakow, Grenadierstrasse was a place of melancholy and nostalgia that he was both attracted to and repelled by: attracted to the familiar atmosphere, which reminded him of his home village and helped to soothe his feelings of loneliness and abandonment; repelled by the bitter poverty and desolation among the Galician, Russian, and Romanian Jews.

Variations of the Ghetto Motifs and Contested Images of Jewishness, 1918–38

The foregoing prewar constructions all clearly distinguish between big-city Berlin and the small-town Scheunenviertel. This opposition made it easy to represent Eastern European Jews as a group closed in on itself, with little or nothing in common with German culture and modern city life. At the onset of World War I, perceptions of the area changed, and not for the better. The war ushered in growing fear of waves of Jewish immigrants from the dissolving Habsburg monarchy. As early as September 1915, some Prussian authorities considered the influx of Galician Jews to Berlin as an "unwanted increase of the destitute city proletariat."[21] The equation of the quarter's Eastern European population with criminal circles soon became a key notion in public debate, wherein the district was considered their "nesting place."[22] This transformed public discourse and perceptions about the Scheunenviertel from being an enclosed, detached space to an integral part of Berlin, where the desolation and decline of German society were particularly evident.

After the collapse of the Russian and Habsburg Empires and the end of World War I, Berlin quickly became the first place of refuge for tens of thousands of Jews fleeing persecution, pogroms, and the civil war in Russia. A petition submitted in February 1920 by Social Democratic police commissioner Eugen

Ernst to the Prussian minister of the interior exemplifies how widespread anti-Semitic clichés were linked to the growing number of Jewish refugees and migrants temporarily living in the Scheunenviertel. Ernst saw the district as overrun by "masses of dishonest characters"[23] and claimed that the migrants hoarded food purchased by illicit trading in their "filthy homes" and that their "habitual uncleanliness" threatened public health. Jewish migrants, he continued, were also extremely dangerous in political terms for their propagation of bolshevist ideas. In his nightmare scenario, which appeared in the major daily newspapers, fear of foreign infiltration and social decline merged with prewar images of this Jewish quarter. This negative picture played into the hands of radical anti-Semites, who unleashed invective aimed primarily at this district concerning the alleged "*Verjudung*" of Berlin.[24]

In the immediate postwar years, the central issue was not whether and how the Scheunenviertel could be regarded as a Jewish space but how negative public perceptions of it affected the status and self-positioning of German and Eastern European Jews in Weimar Berlin. Therefore, the quarter was crucial in controversies about secular and religious conceptions of Jewishness that unfolded in Berlin's Jewish community during the 1920s. Besides Albert Einstein, writer Joseph Roth was a well-known figure who publicly took Jewish refugees' side. In October 1920, Roth portrayed the district as a space of transit where Eastern European Jews could find initial refuge after the traumatic experiences of pogroms and military conflict in their homelands.[25] Roth refuted the equation of criminals with Jews there and sought to draw public attention to the desperate situation of the destitute migrants.

In a lucid essay, Zionist publicist Heinrich Loewe analyzed the image of Grenadierstrasse as a hotbed of criminal activity and pointed to its specific function in relation to domestic policy: by conducting numerous police checks and raids and making a large number of arrests there, the new government was easily able to present itself as tough on crime without having to fully face social and economic issues of postwar German society.[26]

The migration of tens of thousands of Jews from Eastern Europe through Berlin considerably changed the Scheunenviertel, not least for its established Jewish residents. Meir Gontser, the owner of a bookshop on Grenadierstrasse who had moved from Lithuania to Berlin around 1900, initially welcomed the "increasing Eastern Jewish street life,"[27] feeling that postwar Grenadierstrasse had become a "dry and sandy coastline where thousands of Jewish souls had been washed up." The migrants, he wrote in an article for Berlin's first-ever Yiddish-language weekly, "had been cut off from their original source of life, they were thrown to this infertile place ... without any resources for their spiritual and material life. And if there prevail no happy conditions, it is only because of the general conditions of this beached life."[28] In 1920, just one year later, however, Gontser had changed his tune, heavily criticizing the busy street life of Grenadier- and Dragonerstrasse, with its numerous Jewish traders. Although he understood that the experiences of war and the "*galuth* [diaspora] shelters" had marked the lives of the refugees, he reminded migrants that their task was to "prov[e] themselves worthy of being a member of the Jewish family." He felt a need for a higher level of "European and Jewish culture."[29]

In contrast to Gontser, who argued for refugees to restrain themselves and adapt somewhat to German culture, Lodz-born shoemaker and journalist Robert Eulenfeld interpreted Gontser's critique as familiar slander against Galician Jews. Responding to Gontser, Eulenfeld presented himself as an advocate of Galician migrants against the unjustified attacks of a "Litvak," a Lithuanian, non-Hasidic Jew.[30] The Galician Jew, as Eulenfeld saw it, served as a whipping boy for Jews and Gentiles in Germany; Jewish intellectuals, especially, failed to recognize this figure's true significance on Grenadierstrasse. Without Galician Jews, Eulenfeld was convinced, there would be "no true Jewish life in Berlin"; they were "the Jewish Jews of Berlin, who observe the Sabbath and the Jewish dietary laws and preserve their Jewishness."[31] In this idealized reading, the district became a space of authentic and self-confident Jewishness, contrasting with the image of the acculturated liberal German Jew and recalling the positive stereotype of the Eastern Jew as a representative of authentic and genuine Judaism from the Jewish renaissance movement.[32]

One year later, Eulenfeld lamented that Berlin lacked a vibrant Eastern European Jewish culture. While religious Jews could continue their traditional way of life in the many prayer rooms in the Scheunenviertel, secular Jews had no place to develop their *Yiddishkeyt* (Jewishness).[33] In 1927, the newly founded labor union Perez-Farayn aimed to establish a secular Jewish school. Addressing themselves to the district's Jewish parents, union members described an impending state of uprooted and hopeless life in Berlin. Whereas migrants in Eastern Europe had been living "their own lives as one among many" and had been connected to Yiddish language and culture, they were now "thrown into the foreign cauldron of Berlin, where they cannot feel at home, and live a life under a gray yoke without 'today' and 'tomorrow'!"[34] This perception once again transformed the Scheunenviertel as a space, making it stand for big-city Berlin and Eastern European Jews' rootlessness within it. Far from "Jewish," the area was seen as a foreign urban space where Yiddish culture could not grow and flourish. Berlin was perceived as a solid place that allowed migrants to become part of its culture only at the cost of renouncing their cultural identity.

After the crisis years following World War I, public attention turned away from the Jewish quarter. Although the district had been subject to several construction projects and urban integration schemes, it was never systematically redeveloped until the late 1920s, when the inadequacy of housing conditions and the continued social deprivation elicited renewed discussion around the neighborhood's character.

In 1929, Social Democrat Georg Davidsohn emphasized the significant changes the district had undergone since the early postwar years.[35] He was convinced that its bad postwar reputation was unjustified. Also, he refuted the idea that it was then or had recently been a Jewish ghetto in the strict sense of the word. For Davidsohn, the quarter was foremost a proletarian area: the street trading in the western part of Grenadierstrasse contributed to its "original Berlin character." Although Jewish newcomers could still find reminders of home in Grenadierstrasse, many "modern Jewish East-West children" could be found playing in the streets, whose "language," Davidsohn noted, was a mix "between Yiddish and Berlin jargon."

In contrast to Davidsohn's article, which appeared in *Israelitisches Familienblatt* and focused on Jewish migrants' modernization and acculturation in the area, the Communist-sympathizing periodical *Arbeiter-Illustrierte-Zeitung* portrayed a dying centrally located urban district that failed to meet the necessities of modern infrastructure.[36] As the *Arbeiter-Illustrierte* saw matters, the district's old buildings would soon be replaced with shiny new, modern apartment buildings, but there would be no room for the original inhabitants, the traders, the unemployed, and Jewish proletarians in them; poorer residents would be forced to find new homes in other *Elendsviertel*. In this view, the Eastern European Jews formed an integral part of Berlin's lumpenproletariat and were no longer considered a separate, excluded group.

Similarly, the Communist newspaper *Die rote Fahne* portrayed the quarter's Jewish inhabitants as "the poorest of the poor," "languishing in the deepest misery of this capitalist society … [and] slaves of a medieval retardation,"[37] a perceived worldview that put them in a much more desperate position than their non-Jewish counterparts. Here, the Scheunenviertel served as proof of the perfidy and injustice of the capitalist system and underlined the need for communist education among the workers.

We should read both articles against the backdrop of increasing animosity between communist and fascist groups in late Weimar Berlin, accompanied by intense agitation and propaganda among Berlin's workers.[38] Whereas communists defended the neighborhood's poverty-stricken Jews from the "murderous anti-Semitism" of the "fascist scoundrels"—with whom, however, they shared prejudices against "Jewish bourgeois capitalists"—right-wing and nationalist groups viewed the district as a "hotbed of German citizens of the Jewish faith whose sons [were to] become first-class promoters of German culture as lawyers, judges, editors-in-chief, etc."[39] The Scheunenviertel of anti-Semitic propaganda harbored Jewish criminals, constituting a serious threat to German society. It was a significant space in the anti-Semitic battle against an alleged "Jewish race of criminals" because the anti-Semitic repertoire of images of the Jew as a criminal, revolutionary, and swindler could easily be applied to it.

Moreover, the Scheunenviertel was the metaphorical site of the transformation of "the Jew" in German society. Anti-Semitic propaganda utilized the image of external transformation from the visible, traditionally clothed Eastern European Jew into the invisible, acculturated middle-class German Jew to create a particular sense of threat. Thus, it was no coincidence that an early Nazi anti-Jewish measure targeted the district and its Jewish residents: in 1933, in connection with the state-managed boycott of Jewish-owned shops in April, the Berlin police, supported by SA troops, conducted a so-called *Sonderrazzia* (special raid) there. The Nazi propaganda organ *Berliner Beobachter* welcomed the arrest of many Eastern European Jews as the beginning of the end of the "Galician worms in German Berlin."[40] The destruction of the Jewish quarter, more or less explicitly pre-announced in 1933, became reality only a few years later; in 1938 Polish Jews were expelled from Berlin and the German Reich, and in 1941 deportation of the area's remaining Jewish inhabitants to the concentration camps in the East commenced.

Memories and Fantasies: Imagining the Jewish Scheunenviertel after the Shoah

After 1945, few traces of the formerly vibrant Jewish quarter could be found in its physical structure. During World War II, individual streets had suffered severe destruction. The few reconstruction measures between 1950 and 1990 were conducted with great ignorance and apathy toward its history. The identification of former Jewish places became even more complicated when the Berlin Magistrat renamed Grenadier- and Dragonerstrasse Almstadtstrasse and Max-Beer-Strasse in 1951 and renumbered their houses.

Although the quarter's Jewish history was well known among old-established Berliners, it did not attract much attention in postwar Berlin before the late 1970s. The division of the Jewish community in Berlin into eastern and western sections in 1955, the building of the Berlin Wall in 1961, and the socialist doctrine of East Germany, which focused exclusively on anti-fascist socialist figures and largely ignored the fate of Jews, Sinti and Roma, and other victims of Nazi persecution, also contributed to widespread ignorance in East Berlin of the area's Jewish history.

Only when the volume *Im Scheunenviertel* by West Berlin–based journalist and essayist Eike Geisel was published in 1981 did the quarter's Jewish history return somewhat to the collective memory of East and West Berlin. Using photographs and writings from the 1910s to the 1970s, Geisel retraces the Jewish Scheunenviertel's significance against the background of contemporary German and German-Jewish history.

In his introduction, Geisel stresses that the search for historical traces of former Jewish life in the quarter and the approach to historiography focusing on documenting "history from below" are "woefully inadequate ways to commemorate the *Scheunenviertel*'s murdered Jews."[41] For Geisel, the quarter is not a place where remnants of Jewish life can be traced, but rather a neighborhood that exemplifies German xenophobia and anti-Semitism and the inability of both German states to successfully fight these attitudes.[42] The historical place, Geisel claims, is gone: "Selective nostalgia and extensive optimism rise from the ruins; these are the hallmarks of a new obliviousness, where the [area's] history and its dreadfulness vanishes."[43]

Whereas Geisel argues that the physical remnants—houses, walls, and stones—cannot speak and that contemporary photographs and literary texts represent only some of the various perspectives on the Jewish Scheunenviertel, the memory culture of the 1990s was shaped by the belief that "authentic locations" encouraged people to feel a closer connection to history.[44] The rediscovery of the Jewish quarter thus was intimately linked to contemporary discourse on Berlin as the new capital of reunified Germany and about Jews' place within it.[45] As Katharina Gerstenberger summarized, "Jewish Berlin began to play an important role in Germany's claim to normalcy, which required both acknowledging guilt and bringing the process of atonement into the amalgam of post-Wall identity."[46] The Scheunenviertel being *the* historic Jewish district was central to the dispute about how and to what ends Berlin's Jewish history should be linked to the city's contemporary identities and images.

In 1992–93, visual artist Shimon Attie projected historic photographs of pre-Holocaust Jewish street life onto the district's actual or approximate addresses where the events had taken place. Attie, a child of Holocaust survivors, hoped that externalizing his private acts of remembrance, in which he alone saw the faces and forms of now-absent Jews, would make them part of the memory of a larger public. Once seen, these projections were to "always haunt these sites by haunting those" who had seen them.[47] With his "act of remembrance," Attie exposed the gulf between what happened in the past and how it is now remembered. He used iconic photographs of district *Ostjuden* as simulations, not as historical reconstructions.[48] In doing so, he showed not precisely what was lost but rather that loss itself is integral to the quarter's history.

Several other groups and initiatives, by contrast, engaged in remembering and documenting Jewish life in the Scheunenviertel as it actually was before the Holocaust. In 1993, a group of artists founded the Hackesches Hoftheater to maintain and propagate Yiddish culture with theatrical performances and concerts at the historical site.[49] The Betroffenenvertretung Spandauer Vorstadt, a neighborhood group representing residents' interests in the face of the restoration and gentrification taking place across this part of Berlin-Mitte, sought to establish commonalities between its former and present residents. For this group, portraying the district as "a space destroyed but not contaminated by the Nazis" created "a tradition of defiance, and a model of how to stand firm against the gentrification of Eastern Berlin."[50]

While the Betroffenenvertretung drew a clear line from the outsider status of Jewish migrants in Imperial and Weimar Germany to its members' current status, local historians and local history buffs carried out several projects exploring, describing, and illuminating the realities of Jewish life in the district before World War II. The most influential publication, *Im Scheunenviertel. Spuren eines verlorenen Berlin,* compiled academic essays, historical sources, and personal testimonies focusing on the quarter's peculiarities and "exotic" character.[51] Here, the Scheunenviertel appears as an ordinary neighborhood with people trying to survive, where only its cultural diversity set it apart from the rest of the Prussian capital. In this reading, Eastern European Jews, Gentile street traders, students, artists, procurers, and petty criminals were all united by the "spirit of the Scheunenviertel,"[52] not manifested in political or organized opposition to Prussian or Berlin authorities, but in "simple ignorance of the rulers and their orders." This depiction of the Scheunenviertel as a distinct island village in big-city Berlin was closely tied to contemporary efforts to preserve urban diversity in reunified Berlin and a widespread fear, especially among East Berliners, of the capitalist marketing and selling of historic Alt-Berlin.

Although initially intended as a counternarrative, the image of the multicultural Scheunenviertel converged with Gerhard Schröder's political vision, when he assumed the chancellorship in 1998, of reunified Germany as a *Republik der Neuen Mitte* (republic of a new center) with Berlin at its core.[53] The Neue Mitte took material form in the renewed city district around Hackescher Markt. In the marketing for the new historic center of Berlin, the Scheunenviertel's multicultural image was detached from its original historical location and applied to larger parts of the Spandauer Vorstadt, especially the nearby Hackesche Höfe and the area around Oranienburger Strasse, a former center of (German-)Jewish community life in Berlin. Consequently, the mainly invisible history of German Jews in Berlin's center was conflated with the exotic image of the Jewish Scheunenviertel, whereas the district's historical location remained unmarked, not becoming a memorial space.

At first glance, the relationship between the Jewish Scheunenviertel's real, historic space and its imagined space seems to be dissolving. This might lead us to conclude that the imagined space does not need a real location to exist. But actually, the reverse is true. The migration of the image of the Jewish district to the area around Oranienburger Strasse, where one can still find Jewish markers in the urban landscape, shows that the imagined space needs at least some physical reference points. These, however, do not have to be historically correct but must merely fill a certain need. In this case, marking an urban space as Jewish supported Germany's desire for normalcy and the creation of a post-Wall identity.

Twentieth-Century Incarnations of the Jewish Scheunenviertel

In the course of the twentieth century, the Scheunenviertel was crucial both in the perception of Jews in Germany and in more specific debates about the Jewish minority's overall status. These debates were intrinsically linked to contemporary discourses on modernity and urbanism, often progressing along binary lines such as modern/backward, big-city/small-town, familiar/foreign, and genuine/artificial. Ascribing Jewishness to the district's space had a range of functions and meanings closely connected to those who did so and their status. Since World War I, a clear distinction had prevailed between big-city Berlin and the small-town Jewish quarter, which was perceived either as backward or a reminder of home for its refugee inhabitants. During the Weimar years, the district increasingly became an integral part of the metropolis, "an in-between space where the city's two faces—that of a rapidly modernising, hectic metropolis and that of the urban slum—came into close proximity and partially overlapped."[54] Eastern European Jews were just one part, but a distinct part, of the city's culture.

The rediscovery of the Scheunenviertel in the early 1990s was shaped by various quests to explore and remember Jewish life and culture in German history and to identify the place of Jews in a new Germany. In the memory culture of the 1990s, locals and non-Jews, primarily, searched for authentic places and physical markers that could restore history to consciousness and integrate the past into the fabric of contemporary identity. Whereas some people have criticized xenophobic and anti-Semitic tendencies in German society as part of a German history of forgetting and ignoring, others have been interested in reappraising the Nazi past or in acts of remembrance. Still others incorporated the Jewish history of the quarter into their own counternarrative against the political and economic elite in reunified Germany. Interestingly, the mainstream trend of this culture of memory has primarily referenced the district's image from around the year 1900 as a small-town Jewish shtetl crowded with orthodox Jews. This trend is observable in other European countries, too. Due to the almost complete destruction of Eastern European Jewish life and culture, this highly idealized image of Eastern European Jewry as a premodern, intact space of religious unity is deeply anchored in collective European cultures of memory.[55] The clichéd prewar imaginative landscape has merged with the vibrant culture of the "Golden Twenties" into the powerful image of a district of outsiders representing the "other" democratic and anti-fascist Germany, offering a positive history post-Wall Berliners can identify with.

This strong desire to mark historical places as Jewish and to include parts of Jewish history in the urban narrative corresponded with two key terms of 1990s discourse—diversity and multiculturalism—both of which German politicians and some intellectual elites cited as proof of Germany's new cosmopolitanism. That the image of the (Eastern European) Jewish Scheunenviertel became detached from its historical location and was extended and transferred to the former center of the Jewish community in Berlin's Mitte district hints at a crucial ambivalence in Germany's culture of memory during the 1990s and at other times, wherein German-Jewish history is appropriated by the exoticization and simultaneous disappearance of diverse and complex Jewish life and history, even as it loses its point of anchorage in the city.

Anne-Christin Saß was a research associate for the Chair of Eastern and Central European History at the Institute for Eastern Europe at the Free University of Berlin and a Feodor Lynen Fellow at Wolfson College, Cambridge, UK. Her current areas of research are modern Jewish and European history and history of migration. She has written *Berliner Luftmenschen. Osteuropäisch-jüdische Migranten in der Weimarer Republic* (Göttingen, 2012), and edited Fishl Schneersohn's *Grenadierstraße. Roman jüdischen Lebens in Deutschland* (Göttingen, 2012).

Notes

I gratefully acknowledge the Alexander von Humboldt Foundation for granting me a Feodor Lynen Fellowship in 2014, making this research possible.

1. Metzler, *Tales of Three Cities*, 198.
2. See, e.g., Gerstenberger, *Writing the New Berlin*, 77–108; and Gilman and Remmler, *Reemerging*.
3. Meier, "Ohne Pudelmütze und allein," 91.
4. For a detailed analysis of "real" Jewish urban spaces in the Scheunenviertel, see Saß, "Transnational and Transcultural Spaces."
5. This is particularly evident in recent publications on Jewish urban spaces such as Brauch, Lipphardt, and Nocke, *Jewish Topographies*; Kümper et al., *Makom*. From the rich literature on remembrance and memory cultures, see, for a particular focus on Berlin, Jordan, *Structures of Memory*.
6. Several works address different aspects of the quarter's history. For a bibliographical overview see Helas, *Grenadierstraße*; Saß, *Berliner Luftmenschen*. The memory culture is, among others, covered by Gerstenberger, *Writing the New Berlin*, and Waligórska, *Klezmer's Afterlife*.
7. For a short overview of the history of the area, see Saß, "Scheunenviertel."
8. See the extract from the petition quoted by Helas, *Grenadierstraße*, 18.
9. Ostwald, *Dunkle Winkel in Berlin* (vol. 1 of the series *Großstadtdokumente*), 37–47. The *Großstadtdokumente*, published in 50 volumes from 1904 to 1908, were inspired by the journalistic, literary and sociological urban research at the turn of the century.
10. Ostwald, *Dunkle Winkel*, 41.
11. E. Seiffert [before 1914], "Aus einem Weltstadtwinkel," in Verein Stiftung Scheunenviertel, *Das Scheunenviertel. Spuren eines verlorenen Berlin*, 45–46.
12. H. Ermy, "Das Berliner Ghetto," *Israelitisches Familienblatt*, 14 September 1911.
13. On the image and idea of the term *Ostjude*, see Brenner, *Jüdische Kultur*, 145–69. Of continued value is the classic work by Aschheim, *Brothers and Strangers*. For a recent overview, see Saß "Ostjuden."
14. A. Grabowsky, "Ghettowanderung," *Die Schaubühne*, 3 February 1910, 124–26.
15. Ibid., 125.
16. Aschheim, *Brothers and Strangers*, 45.
17. See Heid, "Achtzehntes Bild: Der Ostjude."
18. "Im Ghetto von Berlin," *Deutsche Zeitung*, reprinted 1 March 1907 in *Jüdische Rundschau*, 88–89.
19. Dembitzer, *Aus engen Gassen*.
20. Ibid., 38.
21. Regierungsrat Dr. Doyé to the Police Commissioner of Berlin, 18 September 1915, Geheimes Staatsarchiv Preußischer Kulturbesitz Berlin, 1. HA Rep. 77 Tit. 1176 Nr. 14.
22. Oberkommando in den Marken to the Minister des Inneren, 16 April 1918, Geheimes Staatsarchiv Preußischer Kulturbesitz Berlin, Rep. 77 (M) Abt. I Sekt. 33 Tit 226b Nr. 66, Bd. 1, 12.
23. "Die Schieber im Scheunenviertel," *Berliner Tageblatt*, 18 February 1920.
24. See Schlör, *Das Ich der Stadt*, 226–41.
25. J. Roth, "Flüchtlinge aus dem Osten," *Neue Berliner Zeitung*, 20 October 1920.

26. Assad [pseudonym of H. Loewe], "Die Grenadierstraße," *Israelitisches Gemeindeblatt,* 12 March 1920, 1–2.
27. M. Gontser, "Klaynes Foyleton" [A small feature], *Der Freytog,* 19 September 1919.
28. Ibid.
29. M. Gontser, "Brivlekh tsum liben lezer," *Der Mizrekh-Yid,* 24 September 1920.
30. L. Shpigel [pseudonym of Robert Eulenfeld], "Galitsianer," *Der Mizrekh-Yid,* 25 February 1921.
31. Ibid.
32. See Brenner, *Jüdische Kultur,* 145–69.
33. L. Shpigel, "Mikoyekh a yudishen teater in Berlin Foyleton" [About a Yiddish theater in Berlin], *Der Mizrekh-Yid,* 1 July 1921.
34. Kultur-komisie baym "Perets"-farayn. [without date] Yidishe tate-mames in Berlin [Cultural commission of the workers' union "Perets." To the Jewish parents in Berlin!], International Institute for Social History Amsterdam, Bund Archive, Folder 298.
35. G. Davidsohn, "Grenadierstraße 1929," *Israelitisches Familienblatt,* 12 September 1929.
36. "Im Berliner Scheunenviertel," *Arbeiter-Illustrierte-Zeitung* (1929), 23.
37. "Die Blutsauger des deutschen Volkes im Scheunenviertel," *Die Rote Fahne,* 19 September 1929.
38. On the street fights between communist and fascist groups in Weimar Berlin, see Schumann, *Politische Gewalt.*
39. "Jüdisches Verbrecherzentrum," *Hammer. Blätter für deutschen Sinn* 26 (1927): 593, 136.
40. H. Seehofer, "Groß-Razzia im Scheunenviertel," *Berliner Beobachter. Tägliches Beiblatt zum Völkischen Beobachter,* 5 April 1933.
41. Gerstenberger, *Writing the New Berlin,* 82.
42. Geisel, *Im Scheunenviertel,* 10–32.
43. Ibid., 10.
44. Gruber, *Virtually Jewish.* A more positive approach to the new authenticities of Jewish life has recently been provided by Waligórska, *Klezmer's Afterlife.*
45. For a critical overview, see Peck, *Being Jewish.*
46. Gerstenberger, *Writing the New Berlin,* 77.
47. Young, *At Memory's Edge,* 64.
48. For a short discussion of the photographs and the intentions of the photographer in the Scheunenviertel, see Pilarczyk, "'Ostjuden' im Scheunenviertel," 65–69.
49. The Hackesches Hoftheater was closed in 2006, but there are still concerts and stage performances held at various locations in Berlin. For the theater's self-description, see the website: http://www.hackesches-hoftheater.de (accessed 4 November 2014).
50. Gerstenberger, *Writing the New Berlin,* 84–85.
51. Verein Stiftung Scheunenviertel, ed., *Das Scheunenviertel. Spuren eines verlorenen Berlin.* Some years later, some of the local historians, journalists, and teachers involved in the original project published a much more nuanced local history of the Mitte district, which did not receive similar levels of public attention. Helas, *Juden in Berlin-Mitte.*
52. Weigert, "Scheunenviertel und Großstadt," in Verein Stiftung Scheunenviertel, *Das Scheunenviertel,* 7–11, here 9.
53. Regierungserklärung Gerhard Schröder, Deutscher Bundestag 14. Wahlperiode, Bonn, 3. Sitzung 1998, 62.
54. Metzler, *Tales of Three Cities,* 196.
55. See Gruber, *Virtually Jewish.*

Bibliography

Aschheim, Steven E. *Brothers and Strangers: The East European Jew in German and German-Jewish Consciousness.* Madison: University of Wisconsin, 1982.

Brauch, Julia, Anna Lipphardt, and Alexandra Nocke, eds. *Jewish Topographies: Visions of Space, Traditions of Place.* Aldershot: Ashgate, 2008.

Brenner, Michael. *Jüdische Kultur in der Weimarer Republik.* Munich: Beck, 2000.

Dembitzer, Salamon. *Aus engen Gassen.* Berlin: Schwetschke, 1915.

Geisel, Eike. *Im Scheunenviertel. Bilder, Texte und Dokumente.* Berlin: Severin und Siedler, 1981.

Gerstenberger, Katharina. *Writing the New Berlin: The German Capital in Post-Wall Literature.* Rochester, NY: Camden House, 2008.

Gilman, Sander L., and Karen Remmler, eds. *Reemerging Jewish Culture in Germany: Life and Literature since 1989.* New York: University Press, 1994.

Gruber, Ruth E. *Virtually Jewish: Reinventing Jewish Culture in Europe.* Berkeley: University of California, 2002.

Heid, Ludger. "Achtzehntes Bild: Der Ostjude." In *Bilder der Judenfeindschaft. Antisemitismus, Vorurteile und Mythen,* edited by Julius H. Schoeps and Joachim Schlör, 241–51. Augsburg: Bechtermünz, 1999.

Helas, Horst. *Die Grenadierstraße im Berliner Scheunenviertel. Ein Ghetto mit offenen Toren.* Berlin: Hentrich & Hentrich, 2010.

———. *Juden in Berlin-Mitte. Biografien—Orte—Begegnungen.* Berlin: Trafo, 2000.

Jordan, Jennifer A. *Structures of Memory: Understanding Urban Change in Berlin and Beyond.* Stanford, CA: Stanford University Press, 2006.

Kümper, Michal, Barbara Rösch, Ulrike Schneider, and Helen Thein, eds. *Makom. Orte und Räume im Judentum.* Hildesheim: G. Olms, 2007.

Meier, André. "'Ohne Pudelmütze und allein'—Gedanken eines Passanten." In *Die Hackeschen Höfe. Geschichte und Geschichten einer Lebenswelt in der Mitte Berlins,* edited by Peter Schubert, 87–91. Berlin: Argon, 1993.

Metzler, Tobias. *Tales of Three Cities: Urban Jewish Cultures in London, Berlin and Paris (c. 1880–1940).* Wiesbaden: Harrassowitz, 2014.

Ostwald, Hans. *Dunkle Winkel in Berlin.* Berlin: Seemann, 1904.

Peck, Jeffrey M. *Being Jewish in the New Germany.* New Brunswick, NJ: Rutgers University, 2006.

Pilarczyk, Ulrike. "'Ostjuden' im Scheunenviertel. Eine bildanalytische Recherche." In *Berlin Transit. Jüdische Migranten aus Osteuropa in den 1920er Jahren,* edited by Stiftung Jüdisches Museum, 65–69. Göttingen: Wallstein, 2012.

Saß, Anne-Christin. *Berliner Luftmenschen. Osteuropäisch-jüdische Migranten in der Weimarer Republik.* Göttingen: Wallstein, 2012.

———. "Ostjuden." In *Enzyklopädie jüdischer Geschichte und Kultur,* vol. 4, edited by D. Diner, 459–64. Stuttgart: Metzler, 2013.

———. "Scheunenviertel." In *Enzyklopädie jüdischer Geschichte und Kultur,* vol. 5, edited by D. Diner, 352–58. Stuttgart: Metzler, 2014.

———. "Transnational and Transcultural spaces in the Diaspora: The Case of Berlin 1900–1933." In *Jewish and Non-Jewish Spaces in Urban Context,* edited by A. Gromova, F. Heinert, and S. Voigt. Berlin: Neofelis, 2015.

Schlör, Joachim. *Das Ich der Stadt. Debatten über Judentum und Urbanität 1822–1932.* Göttingen: Vandenhoeck & Ruprecht, 2005.

Schumann, Dirk. *Politische Gewalt in der Weimarer Republik, 1918–1933.* Essen: Klartext, 2001.

Verein Stiftung Scheunenviertel, ed. *Das Scheunenviertel. Spuren eines verlorenen Berlin.* Berlin: Haude & Spener, 1994.

Waligórska, Magdalena. 2013. *Klezmer's Afterlife: An Ethnography of the Jewish Music Revival in Poland and Germany.* New York: Oxford University Press.

Young, James E. 2000. *At Memory's Edge: After-Images of the Holocaust in Contemporary Art and Architecture.* New Haven, CT: Yale University Press.

PART III

Practices

Negotiating, Experiencing,
and Appropriating Spaces and Boundaries

12

A Hybrid Space of Knowledge and Communication

Hebrew Printing in Jessnitz, 1718–1745

Dirk Sadowski

Jewish print shops or printing presses—*bet ha-defus* or *makhbesh ha-defus*—represent one type of Jewish space in existence in Europe in the early modern period, alongside the institutions run by autonomous Jewish communities. Hebrew print shops were not necessarily constituent parts of "Jewish space" as defined in a ritual or administrative sense; they were not even owned by Jews in all cases—a fact to which the famed examples of Daniel Bomberg (c. 1475–1549) in Venice and Ambrosius Froben (1537–95) in Basel bear witness.[1] It is true that they almost always required a Jewish readership as a market for their products, a community that needed prayer books for synagogue services and halachic literature for learning and interpreting and applying the law and that enjoyed reading printed Yiddish literature for purposes of recreation and edification. Nevertheless, some learned Christians also read printed books in Hebrew. Running a Hebrew print shop called for staff with the right training, literacy, and technical skills of type-setters, printers, and proofreaders. While it might have been sufficient for those working the presses to know very little Hebrew—to perhaps be Christian with thorough training in the book printing trade[2]—typesetters and especially proof-readers needed a distinct level of "Jewish" knowledge. Many Christian printers had some Hebraist education yet often employed or partnered with a Jew or a Jewish convert in order to have access to a thorough knowledge of Jewish literature and thus be fully aware of changing demand on the market.

Historian Elizabeth L. Eisenstein, who extensively studied the profound processes of change that accompanied the emergence of the printing press, holds that the early modern print shop represented not only a space of commercial calculation and technical skill but also of intellectual discussion and the communication of ideas; she believes that print shop owners, as businessmen, actively contributed to the creation of the early modern canon of knowledge.[3] Undoubtedly, this also applies to Jewish book printing in early modern Europe. It is almost impossible for us to overestimate the significance of Hebrew book printing, and the communication of knowledge and ideas that took place in and around print shops, to the development of the Jewish body of knowledge from the sixteenth century to the Enlightenment; they represented what we might refer to as a hybrid space of communication, a space for the negotiation of knowledge from their actors' "own" and "other" spheres of reference.

As Amnon Raz-Krakotzkin demonstrated for the sixteenth century, the constant impact of the church's censorship of Hebrew works effected slow yet enduring change in the extant corpus of printed Jewish knowledge. As an element of "Hebraist discourse," censorship brought together the censor, who was usually a convert, the Christian printer with Hebraist interests, and Jewish editors and proofreaders; this diverse collection of individuals negotiated what was to be removed from or remain in the text, thus giving it its final form. In this process, the transformation and preservation of tradition went hand in hand.[4] Although the censorship discourse lost some of its normative power over time and, particularly in Germany, was drowned out and somewhat edged from the scene by territorial nobles' commercial interests, elements of it continued to have an impact in the mid-eighteenth century and beyond. Nonetheless, learned discourse moved increasingly into the foreground, spurred by a growing readership of theologians and philologists at universities and in church institutions who had Hebraist research interests—albeit not entirely free of missionary and indeed repressive undertones.[5]

Having identified Hebrew book printing as a hybrid space of knowledge and communication, in this chapter we will examine the exemplary case of Jessnitz, a town in central Germany, whose printing activities, despite their small scale, were immensely significant to the transformation of the Jewish body of knowledge prior to the Jewish Enlightenment. Focusing primarily on the extraordinary character of the knowledge in the publications issued there and then asking about commercial, censorial, and other "external" factors of Hebrew printing, as well as about Jewish-Christian communication and cooperation in this special case, we will finally identify the "mingled identity"[6] of the printer himself as the key factor leading to the accumulation of printed works in the so-called external disciplines in that town during the first half of the eighteenth century.

Israel bar Avraham and His Hebrew Printing Press

In 1742, Maimonides's *Guide for the Perplexed* (*Moreh Nevukhim*) was reprinted in Jessnitz, a small town on the banks of the river Mulde, some fifteen miles away from Dessau in the principality of Anhalt. This reprint is widely acknowledged to have been of the utmost importance to the intellectual awakening of European Jewry in the eighteenth century. "A literary event of the first order,"[7] the printing of the guide made this key work of medieval philosophy much more accessible. In a well-known story, for example, twelve- or thirteen-year-old Moses Mendelssohn read it in an ecstatic state of fear and fascination and thus became acquainted with the world of medieval Jewish, that is, Aristotelian, philosophy. The last edition of the *Moreh Nevukhim* printed before the Jessnitz edition appeared in 1553 in Sabbioneta, Italy; prior to this, this key work of medieval philosophy had been printed only twice, in Venice in 1550 and previously—as one of the very first printed Hebrew books—in Rome in 1480.[8] Both Jews and Christian Hebraists who wanted to study it in Hebrew before 1742 either had to resort to one of the rare Italian editions or read one of the handwritten copies in circulation.

Even though Maimonides's philosophical magnum opus had been studied occasionally in some Central European yeshivot during the sixteenth century, traditional suspicion of the *Guide* still lingered up to Moses Mendelssohn's time.[9] The question of whether the study of Greek metaphysics in a Jewish guise should be permitted had divided Maimonides's adherents and opponents since the rabbinic ban pronounced in Barcelona in 1305 had forbidden the study of Aristotelian philosophy, and hence the *Moreh Nevukhim,* before the age of twenty-five. Mendelssohn himself, when editing Maimonides's *Logic* (*Milot ha-Higayon*) in 1765, still included an apologetic preface, stating that the Greek philosophy incorporated in it was not really "Greek" but rather the product of one of the most Jewish thinkers of all time.[10]

What printer dared to publish this controversial book in 1742, and why was this most important work of medieval Jewish philosophy printed in a provincial backwater of Germany? Diverse structural and economic factors, alongside personal matters relating to the printer, led to the establishment of a Hebrew printing press in Jessnitz. In 1694, Moses Benjamin Wulff (1661–1729), a court Jew, was granted a license to run a press and to print in Hebrew characters "the Holy scriptures of the Old Testament and various other books and prayer books which are common among the Jews and admissible under the rule of the Christian princes."[11] It began operations in 1697, with Wulff himself operating the press in Dessau until 1704, whereupon it was sold or lent to a printer called Moses ben Avraham, who printed in Halle until 1714.[12] In 1717, a man named Israel bar Avraham purchased the printing equipment, including the famous *otiyot Amsterdam,* Hebrew letters cast in the Dutch capital, which were highly valued among printers.[13] It seems that Israel bar Avraham himself had come to Germany from Amsterdam, where he had learned the craft of printing. Like his predecessor Moses ben Avraham, Israel was a convert to Judaism, most likely a former clergyman or monk. When he applied to the Prussian authorities in 1739 for a printer's license in Halberstadt, Prussian minister Dietloff von Arnim (1679–1753) declined the request, arguing that there was evidence that "the entrepreneur is a born Christian of the Catholic religion and a former Capuchin" and that the king should refrain from granting residence to such persons, who were reputed to be the most acrimonious blasphemers of the name of Christ.[14]

The story of Israel bar Avraham's printing activities starts in the year 1717, when he established his press in the town of Koethen, the same year Johann Sebastian Bach became the director of music there in the court of Leopold, prince of Anhalt-Koethen. In 1717 and 1718, Israel bar Avraham printed four books in Koethen before leaving the town. After a short-lived and unsuccessful attempt to establish his printing press in Blankenburg in the Harz mountains, he moved to Jessnitz. Unlike Koethen, Jessnitz was home to a small Jewish community; this provided the necessary religious infrastructure for Jewish typesetters, printers, and proofreaders, who found work there for short periods. Workers who had learned their craft in Prague or Kraków, in Dyhernfurth in Silesia, or in Wilhermsdorf in Franconia traveled around the Holy Roman Empire searching for a place to stay for a while and work in one of the Hebrew printing offices. Israel bar Avraham employed sixteen such workers during the first period he spent printing in Jessnitz, between 1718 and 1726.[15] Two other factors favored Jessnitz when Israel bar Avraham had to leave Koethen and resettle his press in 1718: its greater proximity to Leipzig as a center of the book trade and the paper mill at the river Mulde, which was established in 1675.[16]

In this mercantile age, Hebrew printing was considered a source of income for the treasury of the absolutist state. In some cases, Hebrew printing presses were established to support the work of local paper mills, as in Wilhermsdorf and, we might assume, in Jessnitz.[17] In some instances, Christian Hebraists and mystics promoted the establishment of a Hebrew print shop in a certain territory, as happened in Sulzbach in Northern Bavaria.[18] From the Jewish side, strong demand for Hebrew books drove the establishment of Hebrew printing presses in Germany after the Cossack uprisings of the 1640s had not only devastated many Jewish communities in Eastern Europe but had also halted the Hebrew printing presses in Kraków, Lublin, and elsewhere. The final decades of the seventeenth and the opening decades of the eighteenth century saw the establishment of Hebrew presses in many places in Germany; the most significant were Wilhermsdorf and Sulzbach in 1669, Fürth in 1691, Berlin in 1697, Dyhernfurth in 1689, and Offenbach in 1714. The opening of the presses in Dessau in 1697 and in Jessnitz twenty years later formed part of this revival of Hebrew printing.[19]

Between 1718 and 1726, a total of forty books left the printing press of Israel bar Avraham in Jessnitz.[20] One-third were prayer books of all kinds—*siddurim, machzorim, tikkunim,* and Psalm books, sometimes with a Yiddish translation. The second largest part of its production consisted of Halakhic literature; biblical commentary and other midrashic or Masoretic works were another important genre. Thus, in 1720 Israel bar Avraham began printing the commentaries of the sixteenth-century Sephardic scholar Moses Alsheikh (1508–1600) on the Prophets and Writings, which gained great popularity within Ashkenazic Jewry in the eighteenth century.

At first glance, the repertory of the first Jessnitz period does not appear strikingly different from Hebrew book production as a whole in that era. One-third of all Hebrew books printed in Germany between 1670 and 1770 were works of liturgical literature, that is, prayer books and Pesach Haggadot. Halakhic literature accounted for another third, while Bibles and biblical commentaries constituted a somewhat smaller proportion.[21]

Some of the books printed by Israel bar Avraham, however, were distinctly unusual in the Hebrew and Yiddish literature printed in Central and Eastern Europe in the eighteenth century. There was a group of grammatical works written by contemporary scholars. The first of these, *Derekh ha-Kodesh,* had already been printed in 1717, during bar Avraham's Koethen period. A second, *Ohalei Yehuda,* was a dictionary of biblical terms compiled by Yehuda Arie Loeb of Carpentras, who, in his preface, reflected on the origins of the various languages of the world. In 1725, Israel bar Avraham printed two books by the grammarian Salomon Hanau (1676–1746), considered to be one of the early *maskilim* greatly concerned with the Hebrew language, who lamented its neglect and the widespread ignorance of its grammatical principles.[22] One, *Bet Tefilah,* was a prayer book, with the text revised according to grammatical logic. The second, *Sha'arei Tefilah,* was a philological commentary on the prayer book. Both were antithetical to the traditional prayer book, and their publication sparked a controversy between Hanau and more traditionalist scholars.[23]

Encyclopedical and Philosophical Works Printed in Jessnitz

It is of little surprise that we can find some works from the *chokhmot chitzoni-yot,* the so-called external (non-rabbinic) disciplines, among the books Israel bar Avraham printed during his first Jessnitz period. Although Ashkenazic culture held strong reservations toward the scholarly study of external disciplines—philosophy, natural sciences, history, grammar—some books in these genres had been printed in Germany in the eighteenth century. Historical works such as Salomon Ibn Verga's (1460–1554) *Shevet Yehuda* about the expulsion of Jews from Spain and Portugal had been reprinted several times in Hebrew and in Yiddish translation. *Chovot ha-Levavot* by Bachiya Ibn Paquda (c. 1050–1120), of which many editions were published in the course of the century, represented medieval Jewish philosophy, although this book was considered a work of ethical literature (*musar*) rather than a philosophical work in the Greek or Aristotelian sense.

In 1721, Israel bar Avraham printed the work *Ma'aseh Tuviya* by Tuviya ha-Cohen (1652–1729), a former student of the university at Frankfurt/Oder and later court physician of the Ottoman rulers in Constantinople.[24] This work compiled knowledge from the external disciplines to provide a Jewish readership with knowledge of the contemporary natural sciences, enabling them to restore "lost Jewish honor" in scholarly discussions with Christians.[25] The book presented knowledge from astronomy, meteorology, geography, botany, mechanics, and especially medicine. Many Jewish readers in Ashkenaz probably gained their first knowledge in the natural sciences through this work.

Evidently for commercial reasons, Israel bar Avraham had to leave Jessnitz in 1726; he headed to Wandsbek near Hamburg, where he remained until 1733 and printed some two dozen Hebrew and Yiddish books.[26] Among these were many of the commentaries, responses, and ethical works of the anti-Sabbatean rabbi Moses Chagis (1672–1751), one of the heads of the Sephardic community of Altona. Israel bar Avraham did not print any works from the external disciplines during the Wandsbek period.

Israel bar Avraham's printing press apparently came to a standstill between 1733 and 1738, when he resettled in Jessnitz, re-established his press, and hired typesetters, printers, and proofreaders from all over Germany and Poland in order to launch one of the most ambitious undertakings in Hebrew printing in Germany in the first half of the eighteenth century: the printing of *Mishneh Torah*, the voluminous halachic code by Maimonides. The chief rabbi of Anhalt-Dessau, David Fränkel (1707–62), apparently convinced Israel bar Avraham, whom he had come to know during his years in Hamburg, to transfer his printing press to Jessnitz again for this remarkable project. The four volumes of the *Mishneh Torah*, totaling more than twenty-six hundred pages, were printed between 1739 and 1742.[27]

Though ambitious, this undertaking did not genuinely transcend the boundaries of tradition. The reprinting of the *Moreh Nevukhim*, shortly after the printing of the *Mishneh Torah* had been completed, by contrast, was somewhat transgressive and revolutionary, given the rabbinical establishment's reservations and even open opposition to the *Guide*. It seems that the *Moreh Nevukhim*, rather than being a "by-product" of the printing of Maimonides's halachic code, formed the core of a small group of extraordinary works. The short period of renewed Hebrew printing activity in Jessnitz from 1739 to 1745 saw the appearance of only a few books alongside the *Mishneh Torah* and the *Moreh Nevukhim*, almost all of them pertaining to the external disciplines. There was, for instance, *Nehmad ve-Na'im* by David Gans (1541–1613), an early seventeenth-century astronomical and geographical work that had never been printed before.[28] In 1744, Israel bar Avraham printed the ethical-philosophical text *Duties of the Heart* (*Chovot ha-Levavot*) by Bachiya Ibn Paquda, shortly after having published another remarkable work of medieval philosophy, *Ruach Chen*, an introduction to the terminology of metaphysics that had been attributed to Jehuda Ibn Tibbon (1120–90) and regarded as a sort of introduction to the *Guide for the Perplexed*. Even more important than the original text was the commentary by Israel Zamosc (1700–1773), an early *maskil* who provided the reader with profound knowledge on the latest achievements in the natural sciences of his day.[29] Another book Israel bar Avraham had announced in his preface to the *Moreh Nevukhim* had remained unprinted: the Hebrew dictionary *Sefer ha-Shorashim* by medieval grammarian and Maimonides supporter David Kimchi (1160–1235). After 1745 Israel bar

Avraham apparently stopped printing and publishing altogether, as no known works left his printing press in Jessnitz or elsewhere after this date.

The philosophical and scholarly works printed in Jessnitz between 1742 and 1744, taken together, seem to be more than a coincidental collection. If we include Kimchi's grammar in this group, although the plan was never realized, and link the works that appeared during these years to the grammatical texts printed during Israel bar Avraham's first period in Jessnitz and, in particular, to the encyclopedia by Tuviya ha-Cohen printed in 1721, a pattern begins to emerge. Israel bar Avraham's publishing program displays an emergent interest in an old knowledge that had long been outside the mainstream of Jewish knowledge in Ashkenaz, alongside new elements of world knowledge still unfamiliar to his Jewish audience and not necessarily reconcilable with extant tradition. We might refer to this knowledge, as well as the space from which it proceeded, as highly hybrid. It consisted, on the one hand, of medieval knowledge with a Sephardic character, which had come into being in a space of contact between Jewish tradition and an Islamic learned culture that had translated elements of ancient Greek thought into its own terms. This knowledge was not untouched by controversy, yet it had found its way into the body of Jewish learned thought.[30] Its other constituent part encompassed new ideas derived neither from Ashkenazic nor from medieval Sephardic learned culture, including reflection on the origins of languages, discussion of the discoveries made by Copernicus, Galileo Galilei, and Tycho Brahe, descriptions of newly discovered continents, and details of inventions such as the air pump and the quadrant. Works like *Nehmad ve-Na'im* and Israel Zamosc's commentary on *Ruach Chen,* in seeking to familiarize Jewish readers with recent and current research on natural phenomena, both set themselves apart from the Aristotelean and Ptolemaic premises of medieval Sephardic learning and, evidently, propagated new, unfamiliar knowledge that had yet to be transferred and translated into contexts that operated within the framework of Jewish tradition. Thus, Israel bar Avraham's printing activities, particularly in his final Jessnitz phase, point to an editorial agenda reminiscent of an early Haskalah manifesto. On a small scale, this agenda foreshadowed the range of publications issued by the central printing press of the Haskalah, the Orientalische Buchdruckerey in Berlin, headed by the *maskil* Isaak Satanow (1732–1805).[31]

Censorship and Commerce: External Factors for Editorial Decision-Making

If we compare the various periods and places of Israel bar Avraham's printing activities, some notable differences come to light. The range of works he produced between 1718 and 1726 did not differ substantially from general Hebrew printing in Germany at that time, except for the Hebrew grammar and the extraordinary *Ma'aseh Tuviya.* His second Jessnitz printing period, however, was exceptional, perhaps marking a revolutionary outburst of nonconventional and partly very modern knowledge. Moreover, between the two Jessnitz printing periods, bar Avraham's activities in Wandsbek between 1726 and 1733 appear to have been rather conventional and traditional. How can we explain these differences in one printer's output during the different stages of his printing activities between 1718 and 1744?

Two important factors in printers' and publishers' editorial decisions in the early modern period were the needs of the book market, in combination with everyday financial and commercial considerations (often the most decisive), and the restraints imposed by censorship. Rich patrons were prepared to fund prestigious works that promised to give them a great reputation; in most cases, these works were either central to the Jewish canon or had not been printed for decades, generating a great need among readers. For example, the court Jew Berend Lehmann of Halberstadt (1661–1730) financed the printing of Alsheikh's Bible commentaries by Israel bar Avraham. The printing of Maimonides's *Mishneh Torah* and also of the *Moreh Nevukhim* was sponsored by Jewish merchants from Berlin.

However, a Hebrew printer could not earn his living from printing prestigious and externally sponsored books alone. In order to guarantee a steady cash flow and consistently employ and pay typesetters and printers throughout less profitable periods, the printer-entrepeneur had to resort to small, ephemeral products, "little jobs"—broadsides, advertisements, calendars, and similar products.[32] Calendars proved particularly lucrative in Hebrew printing, and Jewish and Christian printers fought hard to acquire the special licenses for printing them.[33] We have one calendar printed by Israel bar Avraham in Jessnitz for the year 5500/1740; beyond this, we do not know whether calendars were regularly printed in Jessnitz or if this was the only one, printed to celebrate the half millennium. This small calendar, containing both the Jewish and Christian systems, featured dates of important markets and fairs for everyday commerce between Jews and Christians, alongside the Jewish holidays.

Marketing and clientele were, of course, crucial to the publishing strategies of early modern printers. These concerns help explain, to a degree, the differences in bar Avraham's output during the various stages of his printing activities. In Wandsbek, it seems, he tried to satisfy the great demand for traditional literature arising from the large and conservative Jewish community of Altona-Hamburg-Wandsbek. Here, he also encountered a strict intra-Jewish regime of censorship in the anti-Sabbatean Sephardic scholar Moses Chagis. In Jessnitz, there were no such sophisticated Jewish control mechanisms. The Jewish community was much smaller, and the rabbinical institutions in Dessau were much more lenient concerning unconventional topics, as evidenced by the Dessau rabbi's approval of the work *Ma'aseh Tuviya*. The Jewish communities in Anhalt did not constitute a market for Israel bar Avraham's products. His business practices seemed to be more aligned to a supraregional market, with the key commercial factor being the proximity of Leipzig as a venue for large-scale trade fairs in books; a considerable proportion of Hebrew books printed in Jessnitz were exported to Eastern Europe.

In Jessnitz, bar Avraham did not print for institutions of Jewish learning; only two small-format editions of Talmudic tractates, and none of the great codices of Jewish law, such as the *Arba'a Turim,* appeared in his first Jessnitz printing period. No strong demand for such works existed in the Jewish communities nearby, and other, long-established printing offices in Berlin and Frankfurt/Oder apparently satisfied the needs of the communities in other central and eastern German territories. With his Jessnitz products, bar Avraham addressed a learned, rabbinical readership, perhaps at least partially inclined to seek rather unconventional knowledge. Another sizable segment of his clientele was Christian scholars interested in Hebrew literature, mostly professors of theology or Oriental languages in Halle, Leipzig, or other universities in the region, for whom he published advertisements in the *Acta eruditorum,* one of the leading academic journals in Germany.[34] Leipzig theologian Johann Christian Hebenstreit (1686–1756) contributed a lengthy book review in Latin to the 1743 *Sefer Nehmad ve-Na'im* by David Gans.[35]

Israel bar Avraham seems to have been spared one issue that could cause immense printing delays, if not halt it completely, at least in his Jessnitz years: censorship by governmental or ecclesiastical authorities. The broad license Princess Henriette Catharina von Anhalt granted her court Jew Moses Benjamin Wulff in 1694 to set up a Hebrew printing office did not stipulate specific censorship matters, instead providing a more general guideline: "That in all books and scriptures that are printed and published in the above-mentioned printing office there must be no content that runs counter to the Christian religion, the respect and authority of our princes, the common interest, the laws of the empire and morality."[36] Wulff's print shop apparently operated without being subject to specific mechanisms of censorship, with no censor installed. The license granted Wulff the right to "sell this print shop justly and freely, either in part or as a whole, or to place it in another way in the hands of another, who shall benefit from the same freedom."[37] We may therefore assume that Israel bar Avraham, who took over Wulff's print shop in 1717, likewise enjoyed the liberty granted by this decree.

In contrast to Blankenburg, where bar Avraham had faced the strict censorial regime of Protestant clerics and theologians during his short stay in 1718,[38] it appears that in Anhalt, those in power believed the renown and economic boost a Hebrew printing press in their territory would bring outweighed any religious concerns or reservations. Indeed, the license granted Wulff had explicitly referred to the "commerce to be hoped for with tradesmen and booksellers from elsewhere, as well as the work, sustenance and earnings [to be generated] for the city [Dessau] and [to benefit] various people."[39] In 1739, delays in the printing of Maimonides's *Mishneh Torah* in Jessnitz led to a dispute between the typesetters and printers contracted to carry out the work and its editor David Fränkel and printer Israel bar Avraham. The contracted workers sought compensation for lost earnings and, their request denied, turned to the authorities for redress. The letters that the state representatives wrote about the matter frequently emphasized an overarching economic interest in settling the issue and cautioned against allowing the court case to negatively impact printing activities. As state minister Johann Georg von Raumer (1671–1747) decreed, the court was required to find a compromise so that "the printing should not stagnate under any circumstances."[40]

Cooperation with Christian Printers

Beyond the "Hebraist discourse"[41] connected to censorship and the theological-philological interest of Christian scholars and clerics in Jewish literature, another area of communication between Jews and Christians was involved in the making of the Hebrew book: cooperation between Jewish and Christian printers, which often manifested itself in "hybrid" products of knowledge. Among the six printers and typesetters bar Avraham and Fränkel hired for the *Mishneh Torah* project was a Christian printer, George Friedrich Klesser, originally from Leipzig.[42] He is first mentioned in the early 1720s as a self-employed printer in Jessnitz. The publications Klesser's Jessnitz press issued up to the second half of the 1730s were principally works printed to order, occasional publications, eulogies, and panegyrics. The most substantial work he produced was the *Vollständiges Bitterfeldisches Gesangbuch,* published in Jessnitz in 1734. At the end of 1738, a Johann Ehrenfried Klesser—perhaps George's brother or son—submitted a request to Leopold II to move "from Jessnitz to Dessau with my book printing business due to poor earnings"; his intention was to become court printer to the prince in Dessau, and he did indeed receive the required license shortly afterward.[43] George Klesser, however, appears to have remained in Jessnitz and begun working for Israel bar Avraham.

The cooperation between these two printers of Jessnitz, a convert to Judaism—Israel bar Avraham—and a Christian—George Klesser—in the early 1720s is of considerable interest for the concept of hybrid space. We can find Klesser's name or initials in Latin characters in the colophons of several Hebrew works printed in Jessnitz. His first mention is as "Georgius Klesserii, Lipsiensis," with the addition in Hebrew of *ha-druker* (the printer), in a commentary on the Prophets entitled *Mar'ot ha-Tzov'ot,* written by Moses Alsheikh and printed by Israel bar Avraham in 1720.[44] On occasion, his own works feature the same floral illuminations bar Avraham used to decorate many of his books. Their cooperation in technical matters, likely inspired by purely pragmatic considerations, appears to have been accompanied by shared interests in specific areas of knowledge. In Jessnitz in 1720, and again in 1722, Klesser printed *Eine gründliche Verfassung der jüdischen Lehre,* the German version of the catechism *Lekach Tov,* produced on the basis of a 1704 translation from the Hebrew original into Latin by the Helmstedt-based Orientalist Hermann von der Hardt (1660–1746). *Lekach Tov* (Good Teachings), written by the Italian Jew Abraham Yagel (1553–1623) in the style of a Christian catechism and first printed in Venice in 1595, experienced an astonishing trajectory. After its initial publication, it was widely translated for a multitude of purposes. Yagel had based his volume on the thirteen principles of the Jewish faith articulated by Maimonides and dedicated it to the education of Jewish youth. After being discovered by Christian theologians in the seventeenth century and translated into Latin and English, and then, in the eighteenth century, into German, *Lekach Tov* became a key text of the contemporary Hebraist discourse and was reprinted time and again.[45] The text prompted Christian expectations centering around the idea of a refinement of the Jewish faith that were linked to missionary intentions. Alongside these developments, translations of *Lekach Tov* into Yiddish in the first half of the eighteenth century also shaped a part of its contemporary reception. These translations, which emerged in the context of early maskilic discourse, related to the then-prevalent criticism of the traditional curriculum and, specifically, of the practice of neglecting the study of the Torah to focus exclusively on the Talmud.

A year before Klesser's first edition of the *Gründliche Verfassung der jüdischen Lehre,* bar Avraham had printed a bilingual Hebrew/Yiddish version of *Lekach Tov.* Klesser may well have gotten the idea of issuing a German edition of the Jewish catechism from bar Avraham, who evidently had real interest in this material. In his Yiddish foreword to his own edition of 1719, bar Avraham, writing very much in the spirit of the early *maskilim,* expressed unveiled criticism of the traditional educational practices of the Ashkenazic Jews. He denounced the exclusivity of the use of the *pilpul* method when studying the Talmud, claiming it led to even learned Jews barely being acquainted with the fundamentals of the Torah as formulated in Maimonides's Thirteen Principles; it left the common man without true faith.[46] Evidently, bar Avraham considered it important to assert and demonstrate the rationality of the Jewish faith as revealed in Yagel's catechism to Christian and Jewish readers. Consequently, he became enthusiastic about a German version of the work, which, however, he could not print himself, as his license extended only to printing with Hebrew characters.

The Printer's Personality as a Secondary Agent of Innovation

All of this brings us to the concluding question of how the range of extraordinary texts published by the Jessnitz printing press was interlinked with the hybrid, "mingled" identity of the printer. We may find it helpful here to return to Elizabeth Eisenstein's idea of the early modern printer not only as an entrepreneur driven by economic motives, but also as a secondary agent in the generation and dissemination of knowledge. Prompted by a thirst for knowledge and intellectually inclined toward certain fields, such a printer collects, chooses, and contributes to the body of knowledge and helps establish new ones.[47] This was especially true of Israel bar Avraham. If indeed he was a cleric prior to his conversion to Judaism, or even a monk, as the Prussian minister von Arnim suspected,[48] then his inclination toward certain fields of Jewish knowledge probably derived from his clerical past and was formed by Hebraist interests and knowledge and a critical attitude toward texts, acquired during his theological studies or monastic career. As we have seen, bar Avraham printed a significant number of grammatical works and Bible commentaries in his early years in Koethen and Jessnitz—works from fields typically among Christian Hebraists' readings. For Moses Alsheikh's Bible commentaries and some of the other Jessnitz works—most significantly the *Moreh Nevukhim*—bar Avraham himself was the editor and apparent initiator of their printing.

It is tempting to surmise that bar Avraham's non-Jewish origin and his socialization in a learned Hebraist culture was what actually later influenced the output of his Jessnitz printing press. Grammarians, such as the controversial Salomon Hanau, whose works were printed in Jessnitz, could have been attracted by bar Avraham's open-minded attitude toward knowledge on the periphery of the Jewish canon. And the relationship was reciprocal: these encounters with moderately enlightened authors would have provided intellectual stimulation for the printer himself. As was usual in early modern printing, authors and editors were often present during the typesetting, printing, and proofreading. Hanau spent some weeks in nearby Halle to supervise the printing and intervene if necessary.[49] As Elizabeth Eisenstein showed, the early modern printing office was a place of communication between printers, authors, and editors, a forum for the sharing of knowledge and opinions and, sometimes, for initiating important editorial projects. Thus, the print shop in Jessnitz was a space of encounter between a Christian convert to Judaism with a Hebraist past and Jewish scholars inclined to external disciplines like the grammarian Hanau or the editor of the *Nehmad ve-Na'im* of 1743, Yo'el ben Yekutiel Glogau, and last but not least, Moses Mendelssohn's teacher David Fränkel. In these encounters within or outside the space of the Hebrew print shop lay the seeds for many of the works of hybrid knowledge that bar Avraham printed in Jessnitz.

Research into this printing history has long presented David Fränkel as the driving force behind the printing of both the *Mishneh Torah* and the *Moreh Nevukhim,* with their printer taking a more secondary role.[50] Particular comments bar Avraham made in his forewords to works he published in Jessnitz, however, indicate that he played a very active role in selecting material produced by his press. The strongest example of these is in fact his courageous and unusual preface to the *Moreh Nevukhim,* which appears on its first page, where rabbinic certificates of approval, missing here, would ordinarily have been. He highlights the warnings against printing the book he had received from friends and foes alike. Further, he explains that after his thirst for wisdom had been aroused by the *Mishneh Torah,* he had found that he could satisfy this thirst only by further studying Maimonides's writings, prompting him to publish the *Guide.* He thus presents his own interest in this significant medieval text as his principal reason for printing it. Even his opponents' threat that he would suffer death at God's hands for eating of the forbidden fruit (of philosophy) did not distress him. If God had wished to kill him, bar Avraham continued, then He would not have shown him these things at all. For bar Avraham, the *Moreh Nevukhim* was the "tree of life" (*etz chayim*), and those eating of its fruit would gain knowledge and salvation.[51] If his passionate words are true, then it would seem that the Jessnitz printer played a much more important role in publishing Maimonides's philosophical work than has generally been attributed to him.

Dirk Sadowski is a research fellow and the coordinator of the German-Israeli textbook commission at the Georg Eckert Institute–Leibniz Institute for International Textbook Research in Braunschweig. His book publications include *Haskala und Lebenswelt. Herz Homberg und die jüdischen deutschen Schulen in Galizien 1782–1806* (Göttingen, 2010), and the edited collection (with M. Liepach) *Jüdische Geschichte im Schulbuch. Eine Bestandsaufnahme anhand aktueller Lehrwerke* (Göttingen, 2014). His research interests are in the fields of Haskalah and enlightened Jewish pedagogy, early modern Jewish lifeworlds, and Hebrew printing.

Notes

1. Burnett, "Christian Hebrew Printing"; Heller, *Sixteenth Century Hebrew Book*; Heller, *Seventeenth Century Hebrew Book*.
2. Heller, "And the Work, the Work of Heaven, Was Performed on Shabbat," in Heller, *Studies*, 266–77.
3. Eisenstein, *Printing Revolution*, 46–70, esp. 49–50.
4. Raz-Krakotzkin, *The Censor, the Editor, and the Text*, 95–119.
5. Burnett, *Christian Hebraism*.
6. Ruderman, *Early Modern Jewry*, 159–89.
7. Altmann, *Moses Mendelssohn*, 10.
8. Heller, *Sixteenth Century Hebrew Book*, 1:371; Harvey, "Introductions," 87, with note 8.
9. Reiner, "Attitude," here particularly 593–96; Harvey, "Introductions," 87–88.
10. Maimon, *Be'ur milot ha-higayon*. See Mendelssohn's preface to this edition in Mendelssohn, *Hebräische Schriften* I, 43–46 and 423–27.
11. M. Freudenthal, *Aus der Heimat*, 155–74; Heller, "Moses Benjamin Wulff—Court Jew," in Heller, *Studies*, 206–17; license (*Privileg*) issued on 14 December 1694. Landeshauptarchiv Sachsen-Anhalt (LHASA), Abteilung Dessau, C 9e Nr. 22 Bd. I, Bl. 3. The license is also cited in M. Freudenthal, *Aus der Heimat*, 157–60, here 158.
12. M. Freudenthal, *Aus der Heimat*, 175–88; Heller, "Moses ben Abraham Avinu and His Printing-Presses," in Heller, *Studies*, 218–28 (first published in *European Judaism* 31, no. 2 [1998]: 123–32).
13. Sadowski, "'Gedruckt in der heiligen Gemeinde Jeßnitz'"; M. Freudenthal, *Aus der Heimat*, 188–231.
14. Report by State Minister Dietloff von Arnim for the Preußische General-Oberdirektorium, dated 12 October 1741, Geheimes Staatsarchiv Preußischer Kulturbesitz Berlin, I. HA Rep. 33 Nr. 120 (1740-1807), 482, and 496–97, reproduced in Strobach, *Privilegiert in engen Grenzen*, 2:72–73.
15. M. Freudenthal, *Aus der Heimat*, 273–75.

16. Werner, *Geschichte der Stadt Jeßnitz,* 56, 105; Beckmann, *Historie des Fürstenthums Anhalt,* vol. 1/1, 113.
17. Freimann, "Annalen," 100–101.
18. Weinberg, "Die hebräischen Druckereien."
19. Schmelzer, "Hebrew Printing."
20. M. Freudenthal, *Aus der Heimat,* 194–213, 251–70. Vinograd, *Thesaurus,* 2:352–53, gives the number of works appearing up to 1726 as forty-four.
21. This figure is an estimate, based on the bibliographic entries for the print shops in the territory of the Holy Roman Empire, given in Vinograd, *Thesaurus.*
22. Schatz, *Sprache in der Zerstreuung,* 140–65.
23. M. Freudenthal, *Aus der Heimat,* 209–12.
24. A first edition of the book appeared in 1707 in Venice. Sadowski, "Jupitermonde"; Ruderman, "On the Diffusion."
25. Feiner, *Haskala,* 35–42.
26. Heller, "Israel ben Abraham, His Hebrew Printing-Press in Wandsbek and the Books He Published," in Heller, *Further Studies,* 169–94; Bamberger, "Wandsbeker Druckperiode."
27. M. Freudenthal, *Aus der Heimat,* 214–19; M. Freudenthal, "R. David Fränkel," 575–76.
28. Neher, *Jewish Thought,* 58–91, 169–250; Efron, "Irenism," 636–44.
29. On Zamosc, see G. Freudenthal, "Hebrew Medieval Science"; G. Freudenthal, "Jewish Traditionalism."
30. Fontaine, Schatz, and Zwiep, *Sepharad in Ashkenaz,* ix–xvii.
31. Lohmann, "'Sustenance for the Learned Soul.'"
32. Stallybrass, "'Little Jobs.'"
33. On the printing of Hebrew calendars in Berlin, see Lohmann, "'Sustenance for the Learned Soul,'" 15–17.
34. M. Freudenthal, *Aus der Heimat,* 194–95, 199.
35. *Sefer Na'im ve-Nehmad* [*sic*] (…) quid contineat, recenset D. Io. Christian Hebenstreit in Academia Lipsiensi, Linguae Sanctae p.p.
36. License (*Privileg*) issued on 14 December 1694, ibid.
37. Ibid.
38. Strobach, *Privilegiert in engen Grenzen,* 1:79–88.
39. License (*Privileg*) issued on 14 December 1694, ibid.
40. Letter dated 12 June 1739, LHASA, DE, Abt. Dessau, C 9e, Nr. 24.
41. Raz-Krakotzkin, *The Censor, the Editor, and the Text,* 180–94.
42. Letter dated 9 June 1739, LHASA, DE, Abt. Dessau, C 9e, Nr. 24.
43. Printed application dated 6 December 1738, LHASA, DE, Abt. Dessau, C 9e, Nr. 22, Bd. I.
44. Further Hebrew books, alongside *Mar'ot ha-Tzov'ot,* which contained Klesser's name or his initials GK, are the Alsheikh commentaries *Romemot El* (printed in 1721) and *Rov Pninim* (1722); additionally, he featured in the colophons of *Ateret Zvi* (1722) and *Damesek Eliezer* (1723).
45. Faierstein, "Abraham Jagel's 'Leqah Tov.'"
46. *Sefer Lekach Tov,* Jessnitz, 1719, *Hitnatzlut shel ba'al ha-madpis,* fol. 10b.
47. Eisenstein, *Printing Revolution,* 46–70, esp. 49–50.
48. Report by State Minister von Arnim, dated 12 October 1741 (see note 14).
49. *Sha'arei Tefila,* Jessnitz, 1725, without pagination (last folio, verso).
50. M. Freudenthal, *Aus der Heimat,* 220–21.

51. *Amar ha-madpis*—preface by Israel bar Avrahams, *Moreh Nevuchim.* Jessnitz, 1742, without pagination.

Bibliography

Altmann, Alexander. *Moses Mendelssohn: A Biographical Study.* Tuscaloosa, AL: University of Alabama Press, 1973.

Bamberger, Simon. "Wandsbeker Druckperiode des Israel ben Abraham, 1726–1733." In *Festschrift für Aron Freimann zum 60. Geburtstage,* edited by Alexander Marx and Hermann Meyer, 101–8. Berlin: Soncino Gesellschaft, 1935.

Beckmann, Johann C. *Historie des Fürstenthums Anhalt.* Zerbst: Zimmermann, 1710. Reprint, Dessau, 1993.

Burnett, Stephen G. *Christian Hebraism in the Reformation Era (1500–1660): Authors, Books, and the Transmission of Jewish Learning.* Leiden, Boston: Brill, 2012.

———. "Christian Hebrew Printing in the Sixteenth Century: Printers, Humanism and the Impact of the Reformation." *Helmantica* 51, no. 154 (2000): 13–42.

Efron, Noah J. "Irenism and Natural Philosophy in Rudolfine Prague: The Case of David Gans." *Science in Context* 10 (1997): 627–49.

Eisenstein, Elizabeth. *The Printing Revolution in Early Modern Europe.* Cambridge: Cambridge University Press, 2005.

Faierstein, Morris M. "Abraham Jagel's 'Leqah Tov' and Its History." *Jewish Quarterly Review* 89 (1999): 319–50.

Feiner, Shmuel. *Haskala—Jüdische Aufklärung. Geschichte einer kulturellen Revolution,* Hildesheim: George Olms Verlag, 2007.

Fontaine, Resianne, Andrea Schatz, and Irene E. Zwiep, eds. *Sepharad in Ashkenaz: Medieval Knowledge and Eighteenth-Century Enlightened Jewish Discourse.* Amsterdam: Royal Netherlands Academy of Arts and Sciences, 2007.

Freimann, Aron. "Annalen der hebräischen Druckerei in Wilhermsdorf." In *Festschrift zum siebzigsten Geburtstage A. Berliner's,* edited by Aron Freimann and Meir Hildesheimer, 100–15. Frankfurt am Main: J. Kauffmann, 1903.

Freudenthal, Gad. "Hebrew Medieval Science in Zamość, ca. 1730: The Early Years of Rabbi Israel ben Moses Halevi of Zamość." In Fontaine, Schatz, and Zwiep, *Sepharad in Ashkenaz,* 25–67.

———. "Jewish Traditionalism and Early Modern Science: Rabbi Israel Zamosc's Dialectic of Enlightenment (Berlin, 1744)." In *Thinking Impossibilities: The Intellectual Legacy of Amos Funkenstein,* edited by Robert S. Westman and David Biale, 63–96. Toronto: University of Toronto Press, 2008.

Freudenthal, Max. *Aus der Heimat Mendelssohns. Moses Benjamin Wulff und seine Familie, die Nachkommen des Moses Isserles.* Berlin: F. E. Lederer, 1900. Reprint, Dessau, 2006.

———. "R. David Fränkel." In *Gedenkbuch zur Erinnerung an David Kaufmann,* edited by Marcus Brann and Ferdinand Rosenthal, 569–98. Breslau: Schottlaender, 1900.

Harvey, Steven. "The Introductions of Early Enlightenment Thinkers as Harbingers of the Renewed Interest in the Medieval Jewish Philosophers." In Fontaine, Schatz, and Zwiep, *Sepharad in Ashkenaz,* 85–104.

Heller, Marvin J. *Further Studies in the Making of the Early Hebrew Book.* Leiden: Brill, 2013.

———. *The Seventeenth Century Hebrew Book: An Abridged Thesaurus.* Leiden: Brill, 2010.

————. *The Sixteenth Century Hebrew Book: An Abridged Thesaurus.* 2 vols. Leiden: Brill, 2004.

————. *Studies in the Making of the Early Hebrew Book.* Leiden: Brill, 2008.

Lohmann, Uta. "'Sustenance for the Learned Soul.' The History of the Oriental Printing Press at the Publishing House of the Jewish Free School in Berlin." *Leo Baeck Institute Yearbook* 51 (2006): 11–40.

Maimon, Mose ben. *Be'ur milot ha-higayon.* Edited by Moses Mendelssohn. Berlin, 1765.

Mendelssohn, Moses. *Hebräische Schriften I, Deutsche Übertragung.* Stuttgart: frommann-holzboog, 2004 (Moses Mendelssohn: *Gesammelte Schriften. Jubiläumsausgabe,* vol. 20, 1).

Neher, André. *Jewish Thought and the Scientific Revolution of the Sixteenth Century: David Gans (1541–1613) and His Times.* Oxford: Oxford University Press, 1986.

Raz-Krakotzkin, Amnon. *The Censor, the Editor, and the Text: The Catholic Church and the Shaping of the Jewish Canon in the Sixteenth Century.* Philadelphia: University of Pennsylvania Press, 2007.

Reiner, Elchanan. "The Attitude of Ashkenazi Society to the New Science in the Sixteenth Century." *Science in Context* 10 (1997): 589–603.

————. "Beyond the Realm of Haskalah—Changing Learning Patterns in Jewish Traditional Society." *Jahrbuch des Simon-Dubnow-Instituts* 6 (2007): 123–33.

Ruderman, David B. *Early Modern Jewry: A New Cultural History.* Princeton, Oxford: Princeton University Press, 2010.

————. "On the Diffusion of Scientific Knowledge within the Jewish Community: The Medical Textbook of Tobias Cohen." In *Jewish Thought and Scientific Discovery in Early Modern Europe,* 229–55. Detroit: Wayne State University Press, 2001.

Sadowski, Dirk. "'Gedruckt in der heiligen Gemeinde Jeßnitz.' Der Buchdrucker Israel bar Avraham und sein Werk." *Jahrbuch des Simon-Dubnow-Instituts* 7 (2008): 39–69.

————. "Jupitermonde und 'verschlossene Gärten.' Tuvija Cohens Enzyklopädie der Naturwissenschaften und der Medizin Ma'aseh Tuvija (1707)." *Jahrbuch des Simon-Dubnow-Instituts* 9 (2010): 247–77.

Schatz, Andrea. *Sprache in der Zerstreuung. Die Säkularisierung des Hebräischen im 18. Jahrhundert.* Göttingen: Vandenhoeck & Ruprecht, 2008.

Schmelzer, Menaham. "Hebrew Printing and Publishing in Germany, 1650–1750: On Jewish Book Culture and the Emergence of Modern Jewry." *Leo Baeck Institute Yearbook* 33 (1988): 369–83.

Stallybrass, Peter. "'Little Jobs': Broadsides and the Printing Revolution." In *Agent of Change: Print Culture Studies after Elizabeth L. Eisenstein,* edited by Sabrina Alcorn Baron, Eric N. Lindquist, and Eleanor F. Shevlin, 315–41. Amherst: University of Massachusetts Press, 2007.

Strobach, Bernd. *Privilegiert in engen Grenzen. Neue Beiträge zu Leben, Wirken und Umfeld des Halberstädter Hoffuden Berend Lehmann (1661–1730).* 2 vols. Berlin: epubli, 2011.

Vinograd, Yeshayahu. *Thesaurus of the Hebrew Book* [Otsar ha-sefer ha-ivri]. 2 vols. Jerusalem: Institute for Computerized Hebrew Bibliography, 1993–95.

Weinberg, Magnus. "Die hebräischen Druckereien in Sulzbach." *Jahrbuch der Jüdisch-Literarischen Gesellschaft* 1 (1903): 19–202.

Werner, Ernst. *Geschichte der Stadt Jeßnitz.* Dessau: RUPA-Druck, 1938.

13

Faith in Residence

Jewish Spatial Practice in the Urban Context

Joachim Schlör

Liberal societies today are challenged by diverse fundamentalist movements whose primary preoccupation is often the reinforcement of rigid laws based on religion. The idea of public space as a marketplace where identities and ways of life (*Lebenswelten*[1]) can be negotiated freely—where people may have "faith in residence"—has lost its attraction among substantial parts of our contemporary societies, and these developments need to be investigated. A too narrow focus on violence and exclusion, though, might produce too dark a picture. Generally, we can observe that the secularization paradigm—a continuous loss of significance for religious practice since the Enlightenment era and throughout the nineteenth and twentieth centuries—seems to have overlooked how very lively the encounter between private and public spaces, in terms of religious practices, has remained.[2] This circumstance becomes visible when established knowledge about the cultural practices related to the construction and perception of Jewish homes and settlements is challenged by new research in mobility and migration studies.

This chapter picks up on this challenge and attempts to broaden our understanding of Jewish spatial practices by looking at the importance of movement and mobility in this context. It is concerned with various forms and aspects of a very specific place: the doorstep. I explore points of entry and exit, intermediate and mediating joints that lie between outside and inside: people enter and leave private homes, synagogues, libraries, but also streets and whole quarters or cities—and so, metaphorically, worlds of belonging—*through* such places, from private to public and back. These acts of "going through" can be regarded as performative acts of cultural practice, as exercises in liminality that both relate to and create space—not an all-too-easily identified "Jewish space," but rather different "social, cultural, and political spaces, places, and (symbolic) boundaries as they relate to Jews and Jewishness."[3] Such acts add meaning, beyond the practical, to the complex imaginary that urban dwellers call a home.

From this vantage point, my two main areas of research, urban studies and migration studies, only appear to cover two different spheres; in fact, they are very closely related. In brief, urban studies seem to focus on those who *stay* (live, reside, dwell) somewhere, while migration studies appear to revolve around those who *leave* (move, migrate) elsewhere.[4] However, for a considerable number of cultural situations and historical events, the staying and the leaving cannot be easily separated from each other, but are dialectically connected and need to be researched from both perspectives: Some move within a city or an urban quarter and break with actual or unwritten rules around their way of life; others might instead move to places where they will be able to live a life according to such rules. Some decide to emigrate but leave a part of their home (relatives, property, memories) behind, while taking other parts along. Perhaps a group of people (a family, a community) was expelled from a place they used to call their home and attempt to settle somewhere else. In all these cases, there is a "doorstep," a *limen*, on which farewells and greetings, friendly or unfriendly, are exchanged. Many laws and practices regarding or ruling Jewish life, especially Jewish religious and cultural life within non-Jewish environments and neighborhoods, refer to such points of farewell and welcome, entry and exit, settlement and movement. Settlement contains memories of before or fears of further movement; movement contains memories of before or hopes of further settlement. Movement takes place within settlement(s), and forms of intermediate settlement have even characterized movement, as recent research on the transmigrancy of Eastern European Jews through Europe toward the Americas in the period between 1881 and 1924 (or 1938 or 1948) has shown.[5] Interdisciplinary research on "Jewish space(s)" has developed well in recent years.[6] *Jewish Cultural Studies (JCS)*, created by Simon J. Bronner, an important forum for discussing new and interdisciplinary developments in the field, dedicated its second issue to "Jews at Home: The Domestication of Identity" and defined home as follows: "Calling a house a home suggests an emotional connection to it that has been constituted by the shared history of the building and its occupants."[7]

The sources I use here are partly historical, partly contemporary. I am well aware that not all Jews "do" the things I consider here and that there has been a great variety regarding such practices over time and space. I do not construct any kind of essentialism. The spaces I reflect on here emerge through the interaction of (many) Jews and (many different) "others," across historical periods and geographical locations. Much of this interaction—more than traditionally thought—has arisen, and still does, *through* cultural practices based on religious law confronted with public space. "One begins to see," says Martin Marty, "that in life the distinct elements of what is 'private' and what is 'public' intersect, interact, and interpenetrate each other."[8] The following sections focus on the function of the mezuzah for the private Jewish home, on inscriptions above synagogue entrances as visible signals of a Jewish presence on city streets, and on the function of the *eruv*, the Sabbath boundary, for the urban Jewish quarter, respectively. I conclude with an outlook at the sea voyage and the function of the ship in migration processes.

From the Inside of the Home to the Doorstep

The notion of sacred space, in the Jewish context, that is, in a history of diasporas and displacements, can only be understood through the tension between long-ing and belonging: "The conventional Jewish historical narrative of expulsion, dispersion and consequent exile provokes comment not only about where an emotional home is located, but also about whether one can be 'at home' in the diaspora."[9] In anthropological and ethnographical research, an ideal (but not nec-essarily any historical) Jewish home that offers comfort and protection and an *in-side,* which can only be understood in its relationship to an *outside*—the staircase, the neighbor's flat, the neighbor's house, the street, the village, or the city—is one that contains markers of identity and difference. Ritual objects—a menorah (or rather *hanukiah*), a seder plate, *besamim* (spice box), or a box for *tzedakah* (charity)—are often displayed on shelves or tables and serve as reminders of their function on holidays. Popular guides to the decoration of a "good Jewish home" often contain advice to compile (and display) a Jewish library, with a Hebrew Bible, a complete prayer book, a manual for keeping a Jewish household, works on Jewish history, books on Jewish cuisine, and so forth.[10] The kitchen contains material objects that make visible to an experienced eye whether or to what de-gree a household follows the rules of kashrut. Vanessa Ochs, in her anthropolog-ical study on material culture in the Jewish home, uses material objects and their positions in homes as a medium for understanding the idea of "home": Within Jewish homes, things, people, and even times of day and seasons of the year and of life interact in a fluid process, through which things make the home Jewish, by which things are animated by Jewish life and absorbed by it in specifically Jewish ways.[11] The presence and placement of objects creates "Jewish ways" of doing things; cultural practices create a recognizable spatial environment: "Sometimes Jewish law dictates the placement [of an object]. The mezuzah goes on the right side of the door post; the menorah goes near a window; the Sabbath challah goes on the table."[12] When we translate such findings from anthropological research to the world of German-Jewish history, it is easy to observe that many families have lost, or decided to ignore, such practices since the Enlightenment, as research on the emergence of a Jewish civic and bourgeois lifestyle has illustrated.[13] But I hold that, present or absent, they still constitute an important element of our understanding of the construction of what Jewish families regard as their home.[14]

Private spaces are "products of social negotiation and subject to historical change"; therefore, "the borders between the inside and the outside are perma-nently being contested."[15] It is striking, in this context, that research on German-

Jewish history has not yet made good use of rich and relevant studies in cultural anthropology. The most famous study on doors remains Otto Weinreich's *Türöffnung im Wunder-, Prodigien- und Zauberglauben der Antike, des Judentums und Christentums*. This work examines texts "in which doors are made, or encouraged, spontaneously to open, to admit a divinity or, occasionally, to speed his departure."[16] A new and contemporary version of Weinreich's study, yet unwritten, would offer fascinating insights into urban life: on the threshold.

The many variations in the Jewish community make it possible for families to design their homes in a "Jewish" way. Often throughout Jewish history, however, markers of "Jewishness" were defined by the outside world. Michael Shapiro offers an intriguing approach revolving around how theatrical productions of Shakespeare's play *The Merchant of Venice* characterized Shylock's home as "Jewish."[17] Early modern productions included a number of doors on the set, one of which represented otherness in that Shylock, and Shylock only, entered and exited through it. Later directors "often imagined some of the scenes in Act 2 as taking place inside Shylock's house, and so assumed the additional burden of rendering the space as domestic as well as Jewish."[18]

Shapiro does not cite two productions very interesting in our context: an anti-Semitic one in Berlin, and Leopold Jessner's staging at Tel Aviv's *Habimah* in 1933, which presented Shylock as a proud Jew.[19] In postwar productions, a ritual often marked this otherness: Jessica lighting the candles on Sabbath, Shylock himself wearing a prayer shawl or kissing the mezuzah. This latter cultural practice, urgently prescribed by some and dismissed as superstition by others, is bound up with the importance of entering and leaving a house. The number of "mezuzah stories" published in books or online is remarkable. They mainly appear in the educational framework of "Orthodox" or *haredi* communities, especially of Chabad Lubavitch, an organization known for its efforts to help "return" Jews to a traditional and observant form of life.[20] The storylines are straightforward: forgetting to care for this seemingly small object on the doorpost leads to all kinds of catastrophes, family quarrels, and even loss of life, while a return to the practice brings back happiness and harmony. The manuals on "the Jewish household" describe this practice in detail:

> On the doorposts of Jewish homes a passerby can likely find a small casing … known as a *mezuzah* (Hebrew, "doorpost") because it is placed upon the doorposts of the house. … The commandment to place a *mezuzah* on the doorpost is derived from a passage in the Book of Deuteronomy commonly known as the *Shema*. … The custom became to write the words of the *Shema* on a tiny scroll of parchment, on the back of which the name of G-d is also written, and then roll up the parchment and place it in a case so that the first letter of G-d's name (the Hebrew *Shin*) is visible. … The case and scroll are then nailed or affixed to the right side doorpost on an angle, with a small ceremony called *Channukat Ha-Bayit*. A brief blessing is then recited.[21]

The ritual underscores the importance of passing through threshholds: when Jews pass through doors with mezuzot attached, they kiss their fingers and touch the mezuzah. This is an expression of love and respect for God and God's commandments. By contrast, families are to remove mezuzot when they leave their home.[22] Beyond the observed daily practices, the act of leaving a home (and building a new one elsewhere) forms the topic of many stories, with special emphasis on the experience of loss. Anat Feinberg illustrates this with a story from her own family. Her grandparents Feiwel and Braine Grüngard had moved from Stockholm to Berlin in 1923; however, for them, inspired as they were by the Zionist vision of a return to Eretz Israel (the Land of Israel), the German capital was a transitory home. In a close analysis of several material objects Feinberg grew up with, she reads these private objects as "semiophores,"[23] carriers of memory, in an attempt to understand her family's feelings of home between Germany and Israel:

> Ze ha-Bayit, that's the house, announced Feiwel Grüngard in Hebrew to his wife and his two children as the taxi finally came to a stop. He didn't say Higa 'anu habayta, we're home; for the villa in Freiherr-vom-Stein-Strasse was only intended to serve as a place to live for a short while. The family's real home was thousands of miles from Berlin, at the eastern shores of the Mediterranean. The temporary dwelling in Berlin's district of Schöneberg, however, ended up providing a home to the Grüngards for ten years, until they finally established their home in Eretz Israel in April 1934.[24]

Home can be somewhere other than where one is: lost, lovingly remembered, but also anticipated. The Jewish home Feinberg describes, in the Berlin of 1923, already contains markers of a future home in Tel Aviv. Correspondingly, the home built in Tel Aviv after the family's emigration from Germany contains memories of the earlier place in Berlin. Berlin, Feinberg writes, was "a station in their lives [Lebensstation] of their own choosing, one which enabled Frau Grüngard, my grandmother, to realize her dreams."[25]

The Lebensstation Berlin can be regarded as a doorstep, a limen whence a family's life might take different directions, often depending on external circumstances. In times of persecution, the private home was frequently the last refuge from a hostile environment.[26] German Jews, marginalized under Nazi laws and excluded from public activities such as going to the theater or concerts, turned their private homes into venues for cultural events for a small circle of friends. These last places of refuge were destroyed when persecution was radicalized in 1938. Here is an account of events during the pogrom of November 1938. In one home in the southwest German town of Heilbronn

> incredible acts of destruction took place that bore witness to a spirit of sheer vandalism. In the space of approximately half an hour, all the home's contents and furnishings were smashed to pieces with crowbars, iron bars, and similar implements, even the most "innocuous" objects such as medicine bottles; a number of coat hooks were torn down or smashed to bits one by one; thirty-three windowpanes were broken, as were goblets, glasses, vases, bowls, crockery, furniture, a sewing machine, a typewriter, the wireless, and even a fireplace of Italian marble. In this case, too, the perpetrators had intimidated and threatened the inhabitants, [whom they had] locked in one room.[27]

This terror was clearly directed against the security of the home, with the intention of making those residents feel vulnerable and unprotected. In many cases, the intrusion of Gestapo officers into families' private houses and the destruction or confiscation of private property were final catalysts in the decision to emigrate, coming as they did on top of a series of traumatic losses of profession, social standing, access to social and cultural experiences—a destruction that moved remorselessly from the public into the private sphere.

Making a home, by contrast, is equally important in stories and memoirs of subsequent immigration. Diane Harris, in the preface to *Little White Houses: How the Postwar Home Constructed Race in America,* writes that her maternal grandparents, Rudolf and Eva Weingarten, "obtained citizenship as quickly as possible, and they did everything they could to assimilate, to become as ethnically 'white' and American as possible. Owning a home of their own was an important goal for them, and I think it is fair to speculate that homeownership symbolized much for them, as it did for millions of others."[28] Harris uses this house in Southern California's San Fernando Valley, purchased in 1955, "scrupulously maintained, fastidiously clean, carefully furnished and decorated," as the leitmotif of her book. While the objects and spaces in and around the house appear "as the material dimensions through which racial and class identity and difference are recursively constructed, assumed, and negotiated,"[29] they also serve as reminders of the family's emigration from Germany and the loss of their earlier home. The doorstep is the very point where such constructions, assumptions, negotiations, and representations of identity and difference take place.

On the Street: Inscriptions

On Oranienburger Strasse in Berlin, opposite the partially reconstructed Neue Synagoge, built in 1886, plundered in 1938, destroyed by fire in 1943, left as a ruin until the early 1990s, reconstructed as Centrum Judaicum in 1995, is the restored golden cupola returned to Berlin's panorama in 1995. The inscription above the former main entrance to the synagogue reads: "Open ye the gates, that the righteous nation which keepeth truth may enter."

The "gates" of a synagogue have both a concrete and metaphorical function. They mark the interface between inside and outside, partly private and partly public, and while not much research has yet been done in this field, a short list of inscriptions used on synagogue entrances offers a glimpse at the wealth of information they could provide. Katrin Keßler, of the Bet Tfila-Forschungsstelle für jüdische Architektur in Europa (Brunswick, Germany), has identified the three most common inscriptions in Germany:

> This [is the] gate of the LORD, into which the righteous shall enter. (Psalm 118:20)
>
> This is none other but the house of God, and this is the gate of heaven. (Genesis 28:17)
>
> Blessed shalt thou be when thou comest in, and blessed shalt thou be when thou goest out. (Deutoronomy 28:6)

She adds that during the nineteenth century, when many new synagogues were built in Germany and other European countries, East and West, as visual expressions of arrival and integration, these inscriptions changed in nature. The verses then chosen stress the relationship between the Jewish and Christian religions:

> Thou shalt love thy neighbor as thyself. (Leviticus 19:18)
>
> For mine house shall be called a house of prayer for all people. (Isaiah 56:7).[30]

Going through these examples and trying to make sense of the inside/outside relationship that emerges (and shifts) here, we can see that different (and changing) notions of *house* and *home* are central:

> Krefeld: LORD, I love the habitation of thy house, and the place where thy glory dwelleth. (Psalm 26:8)
>
> Meisenheim, Landkreis Bad Kreuznach, southwestern Germany: How goodly are thy tents, O Jacob, and thy tabernacles, O Israel. (Numbers 24:5)
>
> Budapest: And let them make me a sanctuary; that I may dwell among them. (Exodus 25:8)
>
> Paris, Grand Synagogue: This is the House of God and the Gate of Heaven.[31]

The inscriptions refer to an idea of home. Inscriptions are usually taken from basic scripture: Genesis, Psalms, Prophets—words from the Hebrew Bible that Christian or Muslim neighbors would likely be familiar with. For Jews, the synagogue entrance offers a home away from home in a double sense: away from the Temple in Jerusalem for which the *beit knesset* works as a kind of temporary replacement, but also away from the private home and its security. Passersby are in the public sphere, but not quite; the gates signal the presence of a Jewish neighborhood. Vanessa Ochs speaks of how the thresholds of synagogues "are often poorly 'curated' and those who enter come in with unclear or erroneous impressions about the nature of the space."[32] It is remarkable that Ochs should choose the word "curate" for her description of the cultural and religious activities around synagogue entrances, but the buildings do indeed function as representations, and the invitation to enter in the inscriptions can be read as a call to perform. Rabbi Joel Berger shows what the consequences are, especially for a non-Jewish public, when these calls are "unclear" or even "erroneous."[33]

Research on Jewish forms of settlement, both their historical reconstruction and analysis of their representation in art or literature, appears to have the potential to bridge the two main fields of research in Jewish studies: forms of settlement both reflect the inner discourse around questions of spirituality, of thought and belief, and act as subjects of an outer narrative about the relationship of Jews and Jewish communities to the non-Jewish world. Historical debates about forms of settlement, of living, of movement in space, discussions about the boundaries between "public" and "private" spaces, or about constructs such as the "ghetto," the "Jewish quarter," "Jewish street," or "shtetl," contain and produce information both about the "inner aspects" of laws and the way they are kept—or not—and about the "outer" issues relating to a given place. The variety of Jewish forms of living that have resulted from the ways Jews have attempted to fulfill their need for a life in harmony with Jewish laws, but also in accordance with the laws and conditions of the outside world, means that this field appears to be a mosaic of differences. There is not one single Jewish form of settlement. The outer form of houses, synagogues, streets, and quarters is more than just the everyday manifestation of a bigger, universal, unchanging unit called religion; everyday experiences of a life under so many different conditions have contributed to processes of change in religion and law. Like the private home, synagogues and other community institutions evoke feelings of belonging and a sense of space.

Community Borders: *Eruvim*

The *eruv*, the Sabbath boundary, is a phenomenon discussed in a wide range of disciplines, from rabbinical literature to urban planning. To understand what happens when an *eruv* is established, we need knowledge of both Jewish law and the laws of public space. Charlotte Fonrobert's work on the emergence and development of religious laws and rules concerning the Sabbath and the prohibition on carrying items on that day and Jennifer Cousineau's study on debates and conflicts ensuing from the establishment of *eruvim* in contemporary cities are instructive here.[34] Historical sources document the importance of this institution in stories of inclusion and exclusion, of Jewish life within self-created temporary boundaries, and of destructive events coming from outside. "The Jewish Town in Kazimierz near Krakow" contains a classic description and analysis of the idea of *eruvim*:

> In a proclamation that was published in Krakow in 1892, there was a testimonial to an *eruv* from the times of the *Rama* and the *Bach*. Since Jews were not able to live in the other parts of Krakow, they relied on the walls of the Jewish quarters of Kazimierz. … According to a letter in *HaMagid* (December 22, 1887, number 49) and *HaZefirah* (December 23, 1887, number 274) these walls had gates that were closed on Shabbos to allow carrying.[35]

When a traditional community of observant Jews erects or defines an *eruv*, this act, or cultural practice, creates a home. This home is by no means closed off to others; in fact, it does not intend to *mean* anything to others: it forms an enclosure, functional only on the Sabbath, inside which the law that forbids carrying that day is suspended, because the enclosure takes on, only at this time, the character of a private dwelling. The "mixture" (*eruv*) of private and public space has generated many misunderstandings and, indeed, attacks and pogroms, as indicated in this report from the Polish town of Kalisz in 1878: "In the past the *eruv* led to bitter quarrels between Jews and Christians. … On Sunday, July 3rd, 1878, when the Christians went to Church, the priests delivered sermons of incitement against the Jews, particularly the Rabbi, and the *eruv* served as the excuse. When the Christians left the Church they proceeded to the Jewish Quarter in masses and began rioting."[36]

In both times of construction and destruction, the *eruv* marks a space defined by Jewish identity, a metaphorical gateway. Debates about *eruvim,* as Fonrobert has shown, are crucial in Talmudic and rabbinical discourse.[37] They were also a focus of debate between reform-oriented and Orthodox communities in the decades after the Haskalah.[38] But only in the last two decades has the creation of new *eruvim* in cities in North America, the United Kingdom, and Australia, emerging from efforts to accommodate the needs of growing Orthodox Jewish communities, led to discussions of "Jewish spaces" and "public spaces." The interaction between these spaces has sparked the interest of urban sociologists, urban geographers, town planners, municipal officials, and rabbis, as well as the resistance of some residents.

Often, such debates appear to revolve around the erection of poles connected by wires to symbolically replace and represent the *eruv* borders. But, of course, larger debates loom in the background about what defines a home and what different communities need to belong. Following Talmudic and rabbinical discussions, the creation of an *eruv* requires boundary setting: "Boundaries can include fences, buildings, walls, rivers, even hillsides. Thus, a small island could be said to be an *eruv,* as could a gated neighborhood. In areas where existing structures can't be used, an *eruv* is physically constructed."[39] Some have regarded this as an intrusion into the public lives of those outside of the religious community.

Eruvim create and contest urban space and give rise to discussions around doorsteps, thresholds, *limina,* still today. Britain has seen some of the most heated debates about this, which, interestingly, have attracted the attention of geographers and urban planners rather than of historians. The establishment of an *eruv* has a religious function but also creates a dialogue between Jewish and urban spaces. Opposition has been voiced by non-Jewish neighbors but also by Reform Jews who fear that the construction of *eruvim* will lead to the establishment of new ghettos and, consequently, further marginalize the Jewish community within the wider society.[40]

On the Doorstep of Others

The debate about *eruvim,* as I have shown, revolves around the religious practices in public space. Jewish law is only one of many religiously based coordinating systems in contemporary cities, making observations of how different cultural practices interact possible: Are there "common patterns for integrating the personal practice of religion into public life" between different religious groups? How is the interdependence between culture and religion regulated and discussed among them?[41] Jewish cultural practice has evolved in dialogue with other religions. Only rarely have the results of research into Jewish spatial practices been juxtaposed with debates about Christianity or Islam and the "place-making" practices of these faiths. But this could be highly fruitful, as Hicham Ben Abdallah El Alaoui's remarks on current trends in Muslim culture illustrate: "The public space, then, is increasingly dominated by a cultural norm based on the elaboration of a set of strict rules, a series of dos and don'ts taken from religious texts strictly construed."[42] Can contemporary cities still contain, and create, "pathways between the sacred spaces of religion and the casually secular discourses of profane culture"?[43]

El Alaoui ends on a hopeful note, arguing that a new form of secular culture needs to be developed that does not reject religion but initiates a reconnection with "the Arab and Islamic tradition that built spaces for cultural autonomy over centuries." Conflicts not dissimilar to those surrounding the Orthodox, or rather orthopractical, Jewish *eruvim* arise when Muslim or Christian religious practice interferes (to others) with public space, as discussions about the Islamic call to prayer or about the ringing of church bells show. Have the "differences between religious values and a shared civic culture" become "problematic" or even "irreconcilable," as Halima Begum argues?[44] Or are public urban spaces still places of negotiation?

In his key text on liminality, Victor Turner writes that "the attributes of liminality or of liminal personae ('threshold people') are necessarily ambiguous, since this condition and these persons elude or slip through the network of classifications that normally locate states and positions in cultural space."[45] Religious attitudes and practices have proved to be much more important than theories of secularization have stated: as John Berger states, the world is "as furiously religious as it ever was, and in some places more so."[46] While "doors are permitted within walls," as the laws of *eruvim* say,[47] conflict is as possible as contact or (following Kant) *Beisammensein,* being together. *Wohnen,* dwelling, writes Klaus Duntze, a Berlin-based protestant pastor and activist in the 1970s, is a foundation for the creation of *Heimat.* Drawing on the Jewish diasporic experience, Duntze argues that any reflection upon the exilic situation—in text and in prayer, but also in so-called cultural practice—and upon the consequences of "being called out, embarking upon a journey," requires constant awareness of the idea and the importance of dwelling, of staying in a place and being at home. "Faith as it was lived by Israel needed to become describable in speaking of the city and of [the experience of] residing."[48] This very specific experience has become a matrix for our understanding of the human condition in modern times. For ever more people around the globe, no matter what religious rules they may choose to follow, this condition is an urban one: to consider the city's present and future means to ask how it makes being-in-residence, remaining, dwelling—*Wohnen*—possible[49]—as long, that is, as a family or a community can dwell in peace.

Jewish history has sometimes, and with good reasons, been described as "a journey."[50] Trying to bring together, and into dialogue, motives and questions from migration studies with research on Jewish homes and settlements, we often encounter the idea and practice of rites of passage: processes of real and of symbolic transition, says Philip Schlesinger, involve "stages of separation, marginality and reaggregation"[51]—in Turner's words, of liminality. The line of research suggested here could take us to the port cities of emigration and the following sea voyage as a transitory experience that many migrants, despite the harshness and burdens that came with it, have described as a kind of liberation.[52] The ship itself as a means of transport and reflection—according to Michel Foucault "the heterotopia par excellence"[53]—has been represented as a kind of suspended time/space model that provided an opportunity to look both backward *and* forward, to integrate longing and belonging, if only for a short moment of reflection, in a temporary home away from home.[54]

Joachim Schlör is professor for modern Jewish/non-Jewish relations at the University of Southampton. His contributions to Jewish studies—on the history of the city of Tel Aviv, on the emigration of German Jews to Palestine, or on the "urban character" of modern Jewish culture—are based in ethnography and cultural studies.

Notes

1. See Haumann, "Lebensweltlich."
2. Hitzer and Schlör, "Introduction."
3. Simone Lässig and Miriam Rürup, in the introduction to this volume, p. 9.
4. Universal questions about the process of migration among those who "leave" versus those who "stay" were recently discussed in Dekel-Chen, "East European Jewish Migration."
5. Cf. Brinkmann, *Points of Passage.* Tony Kushner's edited volume *Remembering Transmigrancy: Place and Jewish/Non-Jewish Identities from the 1840s to the Twenty-First Century* will be published in 2017.
6. See Kümper, Schneider, and Thein, *Makom*; Brauch, Lipphardt, and Nocke, *Jewish Topographies*; Mann, *Space and Place in Jewish Studies.*
7. Bronner, "Dualities," 1.
8. Marty, "Religion."
9. Bronner, "Dualities," 3.
10. This reference is from the website of the Union of Progressive Jews in Germany and just serves as an example; of course there are many more (and different) such instructions. Union Progressiver Juden in Deutschland, "Alltag: Den Alltag heiligen," http://www.liberale-juden.de/das-liberale-judentum/leben/alltag/ (accessed 17 Janaury 2017).
11. Ochs, "Jewish Home."
12. Ibid.
13. Lässig, *Jüdische Wege.*
14. See for a very vivid illustration Loewy, *Jüdisches.*
15. Neue Gesellschaft für Bildende Kunst, *Revisiting Home.*
16. McKay, "Door Magic," 1.
17. Shapiro, "Shylock's House."
18. Ibid., 7.
19. Ackermann and Schülting, *Shylock.*

20. "Jewish Practice: Mezuzah Stories," http://www.chabad.org/library/article_cdo/aid/711
    130/jewish/Stories.htm; Aviva Ravel, "A Mezuzah for Mama," http://www.chabad.org/
    library/article_cdo/aid/1750234/jewish/A-Mezuzah-for-Mama.htm; Rabbi Kushi Schus-
    termann, "A Mezuzah Story," Harford Chabad, 3 January 2013, http://www.harfordch
    abad.org/templates/blog/post.asp?aid=1335125&PostID=36080&p=1; Mezuzah Net (Baal
    Shem Tov Foundation 2003-2005), http://www.mezuzah.net/stories.html (all accessed
    July and August 2014).
21. "Jewish Practices and Rituals: The Mezuzah," Jewish Virtual Library, accessed 10 August
    2014, http://www.jewishvirtuallibrary.org/jsource/Judaism/mezuzah1.html.
22. Ibid. For the symbolic implications of the act of removing a mezuzah, cf. Schlör, "'Take
    Down Mezuzahs.'"
23. Pomian, *Der Ursprung des Museums*, 38–54.
24. Feinberg, "Der Fasan," 250 (trans. Katherine Ebisch-Burton).
25. Ibid.
26. Boas, "Shrinking World."
27. Franke, *Geschichte und Schicksal*, 131–32.
28. Harris, *Little White Houses*, ix–x.
29. Ibid., 1.
30. I would like to extend my thanks for this information to Katrin Keßler, Bet Tfila-
    Forschungsstelle für jüdische Architektur in Europa, Brunswick, Germany (e-mail, 18
    August 2014).
31. This refers to Genesis 28:17: "And he was afraid, and said, 'How dreadful is this place!
    This is none other but the house of God, and this is the gate of heaven.'"
32. E-mail to the author, 24 August 2014.
33. J. Berger, "Zweites Bild." The article sums up and analyzes questions and statements by
    visitors to various synagogues where the author served as a rabbi.
34. Fonrobert, "Separatism"; Fonrobert, "Political Symbolism"; Cousineau, "Domestication";
    Cousineau, "Urban Boundaries."
35. "History of City Eruvin: Part 1: The Eruv in Krakow," Eruv Online: Disseminating the
    Truth about Eruvin the World Over, 22 December 2005, accessed 29 August 2014,
    http://eruvonline.blogspot.de/2005/12/history-of-city-eruv-in-part-1-eruv-in.html.
36. "The Jubilee of the Eruv: The Story of a Pogrom," Eruv Online: Disseminating the Truth
    about Eruvin the World Over, 27 December 2005, accessed 29 August 2014, http://eru
    vonline.blogspot.de/2005/12/eruv-stories-sticks-and-stones.html: "The rav of the town of
    Kalish at that time was Harav Chaim Elazar Wachs z'l (1822–1889), the world-renowned
    author of the *Nefesh Chayah*. The account of the pogrom below appeared in *Dos Kalisher
    Leben* 2, July 29, 1927."
37. Fonrobert, "Separatism"; Fonrobert, "Political Symbolism"; cf. also her review of three
    publications relating to the Potsdam Makom project: "New Spatial Turn."
38. Schlör, *Das Ich der Stadt*, 11–27.
39. "The Jewish Religious Landscape: Eruvim (2004)," The Pluralism Project at Harvard Uni-
    versity, accessed 29 August 2014, http://www.pluralism.org/reports/view/160.
40. Watson, *City Publics*, 21.
41. Ibid.
42. El Alaoui, "Split."
43. Ibid.
44. Begum, "Geographies."
45. Turner, "Liminality and Communitas," 1.

46. P. Berger, "Desecularization," 2.
47. "The problem was that it is impractical to build a continuous solid wall around a community. However, the rabbis noticed that doors are permitted within walls, and that a doorway consists of two parts: the vertical members and the lintel on top. In fact, a wall may have quite a few doors, and still be considered to enclose an area." "Shabbat: Eruv," Jewish Virtual Library, accesssed 3 January 2015, http://www.jewishvirtuallibrary.org/jsource/Judaism/eruv.html.
48. Duntze, *Geist,* 138.
49. Cf. Saunders, *Arrival City.*
50. The title of the permanent exhibition at the Jewish Museum in New York has long been "Culture and Continuity: The Jewish Journey."
51. Schlesinger, "W. G. Sebald," 46.
52. A workshop held at Graz University on 20 June 2014 discussed "The Sea Voyage as a Transitory Experience in Migration Processes" (with contributions from Dieter Hecht, Elisabeth Janik, David Jünger, Carolin Matjeka, Philipp Mettauer, Joachim Schlör, Johanna de Schmidt, and Björn Siegel); for a contextualization of the sea voyage in current spatial research, see Rolshoven, "Raumkulturforschung."
53. Foucault, "Of Other Spaces," 336.
54. The first issue of the journal *Mobile Culture Studies* is dedicated to "The Sea Voyage as a Transitory Experience," see http://unipub.uni-graz.at/mcsj (accessed 29 August 2014).

Bibliography

Ackermann, Zeno, and Sabine Schülting. *Shylock nach dem Holocaust: zur Geschichte einer deutschen Erinnerungsfigur.* Berlin: de Gruyter, 2011.

Begum, Halima. "Geographies of Inclusion/Exclusion: British Muslim Women in the East End of London." *Sociological Research Online* 13, no. 5 (2008). Accessed 11 March 2016. http://www.socresonline.org.uk/13/5/10.html.

Berger, Joel. "Zweites Bild: In der Synagoge." In *Antisemitismus. Vorurteile und Mythen,* edited by Julius H. Schoeps and Joachim Schlör, 67–73. Munich: Piper Verlag, 1995.

Berger, Peter. "The Desecularization of the World: A Global Overview." In *The Desecularization of the World: Resurgent Religion and World Politics,* edited by Peter Berger. Washington, DC: Wm. B. Eerdmans, 1999.

Boas, Jacob. "The Shrinking World of German Jewry." *Leo Baeck Institute Yearbook* 29 (1984): 3–25.

Brauch, Julia, Anna Lipphardt, and Alexandra Nocke, eds. *Jewish Topographies: Visions of Space, Traditions of Place.* Aldershot: Ashgate, 2008.

Brinkmann, Tobias, ed. *Points of Passage: Jewish Migrants from Eastern Europe in Scandinavia, Germany, and Britain 1880–1914.* New York, Oxford: Berghahn Books, 2013.

Bronner, Simon J. "The Dualities of House and Home in Jewish Culture." In Bronner, *Jewish Cultural Studies,* vol. 2, 1–40.

———, ed. *Jewish Cultural Studies,* vol. 1, *Jewishness: Expression, Identity and Representation.* Oxford: Littman's Library of Jewish Civilization, 2008.

———, ed. *Jewish Cultural Studies,* vol. 2, *Jews at Home: The Domestication of Identity.* Oxford: Littman's Library of Jewish Civilization, 2010.

Cousineau, Jennifer. "The Domestication of Urban Space and the North-West London Eruv." In Bronner, *Jewish Cultural Studies,* vol. 2, 43–74.

————. "Urban Boundaries, Religious Experience, and the North-West London Eruv." In *Making Place: Space and Embodiment in the City,* edited by Arjit Sen and Lisa Silverman, 126–47. Bloomington: Indiana University Press, 2014.

Dekel-Chen, Jonathan. "East European Jewish Migration: Inside and Outside." *East European Jewish Affairs* 44, nos. 2–3 (2014): 154–70.

Duntze, Klaus. *Der Geist, der Städte baut. Planquadrat—Wohnbereich—Heimat.* Stuttgart: Radius Verlag, 1972.

El Alaoui, Hicham Ben Abdallah. "The Split in Arab Culture." *Midan Masr: An Opinionated Paper.* Accessed 1 August 2014. http://www.midanmasr.com/en/article.aspx?ArticleID=87.

Feinberg, Anat. "Der Fasan, Bialik und die Sehnsucht nach dem Zuhause." In *Exilforschung. Ein internationales Jahrbuch,* vol. 31, *Dinge des Exils,* edited by Doerte Bischoff and Joachim Schlör, 250–67. Munich: Text & Kritik, 2013.

Fonrobert, Charlotte. "From Separatism to Urbanism: The Dead Sea Scrolls and the Origins of the Rabbinic Eruv." *Dead Sea Discoveries* 11, no. 1 (2004): 43–71.

————. "The New Spatial Turn in Jewish Studies." *AJS Review* 33, no. 1 (2009): 155–64.

————. "The Political Symbolism of the Eruv." *Jewish Social Studies* 11, no. 3 (2005): 9–35.

Foucault, Michel. "Of Other Spaces: Utopias and Heterotopias." In *Rethinking Architecture: A Reader in Cultural Theory,* edited by Neil Leach, 330–36. New York: Routledge, 1997.

Franke, Hans. *Geschichte und Schicksal der Juden in Heilbronn.* Heilbronn: Stadtarchiv, 1963. [The volume has been updated and published online by the city archives of Heilbronn: www.stadtarchiv-heilbronn.de/…/03-vr-11-franke-juden-in-heilbronn.pdf.]

Harris, Diane. *Little White Houses: How the Postwar Home Constructed Race in America.* Architecture, Landscape and American Culture Series. Minneapolis, London: University of Minnesota Press, 2013.

Hitzer, Bettina, and Joachim Schlör. "Introduction: God and the City; Religious Topographies in the Age of Urbanization." *Journal of Urban History* 37, no. 6 (2011): 1–6.

Haumann, Heiko. "Lebensweltlich orientierte Geschichtsschreibung in den Jüdischen Studien: Das Basler Beispiel." In *Jüdische Studien. Reflexionen zu Theorie und Praxis eines wissenschaftlichen Feldes,* edited by Klaus Hödl, 105–22. Schriften des Centrums für Jüdische Studien 4. Innsbruck: Studienverlag, 2003.

Kümper, Mihal, Ulrike Schneider, and Helen Thein, eds. *Makom. Orte und Räume im Judentum. Real. Abstrakt. Imaginär.* Hildesheim: Olms, 2007.

Lässig, Simone. *Jüdische Wege ins Bürgertum. Kulturelles Kapital und sozialer Aufstieg im 19. Jahrhundert.* Göttingen: Vandenhoeck and Ruprecht, 2004.

Loewy, Peter. *Jüdisches.* Munich: Kehayoff, 2000.

Mann, Barbara. *Space and Place in Jewish Studies.* Key Words in Jewish Studies. New Brunswick, NJ: Rutgers University Press, 2012.

Marty, Martin E. "Religion: A Private Affair, in Public Affairs." *Religion and American Culture: A Journal of Interpretation* 3, no. 2 (Summer, 1993): 115–27. Accessed 12 March 2016. http://www.jstor.org/stable/1123984.

McKay, K. J. "Door Magic and the Epiphany Hymn." *Classical Quarterly,* n.s., 17, no. 2 (November 1967): 184–94.

Neue Gesellschaft für Bildende Kunst, ed. *Revisiting Home: Dwelling as the Interface between the Individual and the Society.* Published on the occasion of the exhibition Revisiting Home at NGBK, 8 September–15 October 2006. Berlin: NGBK, 2006. [Working group: Ani Corcilius, Birgit Kammerlohr, Iain Pate, Inken Reinert, Janine Sack.]

Ochs, Vanessa Ochs. "What Makes a Jewish Home 'Jewish'?" *Cross Currents* 49, no. 4 (Winter 1999/2000). Accessed 11 March 2016. http://www.crosscurrents.org/ochsv.htm.

Pomian, Krzysztof. *Der Ursprung des Museums: Vom Sammeln.* Berlin: Wagenbach, 2007.

Rolshoven, Johanna. "Raumkulturforschung—Der phänomenologische Raumbegriff der Volkskunde." In *Raum: Konzepte in den Künsten, Kultur- und Naturwissenschaften,* edited by Petra Ernst and Alexandra Strohmaier, 125–40. Baden-Baden: Nomos, 2013.

Saunders, Doug. *Arrival City: How the Largest Migration in History Is Reshaping Our World.* Toronto: Windmill Books, 2011.

Schlesinger, Philip. "W. G. Sebald and the Condition of Exile." *Theory, Culture and Society* 21, no. 2 (April 2004): 43–67.

Schlör, Joachim. *Das Ich der Stadt. Debatten über Judentum und Urbanität, 1822–1938.* Göttingen: Vandenhoeck & Ruprecht, 2005.

———. "'Take Down Mezuzahs, Remove Name-Plates': The Emigration of Material Objects from Germany to Palestine." In Bronner, *Jewish Cultural Studies,* vol. 1, 133–50.

Shapiro, Michael. "Shylock's House: Theatrical Representations of Jewish Space." Paper delivered at the Tenth Congress of the European Association of Jewish Studies, Paris, 22 July 2014.

Turner, Victor. *The Ritual Process: Structure and Anti-Structure.* New Brunswick, NJ: Aldine Transactions, 2008.

Waligórska, Magdalena. *Klezmer's Afterlife: An Ethnography of the Jewish Music Revival in Poland and Germany.* Oxford: Oxford University Press, 2013.

Watson, Sophie. *City Publics: The (Dis)enchantments of Urban Encounters.* London, New York: Routledge, 2006.

14

Photography as Jewish Space

Michael Berkowitz

Before the Nazi onslaught, photography was largely a Jewish space as a field of endeavor in Central Europe and beyond.[1] Although a precise retrospective measure is impossible, it seems that Jews were engaged more intensively in photography than in any other occupational or professional cohort (outside of the religious-Jewish-specific). There was a continuous flow of Hungarian Jews and *Ostjuden* in photography-related occupations into Germany and Austria, so making a clear separation between Central and Eastern Europe is difficult.[2] In a narrow sense, calling photography "Jewish" implies that in addition to photographers per se, studios, photographic laboratories, and photographic equipment stores tended to be owned by Jews. Jews also worked behind the scenes as retouchers and technicians in laboratories developing film, including medical and dental X-rays. They were prominent, and troubling to anti-Semites, as photography editors and agents. Erich Salomon (1886–1944) and Leo Rosenthal (1884–1969) were well-known Jewish press photographers who reached vast national and even international audiences. Jewish women such as Yva (Else Ernestine Neuländer, 1900–1942) and Ilse Bing (1899–1998) were among the most prolific advertising photographers. What might be termed the Jewish space of photography was thus expansive and diverse, comprising working space, living space, commercial premises where Jews interacted with non-Jews, outdoor and indoor public spaces where photographers conducted their work, and spaces of material culture such as billboards, newspapers, and magazines.

Most historiography that focuses on Central European Jewry, which has included a substantial photographic dimension, emphasizes one of two approaches: to show how Jews were represented, stressing their outsider or victim status,[3] or to illustrate how Jews sought to represent themselves.[4] With few exceptions, photography has not been acknowledged as a disproportionately Jewish economic realm.[5] Scholars and curators have nevertheless excavated a notable Jewish presence, followed by an abrupt absence, in photography's history. But attention to the Jewish identities of the principals has been inconsistent. Jewishness is often reduced to a matter of origins, portrayed as significant only as the cause of persecution. Photographers themselves are most readily recalled in this context, while those who facilitated photography have been slower to attract notice. Much greater scholarly attention, however, has been paid to the picturing and representation of Jews "as Jews" in interwar Europe as well as the Holocaust.[6] Notable exceptions include the work of David Shneer on Soviet photojournalists—which brilliantly situates the subject in a larger historical field—and the recent exhibition and catalogue on early Soviet photography and film.[7] Martin Deppner, through individual and collective projects, has been the most dedicated to discerning the myriad connections between photography's development and the involvement of Jews with the media.[8] A great lacuna in contemporary history, generally, due to the Holocaust, is the very memory of the extent to which Jews predominated in continental European photography, which profoundly shaped photographic trends in Britain and the United States,[9] and animated nearly every facet of the application and vocations of photography.

While it is indeed important to discuss the Jews involved in photography who were displaced, dispossessed, and murdered, such as photographers, picture editors, and proprietors of studios and photography stores, I wish in this chapter not only to recall or recognize Jewish roles or contributions in the pre-Nazi socioeconomic order. In an approach consistent with a little-noticed article by Tim Gidal on Jews and photography in the 1987 *Leo Baeck Yearbook,* I wish to underscore the fact that Jews were at the cutting edge of photography and changed the shape of mass visual culture.[10] It is not surprising, therefore, that a particularly treasured and memorable photograph of early post–World War II Germany—for Germans and the world at large—was the work of a formerly German-Jewish photographer.

One of the more unusual books to appear in the aftermath of World War II is *German Faces* (1950), conceived by a Jewish refugee from Hitler's Germany, Henry Ries, and his non-Jewish, American-born wife, Ann Stringer.[11] It was translated and republished in German around the time of the fall of the Berlin Wall, featuring photographs comprising "a portrait of the people, large and small, leaders and led, who must remake Germany in their own image."[12] The text is one of the earliest surveys of postwar Germany to confront the fact that there was, in fact, a small Jewish minority living in the country that faced unprecedented challenges.[13] The book is distinctive in calling attention to the persistence of rabid nationalism and the feeling among many Germans that they were more victims than victimizers.[14]

Henry Ries, born as Heinz in 1917, would later be recognized for having brilliantly captured in an iconic photograph a signal historical moment: the Berlin airlift. The few reviews of *German Faces* mention its portraiture as reminiscent of the work of August Sander (1876–1964) and note the overall quality of the

Figure 14.1. Henry Ries, Berliner Kinder beobachten das Landemanöver eines "Rosinenbombers" am Flughafen Tempelhof, July 1948. © bpk / Henry Ries

pictures.[15] In West Germany, Ries's photography became the subject of at least two exhibitions and another volume,[16] and he published an autobiography in German.[17] As well as acknowledging its artistic value, we might regard *German Faces* as significant in light of German-Jewish history and in relation to Jewish space. The book represents an attempt, however inadvertent, to reclaim photography in Central Europe as a Jewish domain. While Sander and some other leading (pre-Nazi era) photographers were not Jewish, the fact remains that Jews were tremendously overrepresented in photography, generally, before 1933. Helmut Gernsheim (1913–95), himself a refugee from Nazi Germany, a superb photographer, and a foundational historian of photography, believed that there was a compelling Jewish story interwoven in the history of photography, although he never saw this idea through to completion.[18] It is therefore neither an oddity nor coincidence that *German Faces,* a sympathetic collective portrait of the defeated nation, and Ries's scene of the Berlin airlift, one of the more moving images of Germans in the aftermath of World War II, which "reduced the tense showdown between the great powers to the image of children on a hillside watching a plane gliding in,"[19] are the work of a Jewish photographer.

This said, I wish to defuse any impression that the current examination might be an exercise in valorization. One reason there is such a huge disparity between Jews' roles in photography and the notice they have received from non-Jews and Jews—German and otherwise—is that for all the value placed on some photographs, photography itself was not considered the most respectable of vocations. Descendants of photographers are often unaware of their ancestors' profession. Historians had until quite recently barely begun to approach the problem of photographers' obscuration and marginalization. Photography's problematic relationship with respectability provided a critical opening for Jews, while it also helped obfuscate the Jewish inflection on the history of the field. Jewish and non-Jewish historians and theoreticians of photography have paid far too little heed to the social composition and commercial dimensions of the field. Metacritiques of photography as racist, sexist, and beholden to dominant interests indicate a dearth of historical knowledge about its practitioners in general. The Polaroid company of Edwin Land (1909–91), which was spectacularly successful at developing products for the US military during World War II, was in 1944 derided and denounced for greed and disloyalty by the Boston Chamber of Commerce in terms that would have pleased the Nazis.[20] Sir Kenneth Clark, who might have helped photography's institutionalization in the fine arts in Britain, withheld substantial support because he considered the leading expert on photographic history, Helmut Gernsheim, to be so "unattractive."[21]

The production of photographs involved working long hours in darkness, with strange-smelling and hazardous chemicals. Early photography was sometimes seen as inextricable from pornography and unbridled egoism. The main function of studio photography, portraiture, compelled the photographer to touch and position the sitter, which led to accusations that photographers took liberties. Furthermore, photographers were entrusted with retouching prints and negatives in order to please their customers. This dimension of photographers' craft has received little attention, in part due to persistent myths of the neutrality and honesty of photography. I. J. Singer's short story "In der Finster" ("In the Dark") of 1919 reconstructs the dank, dark, and harried working world Singer experienced as a retoucher for Alter Kacyzne in Warsaw,[22] where the photography trade was largely in Jewish hands.[23] (Kacyzne suffered a horrific death at the hands of Ukrainian collaborators with the Nazis in 1941.[24])

Moves toward acknowledging Jews' prominence in photography, especially in Germany and Austria, have been under way since the late 1950s. The rediscovery of Henry Ries in Berlin is a small part of this uncoordinated effort. A vast body of work exists, mostly in the form of exhibition catalogues on individual Jewish photographers—mainly those who escaped Nazi clutches and ended up in England, the United States, Australia, South Africa, South America, and Palestine. There are a few studies of those who worked as partners, and substantial volumes on notable photographers from Hamburg and Munich.[25] Few scholars, though, have observed that beyond outstanding individuals, there were vital networks of Jews in photography that cut across generational and national lines. Rolf Sachsse has written of the multitude of photography studios in Germany in Jewish hands until 1933 and shown how "Aryans" rushed to fill the roles vacated by Jews as agents and photojournalists.[26] Christoph Kreutzmüller features two photographic institutions in his exhibition and catalogue *Final Sale* (2011),[27] and his book *Ausverkauf. Die Vernichtung der jüdischen Gewerbetätigkeit in Berlin 1930–1945* likewise acknowledges the significance of Jews in photography.[28] In 2012–13, an exhibition held at Vienna's Jewish Museum, "Shooting Girls," focused on the women who owned and ran most of the "quality" photographic studios in Vienna before the *Anschluss*.[29] The "Shooting Girls" can be supplemented with Lore Krüger, Aenne Biermann, Yva, Lotte Jacobi, Gertrude Fehr, Ruth Jacobi, Ellen Auerbach, Ilse Bing, Grete Stern, Marianne Breslauer, Eva (Sandberg) Siao, Leni Sonnenfeld, Ursula Wolff-Schneider, Frieda Riess, Lore Lizbeth Waller, Yolla Niclas Sachs, Lisolette Grschebina, Lucia Moholy, and Gisele Freund.

In a recent article in *German History,* Annette Vowinckel argues that Jewish photojournalists fared unusually well as exiles due to their extensive professional networks.[30] While this is an important insight, she underestimates the cases of those who did not succeed in emigrating in the first instance, as well as those who fell under her radar because of having failed to reconstruct their careers. Vowinckel's piece, though, is significant because it is part of a larger effort in Germany to recast Jews and photography and an unusually successful German-Jewish story.[31]

Although many Jewish photography businesses were ruined or taken over shortly after 1933, a number were allowed to remain open and particularly served Jews needing identification and passport photos. Several Jewish photographers, such as Erich Kastan, were able to work during Nazi times. This was due to the National Socialist penchant for documenting the Jewish cultural activities that were allowed and even officially encouraged. Some seventy individuals or families were explicitly assisted by Ernst Leitz and the famed Leica company based in Wetzlar.[32] The Zeiss factories in Dresden and Berlin, like Leica, were reputed to be among the least oppressive sites for Jewish forced laborers. But in 1939, a group of Zeiss employees took it upon themselves to persecute those who were still employed by the firm.[33]

Harvey Fireside's description in an unpublished memoir of his family's business and their fate under the Nazis reflects the experience of a Viennese Jewish photography studio that was not among the "quality" establishments previously mentioned.[34] Since the *Anschluss* the business had been thriving, mainly due to Jews needing identity photos. During the event that came to be known as *Reichs-kristallnacht,* Harvey recalled that two Hitler Youth forced his father to open his shop to them.[35] The men jammed the contents of the store—"cameras, accessories, photo albums, even light furniture"—into their motorcycles' sidecars. In the midst of this, an SS man appeared who challenged the youth to produce orders and halted their plunder—claiming he would take over. Fireside's father recognized the SS man as Gerd Kalmus, "an old customer," and believed that Kalmus might have saved his life. The SS man advised him to flee the country immediately.[36]

In addition to the *Shooting Girls* show, memoirs such as Fireside's, and related artifacts, there is another formidable source of data about Jews and photography in Vienna: the records of the Mauthausen concentration camp. Located near Linz, the infamous granite quarry was the center of the labor camp belonging to the city of Vienna. Close to 200,000 prisoners went through the Mauthausen camp system between August 1938 and May 1945.[37] Some 95,000 died there, among whom at least 14,000 were Jews. Records reveal that at least 125 of these Jews were photographers or worked in studios or laboratories.[38] I would estimate that there might have been between 200 and 400 Jewish photographers who died in or passed through the camp, given the sketchiness of information during different periods and the tendency for inmates to list occupations that they believed might facilitate survival. Many of those declared as photographers were Hungarians who had come through Budapest or elsewhere in Hungary, while others were from Poland.[39]

One of Mauthausen's Jewish photographers was Bernard Gotfryd.[40] Interestingly, Gotfryd did not claim photography as his profession on entry to the camp. He chose instead to call himself a locksmith, because he had heard that this was one of the preferred vocations for work details and therefore a possible factor in one being allowed to live.[41] Gotfryd is well published, but his writing and life story have been underused and perhaps undervalued by scholars.[42] Scant attention has been paid to his work with regard to photography and World War II, which he describes in meticulous detail. Soon after the outbreak of war, Gotfryd explains, he "was hired as an apprentice at a photography studio owned by a friend of the family. (There were no more schools for Jews)."[43] He worked there until August 1942 and then got a position in a remaining studio "owned by a Polish couple of German descent [who] were given permission by the *Polizeifüh-rer* to employ three Jews."[44] Some aspects of Gotfryd's work as a photographer in the Radom ghetto were of particular note. First, he photographed Jews and non-

Jews, and even Nazi officers, in the studio.[45] Indeed, it was very popular with the soldiers, and they also used it to have film developed of the pictures they took of their comrades—typical tourist shots and battlefield scenes—as well as images of grotesque treatment of civilians.[46]

Gotfryd was approached by Alexandra, a non-Jewish woman in the Polish underground, who asked Gotfryd to undertake the dangerous assignment of making duplicate negatives of photographs revealing the brutal nature of the occupation—"atrocities and executions"—so that these could be used to prove the Poles' case of what the Nazis were inflicting on their country. The underground also was keen to have portraits of "high ranking officials of the Gestapo" and SS men, many of whom ignored explicit orders not to have themselves photographed in commercial studios.[47]

In the context of the Holocaust, Gotfryd's relations with the Nazis might seem strange. He was, in many respects, treated "normally," perhaps because he was helping to provide a service that was appreciated. Gotfryd had an unusually close relationship with a Waffen SS man he referred to as "Kurt," who supplied him with bread and extra food, who, Gottfryd suspected, might have had Jewish ancestry. Godfrey also tells a story about how the photographer Orenstein, his wife, and his son were saved by a Nazi, Helmut Reiner, who sent them into hiding during a round-up for transports.[48] Reiner had been

> the photographer at the Gestapo headquarters, and Orenstein, a master in his art, was his negative retoucher. High-ranking Gestapo officials were not to be photographed in privately-owned studios; photography at Gestapo headquarters became solely Reiner's responsibility. As soon as Reiner had heard about the upcoming deportations in the ghetto, he decided to keep Orenstein out of it. He had known him for over a year; he respected him for his superb craftsmanship and punctuality. Reiner knew that he was taking a chance protecting a Jew, but he was willing to do it; it wasn't in his nature to turn his back on people in need of help, particularly an esteemed friend.[49]

Yet there was also a pragmatic dimension. "Reiner wanted to stay in Poland as long as he possibly could. Things weren't too bad for him there. If he could only have Orenstein, his photography operation at the Gestapo might be assured, and he might not have to go to the east."[50] The plan worked: "For almost another year Orenstein was employed by Reiner, retouching the Gestapo negatives. Occasionally Reiner would visit him in the ghetto, bringing him extra food. Late in 1943 the last of the [Radom] ghetto was liquidated, and the Orensteins were shipped to camps."[51] Reiner, a diabetic, was sent home to Vienna, where he stayed until the end of the war. The Orensteins survived the camps, but their son did not.

Although Godfryd does not know for certain, it seems that Reiner had apparently worked as a photographer in Vienna before the *Anschluss* and was not enamored of anti-Semitism.[52] It is likely that he was comfortable with Jews. Departing from Gotfryd's observations, we might delve more deeply into Reiner's sympathy for Orenstein. Why would a retoucher in the Gestapo photo studio be so important? The Nazis found that relatively objective professional and forensic photographs did not necessarily tell the kind of the story they desired. For instance, Nazi press organs avoided using photos of Jews who did not seem to "look Jewish."[53] The Nazis needed an expert retoucher for SS photographs for similar reasons: a lot of the supposed Nazi racial elite had rather big noses and large ears and were not particularly attractive and at the very least needed to be made more presentable and more stereotypically "Aryan" in photographs.[54] Besides the value of a retoucher, there are other reasons Gotfryd and Orenstein may have been treated rather humanely in the context of the Holocaust. At the heart of this lies the notion of photography as a Jewish space: even the Nazis regarded it as normal, and desirable, to have Jews taking care of photography, except for enclaves such as the *Erkennungsdienst* (information service division) at Auschwitz and other camps.[55]

Before coming to power, Nazis were apparently not averse to using the services of a Jew when in need of a professional quality photograph. It is reputed that Lotte Jacobi was offered "honorary Aryan status" by the Nazis in 1935 because her clients included "high-ranking German officials" who "praised her work as 'good examples of Aryan photography.'" This may seem ironic, as Lotte and her family were said to be politically active on the left.[56] Jacobi herself, however, never claimed to have participated in "political activities" or "social causes" beyond conversation. Moreover, she adamantly denied that her work had "political content": "I photographed the politicians of all parties," she asserted. "I didn't select one over the other."[57] Jacobi's comments remind us that Jews in photography did not mainly take pictures of Jews or those they considered sympathetic, but of everybody, including those they might have considered unseemly or who espoused ideologies from which they themselves were excluded.

However, being permitted as a Jew to work under the Nazis as a photographer did not mean that one's life would be spared. The famed Yva, Else Ernestine Neuländer, who had been one of the great advertising and fashion photographers of Central Europe, continued a career in photography of sorts as an X-ray technician in Berlin's Jewish hospital until 1942, when she was deported to her death at Sobibor.[58] One of the most illustrious Holocaust victims was a fellow photographer, Erich Salomon, who is widely credited with establishing enduring trends in photojournalism. Salomon's work was characterized by a combination of impeccable timing and the deft concealment of himself and his camera. He also evinced a distinctly modernist gaze, as in photos shot through car windows or from across a dark street into a building. Salomon's impact was particularly promoted by Stefan Lorant, who became a leading editor of the pictorial press in Britain after his imprisonment by Hitler and escape from Munich. Lorant and his British publications, in turn, played a huge role in shaping the character of *Life* magazine, as did the émigrés on its staff and among its leading contributors, such as Fritz Goro, Alfred Eisenstaedt, and Robert Capa; *Life*'s founder, Henry Luce, had also been exposed to the work of Stefan Lorant.[59]

The case of the photographer Leo Rosenthal also may be instructive for comprehending photography, in Germany and beyond, as "Jewish space." At the time of his death in 1969, Leo Rosenthal was well known and respected, mainly as a photographer at the United Nations, and received a substantial obituary in the *New York Times*,[60] whose author was apparently unaware that Rosenthal had been a distinguished photographer in Germany before his arrest by the Nazis in 1933. Rosenthal's trove of photographs provided the basis for an exhibition at the Berlin Landesarchiv (November 2009 to March 2010) and an accompanying catalogue.[61]

Rosenthal was born in Riga, which was then in Tsarist Russia. He was no traditional, poor *Ostjude*; his father was a well-established jeweler to the city's elite. Leo Rosenthal initially became a lawyer, practicing in Moscow, and engaged in revolutionary politics. Although his sympathies were on the side of the Reds, he did not believe he would fare well enough under the Bolsheviks and subsequently moved to Berlin, becoming a reporter for socialist and "liberal-democratic" newspapers, mainly *Vorwärts*. His photographs also appeared in the *8 Uhr-Abendblatt*, the *Weltspiegel*, the supplements to the *Berliner Tageblatt* and *Vossische Zeitung*, the illustrated *Zeitbilder*, and the *Volksfunk*. He struggled to make a decent livelihood out of his assignments; his living conditions and economically precarious existence, especially in comparison to his opulent family home, were disheartening. In Germany he was a Social Democrat but remained on its "periphery."[62]

Not surprisingly, Rosenthal was prominent enough to find himself taken into "protective custody" upon the Nazi takeover of power. His Latvian citizenship facilitated his release from prison in Berlin.[63] He escaped first to his family's home in Riga, where he participated in a largely Jewish, anti-Nazi German publication, then proceeded west, via Paris and Casablanca.[64] His mother, brother, and three sisters were not so fortunate; all were murdered in the Holocaust. Rosenthal arrived in the United States in 1942. The launch in 1945 of the United Nations with its conference in San Francisco saw him seize the opportunity to return to photography as his main vocation.

Although the exhibition and catalogue discuss Rosenthal's development as a photographer per se and include several comparisons with his better-known colleague Erich Salomon, they reveal a blind spot about the prevalence of Jews in photography. In contrast, the importance of Jewish connections is presented well in Bianca Welzing-Bräutigam's examination of Rosenthal's period in Riga and peripatetic existence after fleeing the Nazis.[65] Erich Salomon was not simply a contemporary and forerunner to Rosenthal; his emergence was due to his situation in a disproportionately Jewish milieu, including agents, editors, picture editors, publishers, and agency heads. Those who are familiar with Salomon would rarely mistake his photographs for those of Rosenthal. Even though Rosenthal took many, if not most, of his courtroom photographs surreptitiously—as did Salomon—there is more of a formal, staged, static quality to them, with some exceptions.[66]

While it is important to focus on Germany and Austria and the dispersion of its émigrés in relation to this field, it is also appropriate to mention lands on which the Nazis imposed themselves. Perhaps the only countries where Jews in photography were not vastly disproportionate to their numbers in the general population were Belgium and Estonia. The figure in France was relatively low, at some 10 to 15 percent, but still significant. We should not neglect in this context to draw attention to Jewish photographers in the countries where Jews were largely obliterated. In Lithuania, to cite one example, 94 to 95 percent of the Jews were murdered, mainly by the Einsatzgruppen, in sites such as the Ninth Fort and pits of Ponar. An attempt to reconstruct and interpret what Jews did in their daily lives in Eastern Europe reveals that photography played a highly significant part in Jewish existence. In Kovno, for instance, there were photography shops throughout the city center, of which perhaps not a single one was owned by a non-Jew.

One reason for including Lithuania in this chapter stems from the phenomenon of the transcendence of conventional boundaries in favor of elective affinities. Jews in photography, throughout Lithuania, were comfortable in the both the German culture to the west and the Russian one to the east. Therefore, in the early twentieth century and through the interwar years, today's city of Kaunas was often presented neither as "Kaunas" nor "Kauen," but as "Kowno." This was the distinctly Jewish way of referring to the city. While by no means "German Jews," Lithuanian Jews' ties to German culture and language were not as tenuous as it might seem. In the western parts of Lithuania in particular, Jews had been part of a German cultural universe, while also being Lithuanian, Polish, and Russian.

There has been at least one attempt to recover the history of a Jewish photographer from Samogitia, the western region of Lithuania near Memel, one Chaimas Kaplanskis, or Chaim Kaplan, through an exhibition and catalogue (2007).[67] Chaimas Kaplanskis has been referred to as "the best known photographer of Telsiai [Telz]"; the family's professional photographs "reflected the cultural and public life of Western Lithuania at that time," and it is claimed that Feitska Kaplan was "the first female photographer in Telsiai and one of the first in Lithuania. The Kaplanskis' photo studio … was open for half a century and owned by the family itself for 47 years (1894–1940).… Even today in almost every house of Telsiai one can find at least one photo made in the Kaplanskis' photo studio."[68]

Perhaps the most important message to be gleaned from this catalogue, however, is that Jews did not exist in a totally separate sphere from Lithuanian non-Jews. Although no one has gone so far as to assert that there was ever something akin to the "German-Jewish dialogue," there were striking points of cordial contact, it seems, between Jews and non-Jews in Lithuania. Photography was an area in which Jews and non-Jews came together with little tension. Chaimas Kaplanskis was regarded as the premier photographer of Telsiai by the entire population, as his clients included prominent priests as well as rabbis, and he made portraits of Catholic seminary students as well as yeshiva *bochers*. We also learn, however, that Jewish photographers were among those who suffered discrimination as the notion that non-Jewish Lithuanians needed to wrest business from so-called "foreign" hands gained ground in the 1930s.[69] Many lost their livelihoods even before 1939, owing not to market forces but to ethnic-national conflict.

Although the catalogue treats Chaimas Kaplanskis and his heirs as special and distinctive, it intimates that there are troves of pictures by similar Jewish photographers whose histories remain to be written. It also tells us a great deal about the practice of photography in an area of Europe that is typically seen as backward in comparison to Central and Western Europe. In Western Lithuania too, photography was a vital and highly prized part of life, and Jews were at its center. There were many photographers like the Kaplanskis operating in towns such as Kelme and Jurbarkas. One of the notable things about the family's work, reminiscent of Jacobi shooting political figures of every stripe, is the extent to which Jewish photographers were employed to photograph nationalist and religious groups and ceremonies. There was an aspect of the trade, though, that is now seen as bizarre where it is noticed at all, but was apparently something of a Jewish specialty: the photographing of the dead. Photographers who engaged in this often had an unsavory reputation for taking advantage of the bereaved, as they would use the occasion to entice the family to make reproductions of existing photos. This aspect is not discussed in the catalogue, nor has it been well explained in the history of photography.[70]

Interestingly, but perhaps not surprisingly, photographs of Jews by Jewish photographers, and the work of Jewish photographers generally, in pre-Nazi Europe seem initially to have become highly valued in Germany's photography marketplace. This is evident in the valuations of the photographic collection of Josef Breitenbach, which was first loaned to and later bought by the Munich City Museum.[71] In postwar Germany and Austria, beginning in the late 1950s, some individuals active in photography began to pay attention to the historical eradication of Jews who had been at the heart or the cutting edge of the field. Therefore, photojournalists such as Alfred Eisenstaedt, Gisele Freund, Gerti Deutsch, Simon Gutmann, Josef Breitenbach, and Lisl Steiner were "welcomed" back to Germany and Austria, as was the pioneering collector and historian Helmut Gernsheim. Erich Salomon was posthumously made part of a German cultural canon. All of this served a twofold purpose: it was part of the process of *Vergangenheitsbewältigung,* and it acted as a means of creating the myth of Weimar Germany, especially Berlin, and pre–World War II Vienna as precursors to multiculturalism and as the greatest incubators of avant-garde trends. The selective appropriation of Jews and photography, as part of *Vergangenheitsbewältigung,* however laudatory in intent, has been fragmentary and occasionally distorted in historical reflections.

As personalities and as important collectors of historical photography, Helmut Gernsheim and Josef Breitenbach have been integrated into what may be seen as a German discourse on photography. Toward the end of his life, Helmut Gernsheim arranged to sell and donate the portion of his collection not given to the initial repository of his trove, the University of Texas, to the Reiss-Engelhorn Museum in Mannheim.[72] While the Reiss-Engelhorn Gernsheim Collection, officially installed in 2002, is certainly impressive, it is not particularly well treated in the grand scheme of German archives and museums.

Josef Breitenbach's collection of historical photography in Munich is less known and heralded than that of Gernsheim; it is much smaller and not nearly as comprehensive, although it contains a number of fabulous pieces. Breitenbach is mainly remembered as a photographer for his portraits of Bertolt Brecht and Max Ernst and as a photography teacher of diverse and rather eclectic interests. There seems no doubt, though, that his invitation back to Germany generally, and Munich specifically, as a native son in the late 1970s was an act of restitution.[73] The first stirrings began in the late 1950s. He was paid well, by Munich authorities, for his "donation," which he had apparently loaned to the city museum before its permanent fate was settled. There were a few grand events, publications, and exhibitions to mark the "return" of Breitenbach to his home country. As early as 1967, we can see in the light of Breitenbach's efforts to sell off his collection that the works of German Jews and refugees, including Erwin Blumenfeld, Alfred Eisenstaedt, and Roman Vishniac, were being accorded relatively high value.[74]

To conclude this survey of photography as a Jewish space, we turn to the most artificial yet lethal of Jewish spaces: the Nazi ghettos. There are a number of pictures revealing the persistence or re-creation of Jewish photo studios in ghettos.[75] Photographic equipment was not systematically seized in the Lodz ghetto until 7 November 1941 and remained in use in numerous ghettos for relatively long periods. Rather than stressing resilience or defiance in the face of adversity, such photographs may be seen as evidence of continuity between Jewish life before the Nazis and that which carried on into the Holocaust. The studios and their patronage were, then, in this view, a piece of "normal" life, and may even have been largely taken for granted. Just as importantly, and as we have seen in this chapter, photography was crucial to Jews' relations with non-Jews as well as intra-Jewish relationships. A better understanding of photography as Jewish space will help us to comprehend the worlds that Jews and non-Jews made together, as well as to recapture, more sharply, the range of Jewish vocations and identities that were part of their lost world.

Michael Berkowitz is professor of modern Jewish history at University College London. His most recent book is *Jews and Photography in Britain* (University of Texas Press, 2015). For 2016–17 he holds research fellowships at the United States Holocaust Memorial Museum (Washington, DC) and the Remarque Institute, New York University.

Notes

1. Berkowitz, "'Jews in Photography'"; Berkowitz, "Photography as a Jewish Business."
2. See Aschheim, *Brothers and Strangers*; Maurer, *Ostjuden in Deutschland.*
3. Zelizer, *Visual Culture.*
4. Berkowitz, *Jewish Self-Image.*

5. For superb analysis of how photography can be conceived as a secular Jewish phenomenon, see Silverman, "Reconsidering the Margins."

6. Shandler, "What Does It Mean to be Photographed as a Jew?"

7. Shneer, *Through Soviet Jewish Eyes*; Goodman and Hoffmann, "The Power of Pictures."

8. Deppner and Janke, *Die verborgene Spur*; Bezjak and Deppner, *Jüdisches: Fotografische Betrachtungen.*

9. See Fox, *Leaving Home.*

10. Gidal, "Jews in Photography."

11. Stringer and Ries, *German Faces.*

12. Stringer and Ries, *Deutsche, Gedanken und Gesichter.*

13. See Patt and Berkowitz, introduction to *"We Are Here."*

14. See, e.g., the interview with Karl Knoebl, a Catholic priest in Mittenwald, in Stringer and Ries, *German Faces,* 52.

15. Lehrman, "Talking Pictures" (review); Goodman, review of *German Faces.*

16. "Henry Ries. Berlin vor 25 Jahren. Fotos aus der Zeit der Berliner Blockade. Ausstellung der Landesbildstelle Berlin vom 18. Mai bis 8. Juli 1973"; and "Berlin, Photographien 1946–1949," 1998.

17. Ries, *Ich war ein Berliner.*

18. Berkowitz, *Jews and Photography in Britain,* 246–71.

19. Martin, "Henry Ries, 86."

20. *Home Truths* 5, no. 7 (May 1944): 4–5 [newsletter], Boston Chamber of Commerce, in Polaroid Admin Records I.88, "Press clippings—Military Tax, 1944," in Baker Library Historical Collection, Harvard Business School, Boston, MA.

21. Berkowitz, *Jews and Photography in Britain,* 223.

22. Singer, "In der finster."

23. Jagodzińska, "Image and Identity."

24. Web, introduction to *Poyln*; Blitz, "Der kreyts-weg fun Alter Kacyzne."

25. Weinke, *Verdrängt*; see also Auer and Kunsthalle Wien, *Übersee.*

26. Sachsse, "'Dieses Atelier.'"

27. Kühnl-Sager, "Schule Reimann"; and Stange, "Yva Photographic Studio."

28. Kreutzmüller, *Ausverkauf.*

29. Meder and Winkelbauer, *Vienna's Shooting Girls.*

30. Vonwinckel, "German (Jewish) Photojournalists in Exile."

31. An identical point about the significance of professional networks among Jews in photography was made in Berkowitz, "Jews and Photojournalism."

32. Frank Dabba Smith is preparing a PhD dissertation on Leica during World War II; see Honigsbaum, "New Life through a Lens."

33. Gruner, *Jewish Forced Labor,* n48, 41.

34. Just how prevalent the Jews in photography were in Vienna is also apparent in its present-day street markets. It takes little effort to find photographs from formerly Jewish studios.

35. Harvey Fireside, "Delusions and Denials: Viennese Life under the Nazis," ME 1486, Leo Baeck Institute, Center for Jewish History, New York.

36. Ibid.

37. Horwitz, *In the Shadow of Death,* 8–22.

38. Christian Dürr, Archiv der KZ-Gedenkstätte Mauthausen, e-mail with the author, 18 July 2013.

39. Card file data set, Archiv der KZ-Gedenkstätte Mauthausen, Minoritenplatz 9, 1014 Wien.

40. The photographer from Mauthausen who has received most acclaim is Francisco Boix, a prisoner captured in the Spanish Civil War; see Bermejo, *Francisco Boix*.
41. Card file data set, Archiv der KZ-Gedenkstätte Mauthausen, Minoritenplatz 9, 1014 Wien.
42. Gotfryd, *Anton the Dove Fancier*; Gotfryd, *The Intimate Eye*; Gotfryd, *Widuje ich w snach*.
43. Gotfryd, *Widuje ich w snach*, 182.
44. Gotfryd, *Anton*, 65.
45. Bernard Gotfryd, interview with the author, Forrest Hills, New York, 28 June 2013.
46. Ibid.
47. Gotfryd, *Anton*, 53–54.
48. Ibid., 61–68.
49. Ibid., 71.
50. Ibid.
51. Ibid., 75.
52. Bernard Gotfryd, interview.
53. Berkowitz, *Crime of My Very Existence*, 42–43.
54. Photograph of Paul Fuchs, in Sebastian Piatkowski and Arkadiusz Kutkowski, *"Distrikt Radom": Polityka okupacyjnych wladz niemieckich wobec lucnosci polskiej w dystrykcie radomskim* (exhibition), http://ipn.gov.pl/__data/asscts/pdf_file/0010/58519/1-19637.pdf; "Postwar, War criminal Dr Herbert Boettcher, Item 57530, Yad Vashem Photo Archive, http://collections.yadvashem.org/photosarchive/en-us/61552.html; "Google Arts and Culture" site: https://www.google.com/culturalinstitute/asset-viewer/galicia-poland-fritz-katzmann-commander-of-the-ss-and-the-police-in-the-galicia-district-meeting-heinrich-himmler/ZwHM14fhfgQ2fA?hl=en; http://www.dws-xip.pl/reich/biografie/lista2/3065.html; commons.wikimedia.org/wiki/File:Pierre_Laval_and_Carl_Oberg_in_Paris.png.; http://www.deathcamps.org/reinhard/krueger.html (all accessed December 2014).
55. Berkowitz, *Crime of My Very Existence*, 74–111.
56. Marion Levenson Ross, "Lotte Jacobi 1896–1990," Jewish Women's Archive, accessed December 2014, http://jwa.org/encyclopedia/article/jacobi-lotte.
57. Phillips, "Interview with Lotte Jacobi."
58. Stange, "Yva Photographic Studio"; Zika, "Dahinter steckte ein kluger Kopf."
59. Berkowitz, *Jews and Photography in Britain*, 26, 80, 113.
60. "Leo Rosenthal, Photographer at U.N. for 24 Years, Is Dead," *New York Times*, 29 October 1969.
61. Landesarchiv Berlin and Rechtsanwaltskammer Berlin, *Leo Rosenthal*.
62. Ibid., 17–18.
63. Ibid., 19.
64. Ibid., 13.
65. Ibid., 18–20.
66. Ibid., 28.
67. *Fotografas/Photographer Chaimas Kaplanskis*.
68. Petrauskiene, "On the Path of the Photographer," 42, 54.
69. Ibid., 47.
70. Berkowitz, *Jews and Photography in Britain*.
71. Evaluation from Münchner Stadtmuseum, c. 1967, appraisal of Josef Breitenbach photographic collection, AG 90: 5/16, Center for Creative Photography, University of Arizona, Tucson, Arizona.
72. Berkowitz, *Jews and Photography in Britain*.

73. Jürgen Kobe to Josef Breitenbach, 15.6.1978, AG 90 3/3, Center for Creative Photography, University of Arizona, Tucson, Arizona.
74. Evaluation from Münchner Stadtmuseum.
75. "Studio portrait of two Jewish girls in the Bedzin ghetto. Pictured are Maniusia Gipsman (left) and her cousin, Dina. Both girls perished in Auschwitz. Maniusia was 13 years old at the time of her death," USHMM, accessed December 2014, http://digitalassets.ushmm .org/photoarchives/detail.aspx?id=1089263; "Studio portrait of Rozia Merin wearing a Jewish badge in the Bedzin ghetto," USHMM, accessed December 2014, http://digitalas sets.ushmm.org/photoarchives/detail.aspx?id=1083888.

Bibliography

Aschhiem, Steven. *Brothers and Strangers: The East European Jew in German and German Jewish Consciousness 1800–1923*. Madison: University of Wisconsin Press, 1982.
Auer, Anna, and Kunsthalle Wien. *Übersee: Flucht und Emigration österreichischer Fotografen 1920–1940 / Exodus from Austria: Emigration of Austrian Photographers, 1920–1940*. Vienna: Kunsthalle, 2010.
Berkowitz, Michael. *The Crime of My Very Existence: Nazism and the Myth of Jewish Criminality*. Berkeley: University of California Press, 2007.
———. *The Jewish Self-Image*. London: Reaktion, 2000.
———. *Jews and Photography in Britain*. Austin: University of Texas Press, 2015.
———. "Jews and Photojournalism: Between Contempt, Intimacy, and Celebrity." In *Die PRESSA/The PRESSA: Internationale Presseausstellung Köln 1928 und der jüdische Beitrag zum modernen Journalismus/International Press Exhibition Cologne 1928 and the Jewish Contributions to Modern Journalism*, edited by Suzanne Marten-Finnis and Michael Nagel, 2:627–39. Bremen: Edition Lumiere, 2012.
———. "'Jews in Photography': Conceiving a Field in the Papers of Peter Pollack." *Photography & Culture* 4, no. 1 (2011): 7–28.
———. "Photography as a Jewish Business: From High Theory, to Studio, to Snapshot." *East European Jewish Affairs* 39, no. 3 (2009): 389–400.
Bermejo, Benito. "Francisco Boix, der Fotograf von Mauthausen." Special issue, *Mauthausen-Studien*. Vienna: Mandelbaum, 2007.
Bezjak, Roman, and Martin R. Deppner. *Jüdisches: Fotografische Betrachungen der Gegenwart in Deutschland*. Bielefeld: Nicolai, 2006.
Blitz, Nakhman. "Der kreyts-weg fun Alter Kacyzne" [The Martyrdom of Alter Kacyzne]. *Dos naye lebn*, no. 10 (1945): unpaginated.
Deppner, Martin R., and Karl Janke. *Die verborgene Spur: Jüdische Wege durch die Moderne / The Hidden Trace: Jewish Paths though Modernity*. Bramsche: Rasch, 2008.
Fotografas/Photographer Chaimas Kaplanskis, Fotoalbumas-Katalogas/Photoalbum-Catalogue, Vakaru Lietuva XIX a. pab.–XX aa. Vid., Western Lithuania 19th –20th c. Telsias-Vilnius: Vilniaus dailes akademijos leidykla, 2007.
Fox, Gerald, dir. *Leaving Home, Coming Home: A Portrait of Robert Frank*, for the *South Bank Show*. LWT, 2005.
Gidal, Nachum T. "Jews in Photography." *Leo Baeck Institute Yearbook* 32 (1987): 437–53.
Goodman, Anne L. Review of *German Faces*, by Ann Stringer and Henry Ries. *New Republic*, 14 August 1950, 22.

Goodman, Susan T., and Jens Hoffmann. *The Power of Pictures: Early Soviet Photography, Early Soviet Film*. New York: Jewish Museum, 2015.

Gotfryd, Bernard. *Anton the Dove Fancier and Other Tales of the Holocaust*. New York: Washington Square Press, 1990.

———. *The Intimate Eye*. New York: Riverside, 2006.

———. *Widuje ich w snach. Nowe opowiadania / I Can See Them in My Dreams: The New Stories* [Polish and English], ed. Krystyna Kasinska, trans. Jaroslaw Wlodarczyk. Radom: Miejska Biblioteka Publiczna w Radomiu, 2008.

Gruner, Wolf. *Jewish Forced Labor Under the Nazis: Economic Needs and Racial Aims*. New York: Cambridge University Press, 2006.

Honigsbaum, Mark. "New Life through a Lens." *Financial Times*, 2 February 2007.

Horwitz, Gordon. *In the Shadow of Death: Living Outside the Gates of Mauthausen*. New York: Free Press, 1990.

Jagodzińska, Agnieszka. "Image and Identity: Warsaw Jews as Others and Non-Others." Unpublished paper, University of Wrocław, 2010.

Kreutzmüller, Christoph. *Ausverkauf. Die Vernichtung der jüdischen Gewerbetätigkeit in Berlin 1930–1945*. Berlin: Metropol, 2013.

Kreutzmüller, Christoph, Kaspar Nürnberg, and Renate Stein, eds. *Final Sale: The End of Jewish Owned Businesses in Nazi Berlin*. Berlin: Aktives Museum Faschismus und Widerstand in Berlin and Humboldt-Universität zu Berlin; New York: Leo Baeck Institute, 2010.

Kühnl-Sager, Christine. "Schule Reimann, College of Applied Art." In Kreutzmüller, Nürnberg, and Stein, *Final Sale*, 58–51.

Landesarchiv Berlin and Rechtsanwaltskammer Berlin, ed. *Leo Rosenthal. Ein Chronist in der Weimarer Republik. Fotografien 1926–1933*. With text by Bianca Welzing-Bräutigam, Janos Frecot, and Bernd Weise. Berlin and Munich: Shirmer/Mosel, 2011.

Lehrman, H. "Talking Pictures" (review). *Saturday Review*, 7 October 1950, 20.

Martin, Douglas. "Henry Ries, 86, Photographer Who Captured Berlin Airlift." *New York Times*, 26 May 2004.

Maurer, Trude. *Ostjuden in Deutschland, 1918–1933*. Hamburg: H. Christians, 1986.

Meder, Iris, and Andrea Winkelbauer, eds. *Vienna's Shooting Girls: Jüdische Fotografinnen aus Wien / Jewish Women Photographers from Vienna*. Vienna: Jüdisches Museum Wien and IPTS—Institut für Posttayloristische Studien, 2012.

Neef, Tatjana, ed. *Unbelichtet: Muenchner Fotografen im Exil = Unexposed: Munich Photographers in Exile*. Heidelberg: Kehrer, 2010.

Patt, Avinoam J., and Michael Berkowitz. Introduction to *"We Are Here": New Approaches to Jewish Displaced Persons in Postwar Germany*, edited by Avinoam J. Patt and Michael Berkowitz, 1–13. Detroit: Wayne State University Press, 2010.

Petrauskiene, Marina. "On the Path of the Photographer." In *Fotografas/Photographer Chaimas Kaplanskis*, 41–57.

Phillips, Sandra S. "An Interview with Lotte Jacobi." *CENTER Quarterly* [Catskill Center for Photography] 3, no. 1 (1981): unpaginated.

Ries, Henry. *Ich war ein Berliner: Erinnerungen eines New Yorker Fotojournalisten*. Berlin: Parthas, 2001.

Sachsse, Rolf. "'Dieses Atelier ist sofort zu vermieten': Von der 'Entjudung' eines Berufsstandes." In *"Arisierung" im Nationalsozialismus: Volksgemeinschaft, Raub und Gedächtnis: Jahrbuch 2000 zur Geschichte und Wirkung des Holocaust*, edited by Irmtrud Wojak and Peter Hayes on behalf of the Fritz Bauer Institute, 269–86. Frankfurt/New York: Campus Verlag, 2000.

Shandler, Jeffrey. "What Does It Mean to be Photographed as a Jew?" *Jewish Quarterly Review* 94, no. 1 (2004): 8–11.

Shneer, David. *Through Soviet Jewish Eyes: Photography, War, and the Holocaust.* New Brunswick, NJ: Rutgers University Press, 2011.

Silverman, Lisa. "Reconsidering the Margins: Jewishness as an Analytical Framework." *Journal of Modern Jewish Studies* 8, no. 8 (2009): 103–20.

Singer, I. J. "In der finster." In *Perl un andere dertseylungen,* 74–93. Vilne [Vilnius]: B. Kletskin, 1929.

Stange, Heike. "Yva Photographic Studio." In Kreutzmüller, Nürnberg, and Stein, *Final Sale,* 68–71.

Stringer, Ann, and Henry Ries. *Deutsche, Gedanken und Gesichter 1948–1949.* Berlin: Argon, 1988.

———. *German Faces.* New York: Sloane, 1950.

Vonwinckel, Annette. "German (Jewish) Photojournalists in Exile: A Story of Networks and Success." *German History* 31, no. 4 (2013): 453–72.

Web, Marek. Introduction to *Poyln: Jewish Life in the Old Country,* by Alter Kacyzne, xxi–xxii. New York: Henry Holt, 1999.

Weinke, Wilfried. *Verdrängt, vertrieben, aber nicht vergessen. Die Fotografen Emil Bieber, Max Halberstadt, Erich Kastan, Kurt Schallenberg.* Weingarten: Kunstverlag Weingarten, 2003.

Zelizer, Barbie, ed. *Visual Culture and the Holocaust.* London: Athlone, 2001.

Zika, Anna. "Dahinter steckte ein kluger Kopf. Else Neulaender-Simon." In *The Jewish Engagement in Photography,* edited by Martin Deppner and Michael Berkowitz. Oldenburg: Carl von Ossietzky Press, 2017.

15

Jews, Foreigners, and the Space of the Postwar Economy

The Case of Munich's Möhlstrasse

Anna Holian

On 30 June 1949, officers from the Munich police drove down "the Möhlstrasse" taking pictures. Since the end of the war, the street had become a key Jewish social space. After the currency reform of June 1948, it had also become the site of a thriving Jewish-run marketplace. Modest wooden kiosks and larger brick buildings had sprung up in front of the neighborhood's stately villas, creating a new streetscape of commercial activity in what had previously been an exclusive residential area. For the municipal authorities, this was a troubling development: they viewed the marketplace as a key site of criminality in the city. Not surprisingly, the photographs taken by the police focused squarely on the marketplace. The day after their picture-taking expedition, the police launched their first mass raid on Möhlstrasse.

At first glance, the photographs seem to illustrate the police's *authority* over Möhlstrasse. Driving down the street with a camera, the police appear to constitute a controlling presence. In fact, however, the photographs attest to the police's *lack* of control. Taken while the car was moving, the images are blurry, in some cases to the point of being useless for forensic purposes. The sharpest object in the photographs is often the police car itself. While the poor quality of the images may reflect the photographer's lack of skill, it also suggests the police's unwillingness to stop and linger. Despite their official authority over the street, they did not feel secure there.

The Munich police had good reason to be nervous. Cameras were not welcome on Möhlstrasse. At the time the photographs were taken, a de facto "ban" on picture taking had been imposed in the neighborhood. People who contravened the ban faced numerous unpleasant consequences: threats, assaults, seizure of their camera.[1] German police officers were also not welcome in the area. Viewed by the Jews on Möhlstrasse as unreconstructed Nazis, they had difficulty carrying out their duties, especially when this involved arrests. They were regularly assaulted, at times even stripped of their weapons.[2] A month after the photographs were taken, the police's lack of authority was made dramatically clear: on 10 August, police officers were attacked after they tried to stop a Jewish demonstration. Twenty-six police officers were injured, some of them seriously. Casualties on the demonstrators' side were significantly lower, with seven people injured. Ultimately, American military police had to intervene.[3] Examining the photographs in the knowledge of this context, we can perceive an implicit threat in the way the people look at the police. Meeting the camera's gaze, they challenge the police's authority, unsettling the power relations that ordinarily take effect when one party wields a camera—or a gun.

Although the municipal authorities' control over Möhlstrasse was fragile in 1949, their actions that summer were the beginning of a concerted effort to "reclaim" the area. Police control over the street was effectively established by the end of 1951, but the campaign to eliminate the marketplace continued into the early 1960s. Other remnants of the Jewish presence lasted longer—most notably an Orthodox synagogue, which remained in use until the early 1970s.

This essay examines how the German authorities constructed Möhlstrasse as a problematic Jewish and foreign space and how they sought to control and

Figure 15.1. Shops in the Möhlstrasse as seen from a Munich police car, 30 June 1949. Used courtesy of the Stadtarchiv München

ultimately eliminate it. I situate this process against the background of Germany's postwar reconstruction, considering what political and economic imperatives were at stake in the conflict and how these imperatives were intimately intertwined with German conceptions of a "foreigner threat." I also consider how this specifically German case fits into the broader context of state campaigns against the informal economy. The ultimate goal is to understand how and why Jewish spaces in postwar Germany became sites of controversy.

Möhlstrasse is well known to students of postwar German-Jewish history.[4] To the extent that scholars have addressed the campaign against Möhlstrasse, they have generally focused on the anti-Semitic German discourse that framed the street as a site of black-market criminality.[5] There is no doubt that anti-Semitism powerfully informed how Germans viewed Möhlstrasse and other Jewish spaces. However, there are a number of other issues that need to be considered. First, the campaign against Möhlstrasse was driven in large part by economic and political concerns connected to reconstruction. These concerns cannot be reduced to alibis for anti-Semitism, even if they were often intertwined with an anti-Semitic discourse about Jews and the economy. Second, the campaign was aimed not just at Jews; it also targeted non-Jewish foreigners. It was both anti-Semitic and xenophobic. Third, the black market was only a small part of the activities the authorities sought to contain. The main focal point of the campaign was the "informal economy" that emerged after the currency reform and its presumed threat to the success of the new market economy. My aim in this essay is to draw out these varied aspects of the campaign against Möhlstrasse without losing sight of either its anti-Semitic or its xenophobic elements. Indeed, I hope to show that in order to fully appreciate the anti-Semitic and xenophobic discourse about Möhlstrasse and similar Jewish spaces, we need to understand the concrete material considerations that Germans believed were "at stake."

To help establish these connections, I draw on David Harvey's work on space, power, and capital. Building on an approach first articulated by Henri Lefebvre, Harvey has developed a very useful typology of spatial concepts. Of particular significance here is his distinction between *appropriation* and *domination.* Briefly, "appropriation" refers to how people use land and the built environment, in the process creating distinct social spaces. It also refers to the mental maps and symbolic hierarchies that accompany these practices, defining spaces as sites of belonging. "Domination" refers to the ways in which hegemonic political and economic actors establish and maintain authority over space, creating spaces governed by exclusive rights. Legislation and policing play an important role here. It also refers to the representations by which spaces come to be defined as the exclusive property of hegemonic forces.[6]

Harvey's distinction between appropriation and domination is useful for expressing how different social actors in postwar Munich related to Möhlstrasse. It also draws our attention to the political and economic interests at stake in the conflict over the street. To briefly summarize, the actions of the Jewish and non-Jewish foreigners who occupied Möhlstrasse were oriented toward appropriation. For them, the street was a space of petty accumulation, of provisional live-lihoods established, in most cases, without any definite intention of remaining. More generally, it was an autonomous social and political space in which it was possible to operate without much interference from Germans. The actions of the municipal authorities, on the other hand, were oriented toward domination. For them, Möhlstrasse was part of a larger urban space that was supposed to lie within their control. Their efforts to establish control were driven by two imperatives: to reassert German political authority over people and space in the city and to re-establish a functioning market economy. These efforts were informed not only by general conceptions about state power and the rules of the marketplace, but also by more specific ideas about the *disorderly* and *disruptive* nature of the Jews and non-Jews who had appropriated Möhlstrasse and whose unwillingness to "play by the rules," in their eyes, threatened the process of postwar reconstruction.

The history of the city's campaign against Möhlstrasse is thus a story about the political and economic reconstruction of Western Germany. It is also a story about conflict between the state and the "informal economy." Since the 1970s, anthropologists, sociologists, and economists have been debating how best to conceptualize trade and production that exist on the margins of the formal econ-omy. The concept of an informal economy grows out of these debates.[7] In its most basic form, it refers to economic practices not sanctioned by the state. It thus covers a wide variety of activities. The most typical are unsanctioned trade and production; more obviously illegal activities such as black marketeering and criminal acts such as theft, robbery, and extortion sometimes also enter into the picture, not least because operating without legitimacy makes traders and pro-ducers vulnerable to pressure to engage in such activities. However, the key fea-ture of informal economic activity is its adversarial relationship with the state. By definition, such a relationship is likely to develop when state control over the economy is incomplete. As Gracia Clark notes, unsanctioned traders are es-pecially threatening to "relatively new governments struggling to gain control over internal and international linkages.... Whether illegal or simply extralegal, traders' highly visible activities advertise the inadequacy of official distribution channels and the existence of fundamental problems related to migration, em-ployment, and pricing."[8] This is an apt description of the situation in early post-war Western Germany. Framing the campaign against Möhlstrasse as a campaign against the informal economy thus helps us identify more precisely how the effort to combat Möhlstrasse was part of a larger process of establishing control over economic activity.

It also brings the conflict between appropriation and domination into sharper focus. It is not difficult to see how spatial practices might figure in the adversarial relationship between the state and the informal economy. To the extent that an informal economy can be localized, the state's efforts to combat it often involve practices of spatial control.[9] As we will see, in their campaign against Möhlstrasse, the Munich authorities utilized practices typical of state efforts to combat the informal economy, including arrests, raids, demolitions, and deportations.

Postwar Möhlstrasse: A Brief History of Appropriation

Möhlstrasse's development into a Jewish and foreign space began shortly after the end of the war. Known as one of the city's most exclusive residential streets, Möhlstrasse was developed in the late nineteenth and early twentieth centuries; it featured sumptuous villas in a variety of historical styles. During the Nazi era, a number of high-ranking party members, most notably Heinrich Himmler, moved into the neighborhood, while Jewish residents were expropriated and forced out. After the war, the American occupation authorities requisitioned a number of buildings in the area. They used some for their own purposes and turned others over to partner organizations like the United Nations Relief and Rehabilitation Administration and the American Joint Distribution Committee. Through their connections with the Americans and these partner organizations, Jews in Munich also gained access to property in the neighborhood.

By war's end, Munich's Jewish population had largely disappeared, destroyed by emigration, suicide, deportation, and mass murder. The survivors included a small number of German Jews, mainly from mixed marriages, and a larger number of foreign Jews, primarily former concentration and extermination camp prisoners. The foreigners were classified as Jewish "displaced persons" (DPs).[10] According to one very approximate count, there were some 6,000 Jews in Munich in November 1945, 2,000 German and 4,000 foreign.[11] Over the next two years, however, the Jewish population in Munich, and Western Germany more generally, grew exponentially. This growth was the product of a massive East–West migration fueled principally by anti-Semitism in Poland and other Eastern European countries. By July 1947, there were some 153,000 Jewish DPs in the American zone of Germany.[12] Around this same time, Munich and its periphery counted about 75,000 Jews.[13] In 1948, as opportunities for immigration to Israel, the United States, and other countries opened up, the numbers again shifted dramatically. Most Jews departed; between September 1948 and September 1949, the number of Jews in Bavaria dropped from around 69,000 to 31,000. Of those who remained, some 23,000 were Polish, and 1,600 were German.[14] The numbers continued to decline into the 1950s. By March 1951, there were only 10,000 Jews living outside DP camps in Bavaria; since by this point most camps had closed, one can assume the total number was at most 14,000.[15]

As Munich's postwar Jewish population grew in the first years after the war, Möhlstrasse became the main gathering place for the Eastern European Jewish DPs. Two important DP committees, the Munich Jewish Committee (representing Jews in the city) and the Central Committee of Liberated Jews (representing Jews in the US zone), established their headquarters in the area. Each developed an elaborate infrastructure that spread out across the neighborhood. By 1947, the neighborhood had a synagogue, a tracing service, a health clinic, a kindergarten, a primary school, and a *Gymnasium,* an academic secondary school. It was a Jewish social, political, and religious space, attracting people from across the city and beyond.

It was also an *economic* space. Initially, economic activity on Möhlstrasse consisted mostly of small-scale black-market trade in the courtyard of the Munich Jewish Committee building.[16] After the currency reform, it expanded dramatically. On the one hand, the street remained a site of black-market activity, though trade in goods was soon supplanted by trade in currency. On the other, it became the site of an important shopping district. As a market economy began to revive, Jewish traders used the capital they had accumulated on the black market to establish more legitimate businesses.

Möhlstrasse's development into a commercial center began in earnest in the autumn of 1948. It included a substantial physical transformation, as dozens of small and medium-sized buildings went up in front of the villas, sometimes with the property owners' permission, sometimes without. These structures housed a variety of small businesses. Most specialized in luxury goods at discount prices: coffee, tea, alcohol, chocolate, cigarettes, nylons, fabric. Although most of these goods were now widely available in German stores, they were significantly cheaper on Möhlstrasse. This was due in large part to the shopkeepers' unsanctioned trade practices: they obtained many of their supplies through unauthorized channels (e.g., the refugee relief agencies) and thus did not pay customs and taxes. At the height of its development between 1948 and 1950, the street had about one hundred shops and a number of restaurants and cafés.[17]

While the heart of Möhlstrasse's economic life consisted of the shops, the street also became a center of petty criminality: theft, robbery, confidence tricks. These activities formed a kind of secondary sector of the Möhlstrasse economy, targeting shoppers and to some extent shopkeepers. They involved a broader range of actors. While most of the shopkeepers were Polish Jews, the petty criminals represented an alphabet soup of foreigners: in addition to a sizable contingent of Greeks of both Christian and Jewish origin, they included Armenians, Bulgarians, ethnic German refugees, Lithuanians, Poles, Polish Jews, Romanians, Ukrainians, and Yugoslavs. At least a few were concentration camp survivors.

This complex informal economy was not unique to Möhlstrasse. It existed in many parts of Munich. Indeed, it was a general Western German phenomenon whose development was premised on a number of factors. The first was the currency reform of June 1948, the initial step in the reintroduction of a market economy to the Western zones. Overall, the currency reform was disadvantageous to displaced persons: it eliminated the opportunities afforded by the black market without providing substantial new avenues for economic participation. However, it did make it feasible to start a business, especially in retail. DPs who took this route could often parlay their relationships with the American occupation forces and the relief agencies into a steady if not entirely legal supply of marketable goods.

The second factor, also connected to the reconstruction of a market economy, was the introduction of *Gewerbefreiheit* (free enterprise), which occurred in the American zone in January 1949 and was extended to all of West Germany after the founding of the Federal Republic. Part of the broader American effort to

promote neoliberal economic policies, *Gewerbefreiheit* made it possible for anyone to ply a trade or craft through a simple registration process. It thus provided
a legal basis for DP participation in the new market economy.

The third factor, specific to the American zone, was the protected status of
displaced persons. While DPs in all three Western zones received Allied assistance, their status as a privileged category was most fully developed in the American zone. In particular, German police in the US zone had limited authority over
displaced persons. They were allowed to arrest DPs but could not hold them for
more than twenty-four hours or prosecute them. Even stronger restrictions applied to victims of persecution, a category that included all Jews: the police were
supposed to turn them over to the Americans immediately. These restrictions not
only made it difficult for the police to exercise substantive control over DPs; they
also made it possible for DPs to transform German locales into extraterritorial
DP spaces.

This last fact helps us understand why Möhlstrasse became a key site of Munich's informal economy: the sheer number of DPs who congregated there made
it an extraterritorial space.[18] However, this situation did not last indefinitely. After the passing of the Homeless Foreigners Law in April 1951, DPs lost their
status as protected Allied nationals. Even before then, their protected status had
been undermined by the "transfer of authority," the process by which German
authorities were given greater control over governance. This was the context in
which the campaign against Möhlstrasse began.

Over the course of the early 1950s, the Möhlstrasse commercial district
declined. As the campaign against the street intensified, both shopkeepers and
shoppers no longer found it a desirable place to do business. The rise of petty
criminality also drove shoppers away. The stabilization of prices and improved inventory in German shops doubtless also played a role. By early 1953, only about
thirty kiosks remained, many of them boarded up.[19] Nonetheless, the remnants
of the marketplace refused to disappear. In late 1957, there were still more than
twenty-five commercial structures on Möhlstrasse, housing a smattering of businesses, including a kosher butcher and a grocer.[20] The effort to eliminate these
remnants continued into the early 1960s.

Domination: The Campaign against Möhlstrasse

The process of reclaiming Möhlstrasse from the Jews and non-Jews who had
appropriated it involved both spatial representations and spatial practices. The
authorities represented Möhlstrasse as a once beautiful street that had been taken
over by Jews, Greeks, and other dubious foreigners, whose illegal and criminal
activities posed a threat to the new economic and political order. Their efforts
to undo this transformation relied on practices commonly employed by states
against informal economies: raids, arrests, demolitions, and deportations. They
targeted individuals operating on Möhlstrasse as both impediments to economic

and political reconstruction and as undesirable foreigners. More precisely, they made foreigners into scapegoats for the more complex and elusive problems associated with re-establishing economic and political control.

The "problems" on Möhlstrasse are usually put under the heading of the "black market." As I have suggested, however, this term is misleading. The black market as such, defined as trade in rationed or otherwise controlled or unsanctioned goods, declined dramatically after the currency reform, that is, at precisely the moment when Möhlstrasse's marketplace began to take off. By the summer of 1949, true black marketeering was limited primarily to currency. As an internal police report from June 1949 noted, "The term 'black market' no longer accurately describes [the activities in] certain public streets and squares of the city. Rather, the issue is unsanctioned trade with circumvention of state and municipal taxes."[21]

These general observations also applied to Möhlstrasse. As an American Military Government report from the autumn of 1949 indicates, German concerns about the street were diverse; they included "violation of real estate zoning laws, squatting on other people's property in erection of stores and other buildings, selling rationed items without required coupons, evading customs and tax laws, buying and selling American and other foreign currency, stores remaining open on Sunday's [sic] contrary to German law, and large-scale black market activities."[22] To this was added concern about criminal activity proper, such as theft, robbery, and assault. In short, the primary issue was not the black market; it was a complex informal economy that had transformed the character of the neighborhood.

This said, the *idea* of the black market certainly played a role in defining Möhlstrasse as a problematic space. Whether out of habit or as a conscious strategy, German authorities continued to talk about the *Schwarzmarkt* (black market) and *Schwarzhandel* (illicit or black-market trade) well after the phenomena had receded. The German press also frequently used these terms. The news magazine *Der Spiegel*, for example, produced disturbingly anti-Semitic reports about Jewish "smugglers and fences."[23] Even in more moderate form, however, the vocabulary applied to the black market established a continuity between the black market of the war and early postwar years and the gray market that predominated after June 1948, labeling both as not only *illegal* but also *immoral.*

The campaign against Möhlstrasse's informal economy began in the spring of 1949. It involved numerous municipal and state agencies and the American Military Government. It also involved German shopkeepers. Not surprisingly, German traders were angry about what they perceived as unfair competition from Jewish shopkeepers on Möhlstrasse and elsewhere. Their official body in Bavaria, the Landesverband des Bayerischen Einzelhandels (State Association of Bavarian Retailers), consistently lobbied the authorities to act.[24] It approvingly cited the US Military Government's assertion "that the responsibility for the developments on Möhlstrasse lies exclusively with the Bavarian state government and the city of Munich."[25] The local and national press gave such complaints considerable publicity.[26]

City and state officials were thus under pressure to demonstrate their authority over Möhlstrasse. In a period of momentous economic transition, they were concerned about not only the direct costs of the informal economy (e.g., loss of tax revenue) but also public confidence. They needed to show that unlike the black market of the early postwar period, the market economy was governed by morality. As Malte Zierenberg notes, in the German discourse about the market economy "a liberal economic order, one in which the invisible hand alone regulated supply and demand, was not conceivable"—state control was also required.[27] Since the authorities could not control who entered the marketplace, they at least wanted to ensure that everyone "played by the rules."

The local authorities regarded the blossoming of an informal economy on Möhlstrasse as a direct result of the *foreignness* of its population. This attitude was especially strong among the police. In the early postwar period, police forces in Western Germany typically viewed foreigners as the main source of criminality. The Munich police were no exception. For example, Franz Xaver Pitzer, the first postwar head of the Munich police, saw foreigners as the principal threat to the city's security; he used the specter of foreigners "armed to the teeth" to argue before the Allies that the police needed firearms.[28] Within this general framework, however, the police also distinguished between different *types* of foreigners. Most importantly, they distinguished between "Jews," on the one hand, and "Greeks," "Yugoslavs," and other non-Jewish "foreigners," on the other. This dualistic framework echoed the Nazi-era division of the Other into Jews and foreigners. The Jew-foreigner dichotomy was central to the authorities' understanding of Möhlstrasse. "Jews" meant the (predominantly Eastern European Jewish) shopkeepers, while "foreigners" meant the (largely non-Jewish Eastern and Southeastern European) black marketeers and petty criminals.

The police viewed each of these groups as problematic, though not necessarily for the same reasons. With the shopkeepers, the key issues were unsanctioned trade, land use and construction practices, and the imposition of nonofficial "laws" like the ban on picture taking. To the police, these practices undermined the orderly workings of the economy and state. As one police report noted, "Some of these businesspeople pay taxes on a portion of their foreign goods and think this gives them license [to make] all their large-scale shady deals."[29] Such comments were echoed by German retailers. What Möhlstrasse shopkeepers were doing, the State Association of Bavarian Retailers asserted, "no longer has anything to do with free enterprise and healthy competition."[30]

Police discussions about the shopkeepers suggest how anxieties around the reintroduction of a market economy mobilized anti-Semitic thinking. The problems connected with re-establishing state control over domestic and international trade, and even more so those associated with increased competition, were represented as problems created by unscrupulous Jews, especially Eastern European ones. In the eyes of the police, Jewish shopkeepers' unwillingness to play by the rules proved that they were a foreign body without a share in the national interest. The survival of anti-Semitism was often hidden behind generic references to "shopkeepers" and "businesspeople" or behind the new language of philo-Semi-tism, as when the police, in editing reports, replaced the word "Jew" with "Jew-ish fellow citizen" (*jüdischer Mitbürger*)[31] or turned "Israelites" into unspecified "persons."[32]

The anti-Semitic tropes employed to discuss Möhlstrasse shopkeepers were not new. The association of foreign Jews with shady business practices can be seen as a direct continuation of discussions about foreign Jews and the black market during the initial postwar years.[33] It can also be seen as part of an older discourse about Jews and capitalist modernity that originated in the late nineteenth cen-tury and peaked under the Nazis, who used it to justify their murderous cam-paign against the Jews. Here, too, the focus was often on "Eastern" Jews and their ostensibly limitless competitive disposition.[34]

While the police took a dim view of the shopkeepers on Möhlstrasse, their assessment of the petty criminals was even worse. They associated Greeks, Yu-goslavs, and other non-Jews with black marketeering and violent crimes such as robbery and assault, regarding them as a fundamental threat to law and or-der. Discussions about petty criminality relied heavily on comparison between Jews and Greeks, the latter standing metonymically for all non-Jewish foreigners. Nonviolent Jewish illegality (sale of untaxed and duty-unpaid goods, unautho-rized construction) was contrasted with violent Greek criminality (theft, robbery, murder). From this perspective, while the Jews were unscrupulous businesspeo-ple, they were at least not violent criminals. A Munich police report from Sep-tember 1949 provides a textbook example:

> According to the local police station, complaints about improper trade on the part of the Jewish businesses have ceased. What irregularities there are, such as the sale of duty-unpaid goods, pale in comparison with the "serious criminal activity" of the *Greeks* and *Yugoslavs* who hang out on Möhlstrasse and run an "organized criminal industry [*organi-siertes Gaunergewerbe*]" there.[35]

The contrast between Jews and Greeks was drawn even more sharply by Philipp Auerbach, head of the Bavarian State Office for Restitution. Auerbach, a camp survivor from a prominent German-Jewish family, had a complex relation-ship with Jewish DPs; he felt they spurred anti-Semitism, but he also strongly defended their interests. In the conflict over Möhlstrasse, he generally took their side. Writing to the Munich police chief in May 1949, after a series of violent events, he sought to ensure that the campaign against Möhlstrasse would focus solely on non-Jews:

> Everything that happens on Möhlstrasse is blamed on the Jews. Observation has shown, however, that a majority of incidents are exclusively the work of Greeks who have insin-uated themselves [into the area] in the last few weeks. These are non-Jews who endanger the reputation of Jewish fellow citizens.
>
> … I recommend [staging] a mass raid on a Saturday, because on Saturdays Jewish fel-low citizens do not spend time on Möhlstrasse.

In the interests of the security of the law-abiding [*ordnungsliebende*] population, immediate action is urgently necessary.[36]

Auerbach thus supported and encouraged massive police interventions on Möhlstrasse even as he sought to channel these interventions away from Jews. His assertion that Jews and Greeks were separate categories cannot, however, be taken at face value. For one thing, criminality was not limited to non-Jews. For another, some Greeks associated with Möhlstrasse were, in fact, Jewish. Yet the Jewish-Greek dichotomy had broad resonance, subsuming Greek Jews under a general "Greek" category defined by criminality.

The German image of Möhlstrasse as a problematic foreign space was thus constructed around a duality rather than a single theme. Nonetheless, the police also tended to lump together shopkeepers and petty criminals and to view the two groups as a single "conspiratorial community" (*verschworene Gemeinschaft*).[37] According to Siegfried Herrmann, the head of the uniformed police, "When it comes to hiding a criminal and obstructing police operations, Greeks and Jews immediately stand together as one, giving the lie to all the assurances that the Jewry of Möhlstrasse doesn't want to have anything to do with the Greek criminal pack."[38] Such conclusions were informed in part by reports that Greeks acted as middlemen for Jewish black marketeers.[39] The fact that some Greeks were Jewish may also have played a role here. However, the central piece of evidence for a conspiracy was simply a shared opposition to the German police.

The municipal authorities' view of shopkeepers and petty criminals as a single conspiratorial community structured their response to Möhlstrasse. In particular, it encouraged the use of raids. A spatial practice *par excellence* and a frequent feature of modern state campaigns against the informal economy, a raid indiscriminately targets everyone occupying a given space at a given time.[40] As Jose Canoy notes, raids were an important aspect of early postwar policing in Bavaria. Despite their authoritarian and intrusive nature, the police were able to employ them on a wide scale by exploiting the public's perception that the black market and other forms of illegal and criminal activity were caused by foreigners.[41]

Although the overall use of raids diminished as the black market declined, their targeted employment against foreigners did not. Indeed, if Möhlstrasse is any indication, the opposite may have been true. The Munich police conducted multiple raids on Möhlstrasse between the summer of 1949 and the autumn of 1951. They also targeted other key sites of the presumed "black market," though not with the same tenacity.[42] The opening salvo in this campaign, the raid of 1 July 1949, involved 550 police officers.[43] Between October 1950 and November 1951, the police carried out twenty mass raids.[44] The largest took place on 25 October 1950, with 1,000 police officers and customs officials participating.[45] Given the size of the marketplace, these were overwhelming displays of force.

Auerbach's request to spare Jews notwithstanding, raids on Möhlstrasse (and elsewhere) aimed at the informal economy as a whole. As Munich mayor Thomas Wimmer noted shortly before the 1 July raid, the goal was to combat both "illegal sales in the street" and "black marketeering."[46] Although newspaper reports described the raid as an attack on the black market, most of the goods seized were in fact non-rationed items that presumably belonged to the shopkeepers.[47] The raids thus seem to have been more successful at combating unsanctioned trade than they were at tackling black marketeering and petty criminality. This has everything to do with the different spatialities of the two types of activities: the fundamentally stationary nature of unsanctioned trade made it more vulnerable to raids than the largely mobile practices of petty criminality.

Efforts to combat the informal economy also involved attempts to restore the right to completely remove "problematic" foreigners.[48] This authoritarian tactic was premised on the 1938 Ausländerpolizeiverordnung (Police Regulations on Foreigners, or APVO), passed once again in modified form in February 1951. The APVO gave the police broad powers to revoke a foreigner's residency permit; grounds for revocation included the commission of a crime in or outside of Germany; contravention of tax, customs, and other financial regulations; and vagrancy. Foreigners who lost their residency permits could be immediately deported.[49] As Karen Schönwälder notes, the return of the APVO, for the federal and state authorities, represented the "recovery of a familiar power of control and removal over foreigners."[50] As the head of the Munich police's criminal division, Andreas Grasmüller, stated in July 1952, the APVO was "ideally suited to restoring orderly conditions in the streets and fighting illegal trade [*Schwarzhandel*]."[51] It is worth noting that the Munich police had not undergone any serious denazification process.[52]

In the end, the city's was not truly able to implement plans to use the APVO. The 1951 Homeless Foreigners Law, which covered former DPs, severely constrained its powers. "Homeless foreigners" could only be deported if they posed a serious threat to public security or order and even then only under non-refoulement terms.[53] In other words, the APVO could not be used against this group. This may explain why, as early as 1952, Munich sought to pass a law that would give it broader powers to deport "criminal and asocial foreigners," including Jewish DPs who had returned from Israel.[54] The Bavarian state similarly sought to expand its powers, representing an "extreme position" on the national scene.[55]

These efforts continued into the later 1950s and 1960s, culminating in the passing in 1965 of the first West German law on foreigners. By this point, the Möhlstrasse marketplace had ceased to exist. Nonetheless, the 1965 foreigners law, far from responding primarily to issues raised by the new labor migration, emerged out of early postwar debates about "undesirable foreign elements."[56] The campaign against Möhlstrasse must be seen as an important part of this earlier history.

Conclusion

By the end of 1951, police interest in Möhlstrasse had declined significantly. Into the early 1960s, however, the remnants of the marketplace—the small number of remaining businesses and the larger number of remaining structures—continued to frustrate the authorities. While these remnants no longer defined the neighborhood, they prevented Möhlstrasse from returning to its previously exclusive state. They acted as an unpleasant reminder of how "Jews and foreigners" had transformed the neighborhood. The project of eliminating these remnants was no longer carried out via broad spatial practices such as raids and deportations but instead through the displacement of individual shopkeepers and the demolition of individual structures.

The informal economy on Möhlstrasse was part of a more general Western German phenomenon. The spatial transformation of Möhlstrasse was also part of a broader phenomenon, one that was not, in fact, specific to the informal economy but rather addressed the more general problem of wartime destruction.[57] Möhlstrasse was nonetheless unique in its ethno-national profile and extraterritorial status. Protected by American DP policy, Jewish and non-Jewish foreigners were able to lay a strong claim to the area. Through a series of spatial practices—ranging from the construction of new buildings to assaults on German policemen—they transformed it into their "own" space. This process of appropriation, to return to the terms introduced at the beginning of the chapter, generated hostility among the German authorities. Pressured by German shopkeepers, and motivated by their own anti-Semitic and xenophobic prejudices, local and state officials interpreted the freedom Jews and non-Jews enjoyed on Möhlstrasse as a threat to their control over people and space. They also saw it as a threat to the new market economy. The ability to dominate Möhlstrasse thus became a key test of their capacity to deliver on the promises of political autonomy and economic revival.

Anna Holian is associate professor of modern European history at Arizona State University. She is the author of *Between National Socialism and Soviet Communism: Displaced Persons in Postwar Germany* (University of Michigan Press, 2011). She is currently working on a book about Jewish space in postwar Germany, as well as another book about "war children" in postwar European film.

Notes

1. Polizeipräsidium München, Diary No. 1559/49, 20 July 1949; Polizeipräsidium München, Diary No. 2152/49, 26 July 1949; Siegfried Hermann to Philipp Auerbach, 25 August 1949, Staatsarchiv München (StAM), Polizeidirektion (Pol Dir) 11349.
2. Hermann to Auerbach, 26 August 1949, StAM, Pol Dir 11349; Hermann to Auerbach, 7 September 1949; Franz Xaver Pitzer to Auerbach, 3 October 1949, StAM, Pol Dir 11350.
3. I discuss the demonstration at length in Holian, *Between National Socialism*, 198–210.
4. See Schwarz, *Redeemers*, 297–305; Wetzel, *Jüdisches Leben in München*, 338–42; Karl, *Möhlstrasse*, 65–81; Kauders and Lewinsky, "Neuanfang mit Zweifeln," 190–91; Crago-Schneider, "Antisemitism or Competing Interests?'; Rühlemann, "Mir zaynen doh."
5. Crago-Schneider, "Antisemitism or Competing Interests?" See also Jacobmeyer, *Vom Zwangsarbeiter*, 210–15; Berkowitz, *Crime of My Very Existence*, chaps. 6–7.

6. Harvey, *Condition of Postmodernity*, 218–22. See also Lefebvre, *Production of Space*, 33–41, 164–68.

7. Key writings include Hart, "Informal Income Opportunities"; Bromley, "Introduction— The Urban Informal Sector"; Clark, *Traders versus the State*; Portes, Castells, and Benton, *The Informal Economy.*

8. Clark, introduction to *Traders versus the State*, 1.

9. Clark, introduction to *Traders versus the State*; Smart, "Resistance to Relocation"; Smart, "How to Survive in Illegal Street Hawking."

10. A creation of the Allied authorities, the term "displaced persons" referred to "civilians outside the national boundaries of their country by reason of the war."

11. "Annual Historical Report for Military Government SK-LK Munich," 3 July 1946, National Archives and Records Service, College Park, Maryland (NACP), RG 260, OMGB, FOD, General Records, Box 406, Historical Reports.

12. "Status Report, Displaced Persons U.S. Zone, Germany as of 31 July 1947," NACP, RG 260, OMGB, ID, Box 163, c. DPs-Polish.

13. "Quarterly Historical Report," 15 October 1947, NACP, RG 260, OMGB, FOD, General Records, Box 406, Historical Reports.

14. "Die Ausländer in Bayern am 30. September 1948 und am 30. September 1949," *Bayern in Zahlen*, December 1949, 357.

15. Informationsdienst des Bayerischen Statistisches Landesamtes, "Die Ausländer in Bayern," 4 May 1951, Bayerisches Hauptstaatsarchiv (BayHStA), MArb 1574.

16. J. Levenson to Celia Weinberg, 3 November 1947, YIVO Institute for Jewish Research, New York (YIVO), 294.1, folder 206; Schwarz, *Redeemers*, 298.

17. Karl, *Möhlstrasse*, 75.

18. I address this issue in greater detail in Holian, "Ambivalent Exception."

19. "Bitte nicht stören in der Möhlstrasse," *Münchner Merkur*, 19 February 1953.

20. "Übersichts-Schema Behelfsläden Möhlstrasse," October 1957, Stadtarchiv München (StadtA Mü), Lokalbaukommission (LBK) 6671.

21. Polizeipräsidium, Kriminaluntersuchungsabteilung, "Schwarzer Markt (Sammelbericht)," 20 June 1949, StAM, Pol Dir 11367.

22. "The Status of Munich's Moehlstrasse," [circa October 1949], [*Information Control Intelligence Summary?*], 11, NACP, RG 260, OMGB, ID, Box 24, Jewish Demonstration (Möhlstrasse etc.).

23. "Am Caffeehandel beteiligt. Deutschlands Schmuggler," *Der Spiegel* 28 (1950): 19–24. For an excellent analysis of this phenomenon, see Heredia, *"Der Spiegel."*

24. Thomas Wimmer to James H. Kelly, 8 June 1949, NACP, RG 260, OMGB, LD, Box 161, Morals and Conduct Jan-Sep 1949 250.1. See also Crago-Schneider, "Antisemitism or Competing Interests?"

25. "Der Basar in der Möhlstrasse," *Münchner Merkur* [circa 30 June 1949]. See also "Ohne Zoll und Steuer—ist es nicht so teuer," *Süddeutsche Zeitung*, 30 June 1949.

26. "Am Caffeehandel beteiligt." See also Heredia, *"Der Spiegel,"* 87.

27. Zierenberg, *Berlin's Black Market*, 203.

28. Schröder, *Münchner Polizei*, 171.

29. Polizeipräsidium, Kriminaluntersuchungsabteilung, "Schwarzmarkttätigkeit; Zeitraum v.26.9 mit 25.10.1949, für das Stadtgebiet München," 3 November 1949, StAM, Pol Dir 11366.

30. "Der Basar in der Möhlstrasse."

31. Polizeipräsidium, Kriminaluntersuchungsabteilung, "Schwarzer Markt (Sammelbericht)," 24 October 1949, StAM, Pol Dir 11367.

32. "Schwarzer Markt (Sammelbericht)," 19 April 1949.
33. Kauders, *Democratization and the Jews,* 70–73.
34. Mosse, *Crisis of German Ideology*; Heredia, *"Der Spiegel."*
35. Otting, "Schwarzer Markt in der Moehlstrasse," 9 September 1949, NACP, RG 260, OMGB, ID, Box 24, Jewish Demonstration (Möhlstrasse etc.). Emphasis in original.
36. Auerbach to Pitzer, 18 May 1949, BayHStA, Staatskommissar für rassisch, religiös und politisch Verfolgte (Auerbach), Vorläufige Nu: 016b. I am grateful to Ari Joskowicz for sharing this document with me.
37. Polizeipräsidium, "Sicherheitsverhältnisse im Gebiet der Möhlstraße," 7 March 1950, StAM, Pol Dir 11356.
38. Herrmann to Auerbach, 7 September 1949.
39. Wimmer to Kelly, 12 August 1949, NACP, RG 260, OMGB, FOD, General Records, Box 406, Correspondence-Outgoing-1948-1949.
40. On the use of raids against informal economic activity, see Bromley, "Organization, Regulation and Exploitation"; Smart, "How to Survive in Illegal Street Hawking."
41. Canoy, *Discreet Charm of the Police State,* 108.
42. James H. Kelly, "Operation 'Crackdown'," 5 July 1949, NACP, RG 260, OMGB, LD, Box 161, Morals and Conduct Jan-Sep 1949 250.1. Other sites targeted included the central train station, Isartorplatz, and Karlsplatz.
43. Kelly, "Operation 'Crackdown'."
44. "Schwarzhandel läßt nach," *Süddeutsche Zeitung,* 3–4 November 1951.
45. Karl, *Möhlstrasse,* 77.
46. Wimmer to Kelly, 28 June 1949, OMGB, LD, Box 161, Morals and Conduct Jan–Sep 1949 250.1.
47. Murray D. Van Wagoner to Clarence R. Huebner, 11 July 1949, NACP, RG 260, OMGB, LD, Box 161, Morals and Conduct Jan-Sep 1949 250.1; "Aktion in der Möhlstrasse."
48. On efforts at expulsion in other historical settings, see Bromley, "Organization, Regulation and Exploitation," 1164.
49. "Ausländerpolizeiverordnung, vom 22. August 1938."
50. Schönwälder, "'Ist nur Liberalisierung Fortschritt?,'" 129.
51. Grasmüller to Polizeipräsident, 7 July 1952, StAM, Pol Dir 11418.
52. Schröder, *Münchner Polizei.*
53. *Gesetz über die Rechtsstellung heimatloser Ausländer im Bundesgebiet.* See also Bayerisches Staatsministerium des Innern, "Ausweisung von DPs," 2 June 1952, StAM, Pol Dir 11420.
54. Quoted in Schönwälder, "'Ist nur Liberalisierung Fortschritt?,'" 130.
55. Ibid., 131.
56. Ibid., 130.
57. On the use of such provisional structures in both Western and Eastern Germany, see Dossmann, Wenzel, and Wenzel, *Architektur auf Zeit.*

Bibliography

"Ausländerpolizeiverordnung, vom 22. August 1938." *Zeitschrift für ausländisches öffentliches Recht und Völkerrecht* 8 (1938). Accessed 18 September 2014. http://www.zaoerv.de/08_1938/vol8.cfm.

Berkowitz, Michael. *The Crime of My Very Existence: Nazism and the Myth of Jewish Criminality.* Berkeley: University of California Press, 2007.

Bromley, Ray. "Introduction—The Urban Informal Sector: Why Is It Worth Discussing?" *World Development* 6, nos. 9/10 (1978): 1033–39.

———. "Organization, Regulation and Exploitation in the So-Called "Urban Informal Sector": The Street Traders of Cali, Colombia." *World Development* 6, nos. 9/10 (1978): 1161–71.

Canoy, Jose Raymund. *The Discreet Charm of the Police State: The Landpolizei and the Transformation of Bavaria, 1945–1965.* Leiden: Brill, 2007.

Clark, Gracia. Introduction to Clark, *Traders versus the State,* 1–16.

———, ed. *Traders versus the State: Anthropological Approaches to Unofficial Economies.* Boulder: Westview, 1988.

Crago-Schneider, Kierra. "Antisemitism or Competing Interests? An Examination of German and American Perceptions of Jewish Displaced Persons Active on the Black Market in Munich's Möhlstrasse." *Yad Vashem Studies* 38, no. 1 (2010): 167–94.

Dossmann, Axel, Jan Wenzel, and Kai Wenzel. *Architektur auf Zeit. Baracken, Pavillons, Container.* Berlin: B_books, 2006.

Gesetz über die Rechtsstellung heimatloser Ausländer im Bundesgebiet, 1951. Accessed 22 November 2014. http://www.gesetze-im-internet.de/bundesrecht/hauslg/gesamt.pdf.

Hart, Keith. "Informal Income Opportunities and Urban Employment in Ghana." *Journal of Modern African Studies* 11, no. 1 (1973): 61–89.

Harvey, David. *The Condition of Postmodernity: An Enquiry into the Origins of Cultural Change.* Oxford: Blackwell, 1989.

Heredia, David. "*Der Spiegel* and the Image of Jews in Germany: The Early Years, 1947–1956." *Leo Baeck Institute Yearbook* 53 (2008): 77–106.

Holian, Anna. "The Ambivalent Exception: American Occupation Policy in Postwar Germany and the Formation of Jewish Refugee Spaces." *Journal of Refugee Studies* 25, no. 3 (September 2012): 452–73.

———. *Between National Socialism and Soviet Communism: Displaced Persons in Postwar Germany.* Ann Arbor: University of Michigan Press, 2011.

Jacobmeyer, Wolfgang. *Vom Zwangsarbeiter zum Heimatlosen Ausländer. Die Displaced Persons in Westdeutschland 1945–1951.* Göttingen: Vandenhoeck & Ruprecht, 1985.

Karl, Willibald. *Die Möhlstrasse. Keine Strasse wie jede andere.* Munich: Buchendorfer, 1998.

Kauders, Anthony. *Democratization and the Jews: Munich, 1945–1965.* Lincoln: University of Nebraska Press, 2004.

Kauders, Anthony, and Tamar Lewinsky. "Neuanfang mit Zweifeln (1945–1970)." In *Jüdisches München. Vom Mittelalter bis zur Gegenwart,* edited by Richard Bauer and Michael Brenner, 185–208. Munich: C. H. Beck, 2006.

Lefebvre, Henri. *The Production of Space.* Oxford: Blackwell, 1991.

Mosse, George L. *The Crisis of German Ideology: Intellectual Origins of the Third Reich.* New York: Schocken Books, 1981.

Portes, Alejandro, Manuel Castells, and Lauren A. Benton, eds. *The Informal Economy: Studies in Advanced and Less Developed Countries.* Baltimore: Johns Hopkins University Press, 1989.

Rühlemann, Martin W. "'Mir zaynen doh.' Die Möhlstrasse als Schauplatz jüdischer Proteste." In *Auf den Barrikaden: Proteste in München seit 1945,* edited by Zara S. Pfeiffer, 31–38. Munich: Volk Verlag, 2011.

Schönwälder, Karen. ""Ist nur Liberalisierung Fortschritt?" Zur Entstehung des ersten Ausländergesetzes der Bundesrepublik." In *50 Jahre Bundesrepublik—50 Jahre Einwanderung.*

Nachkriegsgeschichte als Migrationsgeschichte, edited by Jan Motte, Rainer Ohliger, and Anne von Oswald, 127–44. Frankfurt am Main: Campus, 1999.

Schröder, Joachim. *Die Münchner Polizei und der Nationalsozialismus.* Essen: Klartext, 2013.

Schwarz, Leo W. *The Redeemers: A Saga of the Years 1945–1952.* New York: Farrar Straus and Young, 1953.

Smart, Alan. "Resistance to Relocation by Shopkeepers in a Hong Kong Squatter Area." In Clark, *Traders versus the State,* 119–38.

Smart, Josephine. "How to Survive in Illegal Street Hawking in Hong Kong." In Clark, *Traders versus the State,* 99–117.

Wetzel, Juliane. *Jüdisches Leben in München 1945–1951. Durchgangsstation oder Wiederaufbau?* Munich: Kommissionsverlag UNI-Druck, 1987.

Zierenberg, Malte. *Berlin's Black Market: 1939–1950.* New York: Palgrave Macmillan, 2015.

16

Creating a Bavarian Space for Rapprochement

The Jewish Museum Munich

Robin Ostow

Fragments of Writing on the Wall

Located on the old and new St. Jakobs Platz, a stone's throw from the Marien-kirche and the marketplace, Munich's Jewish Center consists of three square, sleek travertine structures—the synagogue, the Jewish community center, and the Jewish museum—surrounded by ongoing gentrification: boutiques, cafés, a playground, and the City Museum of Munich across the street.[1] Approaching the museum building, trying to work out which one it is, visitors notice writing on the wall. At the Jewish Museum Munich (JMM), the wall is glass, and the writing, in vinyl letters, consists of selected excerpts from awkward and banal conversations about Jews and Germans. "C: We were both born during the war, and our parents were very nationalist. M: At that time? C: Yes, at that time."[2] The text is in different sizes, the largest being visible from twenty meters away. To read the smaller pieces of text, one has to approach the building; to decipher the very smallest, visitors have to come right up to the wall, where they see the museum's foyer inside.

Munich's Jewish museum is probably the only major Jewish building in Germany where the entrance is marked, in large letters, with a local and traditionally Catholic welcome—"Grüß Gott" (May God bless you!)—and without visible security installations.[3] The foyer, with glass walls on three sides, is full of light and contains the elegant Café Makom ("place" in Hebrew) and a Jewish bookstore. When the weather is warm, the café spills over into the space around the play-

Figure 16.1. View of the northwest corner of the Jewish Museum Munich. Used courtesy of Jüdisches Museum München

ground, bringing the museum into the outdoor life of the neighbourhood. The JMM is chic and welcoming.

Jewish museums are classic examples of Jewish space. They announce the presence of Jews in a city's history and/or contemporary life, exhibiting Jewish religion, art, and history to the public. This chapter explores how, as suggested by the museum's glass outer wall, the writing on it, and the café, the JMM aims to do more than mark the presence of Jews and exhibit culture. It attempts to create a new kind of space where Jewish life flows into the life of the city around it to encourage conversations and encounters—even awkward ones—between Jews and non-Jewish Bavarians and tourists.

This chapter discusses the museum's history, its location in Munich's cityscape, its permanent exhibition, and some of its temporary exhibits. It describes how they mark Jewish geographical and cultural space and how they visualize Jewish life in Munich and Bavaria as a narrative of discontinuity informed by place, displacement, and sometimes replacement. It points to the many ways the museum and its exhibits are designed to bring Jews into Bavarian space and invite non-Jews to enter this Jewish museum space where they can engage in conversations—with other visitors and with Bavarian Jewish history. These, in turn, can become the first step toward a rapprochement.

Embarrassed into Being: The Prehistory of the Museum

Although Munich has the second largest Jewish population in Germany — over nine thousand, second only to Berlin—and a vibrant Jewish culture,[4] the Jewish

Museum Munich now stands on St. Jakobs Platz because the city was embarrassed into building it.[5] A Jewish museum had been planned in Munich in the prewar years, but when the Nazis came to power, its promoters emigrated and the display objects they had collected were most likely destroyed.[6] After the war, both Jews and non-Jews were decidedly uninterested in building Jewish museums in Germany. As Polish-Jewish Holocaust survivors, West Germany's postwar Jews had no interest in German-Jewish history and maintained a hostile relationship to the state.[7] In the mid-1980s, then president of Munich's Jewish community, Hans Lamm, called for a Jewish museum in Munich, but the idea found little resonance within the Jewish community or the municipal administration.

In 1989, though, Richard Grimm, a Munich art dealer active in fostering German-Jewish dialogue, opened a very small Jewish museum (just twenty-eight square meters) in a former servants' apartment on elegant Maximilianstrasse. The city's cultural office reacted by opening a "Jewish Museum" file whose first included item was a newspaper article about this museum's opening; an attached note read: "Suspicious!! ... No further steps taken with regard to this matter."[8] Over the years, though, the ten thousand annual visitors to this private Jewish museum, including dignitaries such as Israeli philosopher Schalom ben Chorin and then mayor of Jerusalem Teddy Kollek, pressured the city into agreeing to take it over in 1998, when it could no longer pay its rent. Soon thereafter, a decision was made to build a new municipal Jewish museum.

In contrast to cities like Berlin and Warsaw, where it took decades between deciding to build and opening a Jewish museum—with concepts and staff repeatedly turned over—the process in Munich was relatively swift and straightforward. In 1999, architects, educators, and journalists discussed ideas for the museum at a two-day international symposium, concluding that Munich's new Jewish museum would be secular and not a Holocaust museum. Moreover, it would not present anything that might encourage stereotyping of Jewish ways of life, Jewish history, or Jewish identity.[9] In 2003, art historian Bernhard Purin was named the museum's founding director. The Jewish Museum Munich, which opened its doors in March 2007, represents his vision.

Jewish Space and Urban Renewal

St. Jakobs Platz, where the museum, synagogue, and Jewish community center are located, carries a complex history of its own. Mostly, it was at the center of a municipal life that excluded Jews. Located close to the center of twenty-first-century Munich, the square was also central to the medieval city. From the twelfth century, it had been the site of a chapel to Saint Jacob and a station on the route to Santiago de Compostela, a highway where the Christianization of Europe (including anti-Semitic outbreaks) routinely took place over centuries.[10] The square also became the site of trade fairs, a fire station, a silk factory, and a haymarket. Although Jewish life in Munich dates back to 1229, Jews were banned from the city for four hundred years, starting in 1442. Synagogues were not allowed in the city center until 1882. The downtown synagogue built that year was vandalized and then destroyed in 1937, on the personal orders of Adolf Hitler. During World War II, a large underground air-raid shelter was built underneath the square. In the postwar years, the space was used mainly as a parking lot.

Munich's postwar synagogues were built far from the center. The city's decision in 1999 to build a Jewish complex on St. Jakobs Platz was not reached easily. Local government officials had originally proposed to build the museum on Munich's periphery; only the persistence of the Jewish community's president, Charlotte Knobloch, secured the central location.[11] The construction of a large Jewish center on St. Jakobs Platz signaled that Jews had once again become accepted in the city center and also that Munich's Jewish community was willing to make itself visible there; it made the square a neighborhood for young professionals and a magnet for tourists.[12] Still, the site drew aggression: in 2003, the police foiled an attempt by neo-Nazis to bomb the construction site, and four hundred police officers protected the cornerstone-laying ceremony.

Symbolically, then, the new synagogue, museum, and community center on this site Judaized, and perhaps redeemed, a space that had hosted muscular Christianizing migration and formed the center of a municipal life from which Jews were banished or marginalized for centuries—and a space of protection during World War II that excluded Jews. Press reports celebrated the Jewish Center's opening as a revival and a rededication of this space to religion and culture, as well as a bulwark against the increasing commercialization and uniformity of the inner city. This claim, or perhaps desire, failed to acknowledge the role of museums in supporting the city's tourist economy and in driving gentrification—processes that precipitate commercialization.[13]

Within this complex, which is a top-down designated Jewish space, the synagogue and community center are places for Jews to come together to worship and socialize with one another. By contrast, the Jewish Museum aims to be an open, inclusive space where Jewish and non-Jewish residents and tourists can meet, encounter Munich's Jewish history, and begin to overcome their social and cultural separation. The building is a simple cube with four levels. The ground floor hosts the bookstore, café, and ticket counter; the two upper floors house temporary exhibits and a multimedia database and small library. The permanent exhibition, *Voices_Places_Times*, is located in the basement.

Fragments of *Voices_Places_Times*: The Permanent Exhibition

This exhibition begins with the writing on the museum's outside wall, part of Sharone Lifschitz's installation *Speaking Germany* (2005). In summer 2005, the London-based, Israeli-born artist placed personal ads in major German newspapers, including *Süddeutsche Zeitung*, *Frankfurter Allgemeine Zeitung*, and *Die Zeit*, that read, "Young Jewish woman visiting Germany would like to have a conversation about nothing in particular with anyone reading this." Lifschitz then traveled through Germany, interviewing 45 of the 180 Germans who responded. The writing on the wall presents short excerpts from these conversations. For several weeks in 2007, they were also posted in public places throughout the city.[14] In this way, an Israeli/Jewish visitor to Germany invited non-Jewish Germans to communicate with her. The welcome next to the museum's entrance, "*Grüß Gott*," was taken from the first e-mail response she received: "*Grüß Gott*, young lady."

As they enter the museum, then, visitors are greeted by the words of a non-Jewish German reaching out to a female Jewish visitor. No full conversations are displayed. The installation consists entirely of banal and uncomfortable fragments ("'And do you like the Germans?' 'I don't know, do you like them?'"), which display the difficulty non-Jewish Germans experience in trying to communicate with Jews and approach their own past. It also reveals that despite these problems and inhibitions, many Germans today are taking up this challenge. *Speaking Germany* was the JMM's opening temporary exhibition, but the writing on the outer wall has remained part of the permanent display. In 2014, the museum hosted *Smiling at You*, a retrospective of Lifschitz's work based on the memories of Jewish refugees, but also including photographs of smiling truck drivers and accounts of traveling through Germany, eating and talking with strangers.[15] These images foreground travel and the way it generates communication between strangers, and between Jews and non-Jewish Germans.

Inside the Museum

One-third of the museum's nine hundred square meters of exhibition space is devoted to the permanent exhibition, which the visitor enters by descending a flight of stairs to the basement. Does the visitor become a postmodern Orpheus? Or is this a descent into Bavarian-Jewish history? Museum scholar Reesa Greenberg cites William Connolly's discussion of "neuropolitics," through which cultural forms can spark subconscious thought processes and reconfigure perception,[16] and she explores associations around underground exhibitions and their relation to buried memories.[17]

In Munich, the underground location of the permanent exhibition suggests that Jewish history and culture are buried beneath the surface of everyday life in this city, in a cavity that once offered wartime protection. Two other major Jewish museums, recently opened in cities where the Jewish population was murdered under the Nazi regime, also begin their permanent exhibits underground: the Jewish Museum Berlin and POLIN in Warsaw. The underground exhibition space in Munich is divided into seven areas.[18] Visitors enter via a bare, narrow audio corridor called *Voices,* which is suggestive of an airport—a space of arriving, departing, and conversations with strangers—with marked stations where visitors can stop and listen to the voices of Jews describing their arrival in Munich.

Some of the audio pieces are taken from memoirs dating back to the eighteenth century, and some from interviews with living Jews.[19] These accounts, like the fragments of conversation on the outer wall, are short and out of context. Voices of people from many walks of life describe the many historical situations that brought Jews to Munich and their first impressions of the city. They also explain that most of Munich's Jews (as, indeed, its non-Jews) came from elsewhere. Like Lifschitz's conversations, these disembodied narratives of encounters with non-Jewish Munich are many and fragmentary; they never allow for sustained engagement with a narrator.[20]

Marking Jewish Space in Munich

The second area, *Places,* is an installation by Renata Stih and Frieder Schnock. It is a forty-square-meter, interactive orange carpet that is also a map of Munich, with fourteen numbered, movable white markers. The Jewish places on the map are also numbered, and when the visitor sets a marker on the space with the corresponding number, a photograph on the wall behind the map lights up. The marker displays a paragraph about the significance of the designated place in Munich's Jewish history, and the photograph illustrates an event involving Jews that took place there.[21] Among the events highlighted in the photographs, which span the years 1910 to 2006, are the national championship celebration of the Bayern München soccer team in 1932, with its Jewish president, Kurt Landauer; the widow of Israeli athlete André Spitzer in the destroyed interior of his room after the massacre at the 1972 Munich Olympics; and a portrait of the five children of Jewish art collector Alfred Pringsheim. This installation, too, presents small, disjointed pieces of Jewish history.

A Jewish history section is a standard component in Jewish museum exhibitions. The JMM places this history in the third section, *Times,* which consists of a horizontal black timeline on a white wall. Fifty-one dates in Jewish history in Munich and Bavaria are marked, beginning with the arrival of Munich's first identified Jew—Abraham of Munich—in 1229, and ending with the completion of the Jewish Center in 2006. The events marked include expulsions and burnings of Jews, edicts granting Jews the right to return, the establishment and demolition of synagogues, the development of Jewish organizational infrastructure, and changes in Jews' legal status and demography.

Grasping the Holidays

Most Jewish museums exhibit Jewish holidays and ritual objects. In Munich, the fourth section is called *Rituals,* and the ceremonial objects are first glimpsed in shadow-box format. Behind the white screen, visitors can touch and pick up reproductions of some of these artifacts. Basic explanations of the objects and the holidays they represent appear on the wall behind them. In addition to major traditional holidays, the curators in Munich have added three new ones celebrated in Israel today, with new ritual objects: Independence Day, Remembrance/Fallen Soldiers Day, and Holocaust Memorial Day. *Rituals* represents a minimalist and interactive variant of the *Schatzkammer* (treasure chamber),[22] the premodern idea of museums as treasure cabinets. Unlike the Jewish museums in Paris, Amsterdam, Prague, and prewar Berlin, which showcased ritual objects as treasure, the display of ceremonial objects in Munich's Jewish Museum visualizes their place in family histories (in the *Objects* gallery) and in religious observance. Encouraging visitors to pick these objects up has the effect of desacralizing them.[23] Instead of merely looking at treasures in vitrines, visitors are invited to engage with the ritual objects, which also breaks down boundaries of unfamiliarity and inhibition.

The heart of the permanent exhibition is the wide gallery *Objects,* the fifth section. Seven members of the museum's staff each selected a favorite object to display in a glass case with an accompanying text identifying it and explaining why it was chosen. Each staff member's text expresses a personal relationship with the object selected. It also communicates ideas about the curatorial process and ways of approaching objects. Among other objects, the vitrines exhibit a late eighteenth-century Sabbath prayer book; *Girl from the Ghetto,* a portrait of a Polish-Jewish girl, painted in Munich in 1915; and an *Aliyah* (emigration to Palestine) board game manufactured in Berlin in 1935. This display is the permanent exhibition's only reference to Zionism, which is presented as a form of displacement. The recently arrived Polish-Jewish girl in the image has also been displaced.

The Holocaust from the Point of View of a Refugee Survivor

Perhaps the most moving installation in the permanent exhibition is the sixth section, *Memories of a Survivor.* It is a real wall cupboard that belonged to Simon Snopkowski (1925–2001), a Polish Holocaust survivor who settled in Munich after the war. He became a distinguished surgeon and also served as president of the Bavarian Association of Jewish Communities. The cupboard, on permanent loan from his family, contains shelves with books in Polish, English, and German and small objects and photographs documenting his childhood and the Holocaust, especially the concentration camps and death camps. The order of the objects on the shelves is neither alphabetical nor thematic. Some of the books are upside down. In Snopkowski's home, these shelves had been behind closed doors: they were private. The curator, Verena Immler, calls it a "personal memorial";[24] it is the museum's internal Holocaust memorial.

Snopkowski's wall cupboard displays trauma—trauma contained in a private bookshelf, behind closed doors, the trauma of a Jewish immigrant to Bavaria. It is another confrontation with Munich's Jewish history created by an outsider and

Figure 16.2. Wall cupboard by Dr. Simon Snopkowski. Used courtesy of Jüdisches Museum München

from an outsider's experiences.[25] Today, Europe's Jewish museums differentiate themselves from Holocaust memorials and aim to display Jewish "life," yet each one has a major Holocaust installation that functions as an internal memorial. In Prague's Jewish Museum, the empty Pinkas Synagogue, with the names of the 77,297 Czech and Slovak Holocaust victims inscribed on the walls, serves this function. In Paris, it is Boltanski's installation with the names of the deported inhabitants of the building currently housing the Jewish Museum; in Vienna, the *Schaudepot* (viewable storage area) on the top floor of the Jewish Museum showcases possessions that Jews of prewar Vienna left behind.

Snopkowski's bookshelves are a personal Holocaust memorial and also a library. Library installations, too, are frequently found in Europe's Jewish museums, where they represent Jews' self-identification as "people of the book."[26] The Jewish museums in Amsterdam, Prague, and Warsaw have library installations. Vienna's prewar Jewish museum featured an installation of a bourgeois library; in 2001, Rachel Whiteread's inside-out library sculpture was unveiled as the city's Holocaust memorial. In Vienna, and now in Munich, the library of death and the Holocaust memorial fuse, becoming one.[27]

Comics and Travel

The permanent exhibition ends with more writing on a wall, this time an internal wall of the museum: the installation *COMICS* by Jordan Gorfinkel, an American Jewish cartoonist. This installation, commissioned by the museum, recounts, in ten comic strips, the visit to Munich of Zayds, a fictional emigrated Holocaust survivor who grew up in Munich,[28] and his American grandson, Bernie. Their journey is punctuated by a series of confrontations and conversations that bring into play hesitations, perceptions, and misperceptions harbored by American, German, and Russian Jews (of different generations) concerning one another and non-Jewish Germans. A second focus is German stereotypes of Jews. *COMICS* and the permanent exhibition end as Bernie asks, "Well, Zayds, our trip to Munich is almost over. Is there anything you missed that you'd like to see?" The final frame shows grandfather and grandson walking off, as Zayds answers, "My family."

Bernhard Purin chose to end the exhibition with a cartoon for strategic, historical, and personal reasons. The cartoon is a genre that can reach out to young visitors; the success of Art Spiegelman's 1986 comic-book-style novel *MAUS* shows the usefulness of this format for communicating difficult material. Because the Nazis corrupted the cartoon by exploiting it for the dissemination of anti-Semitic propaganda, Purin wanted to decontaminate and restore this format by using it to help Germans and Jews overcome the divisions caused by Nazism. Finally, Purin turned to comics to restore a missing part of his own life; his parents, part of the generation of "1968" associated with idealistic rebellion in Germany and elsewhere, did not allow him to read comics when he was a child.[29] Greenberg points to the significance of this installation in breaking the taboo, in place since 1945, around using cartoons in Germany to portray Jews.[30] And the JMM has continued to engage with comics as a form of communication. In 2013, it showed a temporary exhibition, *Dr. Wertham and the Comic Code,* which coincided with Munich's biennial Comicfestival. Dr. Fredric Wertham was a Jewish psychiatrist in Munich who believed that comics presented a dangerous influence on young people. In 1922, he emigrated to the United States, where he became an anti-comic-book crusader.

In the 1940s, when comics were being used in Germany to disseminate anti-Semitic propaganda, in the United States they depicted the contours of nationalist masculinity developed by second-generation Jewish immigrants. In featuring exhibits of and about comics and anti-comic crusaders, Purin reclaims this mode of expression as a cultural space linking Germany with American Jews.[31]

The JMM's permanent exhibition, then, is bookended by two commissioned installations representing German-Jewish encounters occasioned by visits of foreign Jews to Germany: the first by a real, young Israeli woman, and the second by a fictional, older man, a Holocaust survivor, who has the final word. *Speaking Germany* and *COMICS* suggest the transnationality of German-Jewish space—and that much German-Jewish history and culture unfolded abroad. They also invite emigrated Jews and their descendants back to Munich and back into Jewish history and culture in Germany. And they begin to reintegrate Jewish life in Germany, which was isolated through most of the Cold War period, back into world Jewish culture, with its centers in Israel and the United States.

Bavaria and Jewish Displacement

The Jewish Museum Munich has explored Jewish displacement largely through a series of temporary exhibitions, starting with *City without Jews: The Dark Side of Munich History.* In 2008, to mark the 850th anniversary of the city's founding, major cultural institutions were asked to organize commemorative events. In contrast to the mostly celebratory programs produced by other museums, the JMM's contribution focused on the persecution of Jews in Bavaria and the more than four hundred years—half the city's history—when Jews were banned from Munich. *City without Jews* was followed by three temporary exhibitions in 2008–9 about cities in which some Munich Jews found refuge in the 1930s and '40s. Under the overarching title *Places of Exile,* the exhibitions were Münih ve Istanbul, *Minchen ve' Tel Aviv,* and *Munich and Washington Heights.* In 2011–12, two exhibitions, jointly called *Jews 45/90,* showcased Jewish immigration to postwar Bavaria. *From Here to There: Survivors from Eastern Europe* explored the experiences of Jewish displaced persons in Bavaria, and *From Far Away: Immigrants from the Former Soviet Union* featured more recent Jewish immigration. Together, these exhibitions highlighted the expansions and contractions of Jewish space in Munich and Bavaria and of Bavaria's Jewish population. And they extended Bavarian Jewish space to neighborhoods in Istanbul, New York, and Tel Aviv.

Home and Migration

The JMM's emphasis on emigration, expulsion, and immigration references migration museums as a recent and dynamic cultural development on several continents. Migration museums foreground discontinuity and dislocation and interrogate possibilities and conditions for integrating migrants into their receiving societies.[32] By emphasizing migration, the JMM expands the geography of Munich's Jewish history to encompass the areas from which many of the city's Jews originated and areas that absorbed its Jewish emigrants and refugees. It is a postmodern response to the traditional German *Heimat* (hometown) museum that displays a community as a home to its inhabitants and showcases its history as a continuous narrative.

The Jewish museums in Paris, Prague, and Amsterdam are *Heimat* museums. The JMM includes some "hometown" installations in its permanent exhibition: the photograph of Bavaria's soccer team in 1932, a beer mug with the inscription "Reserve Military Hospital—Jewish Nursing Home. Christmas 1917," and a pipe bowl featuring a portrait of Kurt Eisner, the first prime minister of the state of Bavaria. The JMM's temporary exhibition, *Dirndl, Trunks and Edelweiss: The Folk Art of the Wallach Broth*ers, mounted in 2007, displayed the role of a Jewish family in the production of traditional Bavarian costume. Another temporary exhibition in 2016 was about Jews and beer, Bavaria's iconic drink. These exhibitions claim Jewish space at the heart of Bavarian popular culture and reach out to non-Jewish Bavarians. Nevertheless, the JMM's emphasis remains on mobility and migration.[33]

The Foreign Gaze

The Jewish Museum Munich contains many of the standard elements found in Europe's Jewish museums, including a chronology of local Jewish life, Jewish holidays and ceremonial objects, a library installation, and a Holocaust memorial installation. However, in contrast to the Jewish museums in Paris, Amsterdam, and Prague, this museum consistently presents Bavaria's and Munich's Jewish history from an outsider's point of view. Director Purin grew up in Austria.[34] Sharone Lifschitz and Jordan Gorfinkel are Jewish "outsiders," as are the sound-station voices at the entrance to the permanent exhibition. Although objects and geographical spaces that frame the lifeworlds of Munich's long-term Jewish residents appear in the exhibition, the museum's installations do not foreground their voices, their friends, and their experiences in postwar Bavaria. Visitors see no representations of the exclusion Jews felt in postwar Munich, their segregation (with Protestants) into separate school classes, their feelings of shame for living in Germany after what had happened or feelings of community that helped sustain them.[35] By contrast, Hanno Loewy's section on the postwar Germanies at the Jewish Museum Berlin highlighted the experiences and social engagement of Germany's Jewish youth and Jewish communities.[36]

Individuals, Fragments, Decontextualization, and Distance

The voices, perceptions, experiences, and encounters that the JMM features are largely those of individuals. Lifschitz's conversations are with individual Germans, the voices at the entrance are of individual migrants, the Holocaust installation is a private library, and *COMICS,* too, displays a series of individual exchanges. The Jewish and non-Jewish communal structures in Munich and the many subcultures that grew out of the city's Jewish and non-Jewish life remain in the background. Their role in providing meanings and biographical trajectories for individuals is neglected in favor of presenting experiences and values through individual voices, objects, travel, and migration.

The JMM's strategy of displaying decontextualized objects, individuals, and small pieces of history, culture, and conversation challenges traditional displays of Jewish history as an epic narrative as they are found in Jewish museums in Berlin, Paris, and Warsaw. The permanent exhibition catalogue explains, "The museum's own collection, as well as local and regional collections, are often arbitrary and full of gaps, not least as a result of the Shoah. Accordingly, only fragments of Munich's Jewish history can ever be exhibited and elucidated."[37] This foregrounding of fragments emphasizes historical and cultural loss. At the same time, by leaving it to the visitor to piece together the larger story, the museum activates viewers, positioning them as co-narrators rather than passive addressees. Ernst van Alphen discusses this strategy as the "narrative" exhibition model.[38] The story, "co-narrated" by the curator(s) and viewers, incorporates viewers' own creativity and knowledge. It ends in an "effect" that is the cumulative impression left by visitors' confrontation with the sequence of displays.

The presentation of many small parcels of speech and bits of history also contributes to maintaining distance between the visitors and the exhibition. The permanent exhibition offers no heroes, villains, or personalities to identify or engage with, no room installations, no items of clothing, no personal letters or diaries. One journalist complained that the museum is informed by "an extravagant introversion and a frozen intellectuality conveyed in material [objects]."[39] Purin, however, defends curatorial distance as a strategy for avoiding voyeurism.[40]

The Jewish Museum Munich as a Meeting Place

The JMM's mission statement ends with an invitation: "The museum invites everyone—Munich residents, guests from inside and outside the country, students, Jews and non-Jews—to gather at this open forum from which they can also explore Jewish Munich."[41] And the permanent exhibition catalogue features a section entitled "A Meeting Place: The Architecture of the New Jewish Museum in Munich."[42] "Meeting" is a translation of the German word *Begegnung,* which suggests a serious encounter. Sociologist Erhard Stölting described a postwar German *Begegnung* as a "meeting of people who dislike each other and come together to work out their problems so that they can overcome their hostilities and love each other."[43] The awkward bits of conversation recorded by Sharone Lifschitz and the comic strips by Jordan Gorfinkel represent less ambitious exchanges. They do not promise museum visitors a successful German-Jewish love-in or even a reconciliation. Rather, they encourage communication around the tensions between Germans and Jews, while accepting their limitations.

The attractive café, in particular, invites visitors to sit and talk. Director Purin participates regularly in café conversations. By 2014, Café Makom had also become a place where young members of Munich's Jewish community regularly take coffee and e-mail breaks during prayer services next door. The museum's bookshop promotes communication through written words.[44]

Curating and Creating Community

The Jewish Museum Munich also promotes encounter and integration through its practices of collection and curation. With only 150 display objects of its own, this museum probably has the smallest collection of any municipal Jewish museum in a major Western European city. Offe outlines how, starting in the 1980s, as many as one hundred communities in West Germany opened (usually small) Jewish museums, which exhibited Jewish artifacts implicitly or explicitly identified as the rescued remains of local Jewish culture.[45] Purin points out that, today, prewar Jewish artifacts on display are considered more likely to have been looted than rescued.[46] Given the shortage of authentic Jewish display objects—especially in the wake of the recent wave of new Jewish museums—as well as current sensitivities about looted art and the difficulty of ascertaining the provenance of objects, the JMM works most often with borrowed objects and new installations commissioned from young artists, some of them from Israel and the United States. In this way too, the museum reaches out to younger visitors and to Jewish communities abroad.

Is It a Museum of Jewish History, Space, and Culture?

Visitors to the Jewish Museum Munich learn that the city's Jewish presence dates back to the Middle Ages and that it has been discontinuous, punctuated by immigration, emigration, persecution, and periods of integration. They find out that many different kinds of Jews have lived, worked, and come together in various locations in Munich; that Jews have a distinct set of rituals and holidays; and that many Munich Jews have been associated with the arts. But do the museum's innovative installations provide an overview of Munich's Jewish history and culture? Purin rejects what he calls the "didactic encyclopaedic model of representation": "We know that the average visitor spends one hour in the museum. Even with the best didactic processes, if somebody knows nothing about Judaism, he will not know very much more after an hour, either." He calls for "dialogue instead of didactics."[47]

With its focus on the outsider's view, on the many individual Jews who have passed through the city, and on ways to approach objects and people, the JMM displays Jewish history, culture, and space in Munich as fluctuating movements, episodes of cooperation and rejection, and the difficulty Germans experience in overcoming their inhibitions concerning Jews and Germany's past. The display objects are presented largely as witnesses, products, and facilitators of these movements, encounters, and interactions, rather than as treasures valued for the material from which they are created or the workmanship that produced them. The installation *Places* makes visible spaces that were—and in some cases still are—part of Jewish life in Munich, including Jewish history, sports, worship, homes, and workplaces.

Museums extend different kinds of invitations to their visitors, and this is true of Jewish museums, as well. The Jewish museums in Paris and Prague provide visitors an opportunity to gaze at Jewish treasures, largely ceremonial objects and works of art. The Jewish Historical Museum in Amsterdam encourages guests to become intimate with Jewish culture through displays of letters, diaries, and household and personal items. Its children's section invites visitors into a Jewish home installation, and literally into bed.[48] In Munich, by contrast, the JMM beckons to tourists and people in the neighborhood to engage with the fragments of Bavarian Jewish history and culture—to respond to Sharone Lifschitz's invitation at the museum's entrance to "speak Germany."

Today, the Bavarian capital that was once the stronghold of the Nazi movement boasts an innovative, state-of-the-art Jewish museum that is also an upscale meeting place. It does not dazzle visitors with Jewish treasures; it does not profile heroes or villains. It does not present a continuous history or overwhelm its guests with facts or horrific displays of the Nazi years. Rather, the Jewish Museum Munich reveals Jewish life in Bavaria as fluctuating presences and absences represented by fragmentary remains and installations that express the experiences of many kinds of Jews. The final impression—in van Alphen's scheme, the "effect"—is a tension between the integration of Jewish culture into the city center, as evidenced by the new Jewish complex, and the persistence of absence expressed by Zayds, who misses his family, and the exhibition's fragmentary nature. But, most importantly, the JMM provides a space that displays and encourages movement and contact, a place for Jews and non-Jewish Germans and tourists to engage with Jewish history and culture, with Germany, and with one another.

Robin Ostow teaches in the Sociology Department at Wilfrid Laurier University, in Ontario, Canada. She has published extensively on national museums, Jewish museums, and immigration museums in Europe, the Americas, and Australia. Her work has focused on these museums' social histories, their displays, and their relations with the communities around them. Her most recent book, *(Re)Visualizing National History: Museums and National Identities in Europe in* *the New Millennium,* was published by University of Toronto Press, 2008. She is currently writing a comparative ethnography of six human rights museums on four continents.

Notes

Bernhard Purin and the staff of the Jewish Museum Munich were more than generous and helpful and made me feel very much at home in the JMM. Reesa Greenberg contributed several key insights. Important ideas also came from Michael Brenner (University of Munich), Barbara Dietz (University of Regensburg), Oliver Lubrich (University of Bern), and Elie Teicher (Munich). The research for this essay was supported by the Social Sciences and Humanities Research Council of Canada.

1. The Jewish Museum Munich is one of the city's four municipal museums. The others are Villa Stuck, the Lenbach House, and the City Museum of Munich. Unlike the synagogue and the community building, the JMM belongs to the city, not to Munich's Jewish community.
2. All translations are mine.
3. The security precautions in this Jewish museum are elaborate, but visitors are never made aware of them. This contrasts decidedly with the Jewish museums in Berlin and New York, where visitors pass through airport-style security checks.
4. Munich's Jewish population was 9,507 in 2015; Zentralrat der Juden in Deutschland, http://www.zentralratdjuden.de/en/topic/126.local-communities.html?gemeinde=69, accessed 03 February, 2017.
5. Munich's mayor, Christian Ude, said he had "always felt ashamed of the historical indifference of this city." *Rathaus Umschau,* 10 November 2006, cited in Purin and Fleckenstein, *Jüdisches Museum München,* 80.
6. Purin, "Building a Jewish Museum."
7. Kugelmann, "Das Jüdische." Cilly Kugelmann is a historian and former deputy director of Berlin's Jewish museum.
8. Purin and Fleckenstein, *Jüdisches Museum München,* 14.
9. Ibid., 79.
10. These pilgrimages reached their peak in the Middle Ages. They declined in the sixteenth century due to political conflicts, the plague, and the Protestant Reformation.
11. Anne Goebel, "Sieg der Opfer über die Täter," *Die Welt,* 23 March 2007.
12. Welz, *Inszenierungen kultureller Vielfalt,* discusses the gentrification of downtown areas of globalizing cities, partly through the development of ethnic heritage sites.
13. Matthias Kristlbauer, "Triumph der Opfer über die Täter," *Münchner Merkur,* 23 March 2007. Oliver Lubrich brought this contradiction to my attention (e-mail, 6 December 2007). Becker, *Ankommen in Deutschland* (48), discusses the idea, widespread in Germany, that Jewish culture will enhance German culture and restore its prewar prominence.
14. See the artist's website, www.speaking-germany.de.
15. See http://www.youtube.com/watch?v=qPRtwXQ7o6k (accessed 14 October 2014).
16. Greenberg, "Constructing," 195.
17. Greenberg, "Private Collectors."
18. The remains of the air-raid shelter were removed due to the advanced state of their deterioration.

19. The historical texts are read by actors. Most interviewees speak in their own voices.

20. By contrast, visitors to the US Holocaust Memorial Museum in Washington, DC, are given a card to carry through the exhibition with the name and biography of a Holocaust victim. This card is meant to engage them and to provide identification with one victim and continuity throughout their visit.

21. The museum refuses to exhibit photos taken between 1933 and 1945 that represent the Nazis' point of view. Purin and Fleckenstein, *Jüdisches Museum München*, 33.

22. Reesa Greenberg called my attention to this (pers. comm., 16 December 2008).

23. For a more thorough exploration of past and current models for exhibiting Jewish ceremonial objects, see Greenberg, "Displaying Judaica."

24. Cited in Purin and Fleckenstein, *Jüdisches Museum München*, 69.

25. Greenberg, pers. comm., 16 December 2008.

26. Fonrobert and Shemtov, "Introduction," 3, cite a Heinrich Heine quotation on the Jews from 1840: "A book is their fatherland, their possession, their ruler."

27. Greenberg notes the juxtaposition of this "dead library" in the basement with the living library—accessible to visitors—on the museum's second floor (pers. comm., 16 December 2008).

28. "Zayds" is a variant of *zaydie*, the Yiddish word for "grandpa." Gorfinkel told the *Cleveland Jewish News* that his comic character Zayds is based on his father-in-law, a Holocaust survivor. Alan Smason, "New Munich Jewish Museum Features Local Cartoonist's Work," *Cleveland Jewish News*, 5 April 2007.

29. "Reine Harmonie wird es nie geben," *Donaukurier*, 28 March 2007. In 2006, the Jewish Museum in New York hosted a temporary exhibition, *Masters of American Comics*. A similar temporary exhibition, *De Superman au Chat du rabbin. Bande dessinées et mémoires juives*, opened in late November 2007 at the Musée d'Art et d'Histoire du Judaisme in Paris.

30. Greenberg, pers. comm., 16 Dec. 2008.

31. Greenberg, pers. comm., 9 Aug. 2007.

32. Greenberg points to the difference between migration museums and more traditional diaspora museums. Migration museums emphasize flux and instability. They are also more outward-looking, and they tend to challenge traditional narratives found in museums that exhibit rooted, sedentary life.

33. Michael Brenner pointed out that Munich's Jewish history and culture are not as rich as the Jewish cultures of Paris, Prague, and Amsterdam (e-mail, 14 December 2007).

34. Reesa Greenberg reminded me of this (pers. comm., 16 December 2008).

35. TE (his identity is disguised) commented on this (pers. comm., 10 August 2007). For accounts of these experiences and feelings, see Seligmann, *Rubinsteins Versteigerung*; and Sucharewicz, *Israelische Geschichten*.

36. It was removed in 2014.

37. Purin and Fleckenstein, *Jüdisches Museum München*, 26.

38. Alphen, "Exhibition as a Narrative."

39. See Alexander Kluy, "Rückkehr ins Zentrum," *Die Welt*, 23 March 2007.

40. Bernhard Purin, pers. comm., 9 August 2007.

41. "Mission Statement," Jüdisches Museum München, accessed 15 October 2014, http://www.juedisches-museum-muenchen.de/about-the-museum/mission-statement.html?L=1.

42. Purin and Fleckenstein, *Jüdisches Museum München*, 79–89.

43. Erhard Stölting, pers. comm., August 1996.

44. Today, most museums have bookshops and cafés. In this museum, their location in the bright, flow-through, accessible foyer provides a particularly attractive space for sociability. Purin observed that only half the people who enter the museum visit the exhibition. The rest browse through the contents of the bookshop and/or meet for coffee (Bernhard Purin, pers. comm., 9 August 2007).
45. Offe, *Ausstellungen.*
46. Purin, "Building a Jewish Museum," 149.
47. Purin and Fleckenstein, *Jüdisches Museum München,* 89.
48. In my many visits to this museum in the summers of 2002 and 2003, I found the bed installation on the top floor almost always occupied, quite often by adults.

Bibliography

Alphen, Ernst van. *Caught by History: Holocaust Effects in Contemporary Art, Literature and Theory.* Stanford, CA: Stanford University Press, 1997.
———. "Exhibition as a Narrative Work of Art." In *Partners: Haus der Kunst,* edited by Y. Hendeles, C. Dercon, T. Weski, and Haus der Kunst München, 143–85. Munich: Haus der Kunst, 2003.
———. "Playing the Holocaust." In *Mirroring Evil: Nazi Imagery/Recent Art,* edited by N. Kleeblatt, 65–83. New York: Jewish Museum and Rutgers University Press, 2001.
Bal, Mieke. "Exhibition as Film." In Ostow, *(Re)Visualizing,* 15–44.
Becker, Franziska. *Ankommen in Deutschland: Einwanderungspolitik als biographische Erfahrung im Migrationsprozess russischer Juden.* Berlin: Reimer, 2001.
Bennett, Tony. *The Birth of the Museum: History, Theory, Politics.* New York: Routledge, 1995.
Bunzl, Matti. "Of Holograms and Storage Areas: Modernity and Postmodernity at Vienna's Jewish Museum." *Cultural Anthropology* 18, no. 4 (2004): 465–68.
Fonrobert, Charlotte E., and Vered Shemtov. "Introduction: Jewish Conceptions and Practices of Space." *Jewish Social Studies* 11, no. 3 (2005): 1–8.
Greenberg, Reesa. "Constructing the Canadian War Museum/Constructing the Landscape of a Canadian Identity." In Ostow, *(Re)Visualizing,* 183–99.
———. "Displaying Judaica: Secularizing the Sacred." Unpublished manuscript, 2007.
———. "Jews, Museums and National Identities." *Musées/Museums* 24, no. 2 (2002): 125–36.
———. "Private Collectors, Museums and Display: A Post-Holocaust Perspective." *Jong Holland,* no. 1 (2000): 29–41.
Gruber, Ruth Ellen. *Virtually Jewish: Reinventing Jewish Culture in Europe.* Berkeley: University of California Press, 2002.
Jüdisches Museum München. *Pressespiegel: März > April 2007.* Munich: Jüdisches Museum München, 2007.
Kugelmann, Cilly. "Das Jüdische als Exponat der Zeitgeschichte." *Wiener Jahrbuch für Jüdische Geschichte, Kultur und Museumswesen* 2 (1996): 43–56.
Malkki, Liisa. "National Geographic: The Rooting of Peoples and the Territorialization of National Identity among Scholars and Refugees." In *Culture, Power, Place: Explorations in Critical Anthropology,* edited by R. Rouse, J. Ferguson, and A. Gupta. Boulder, CO: Westview Press, 1992.
Offe, Sabine. *Ausstellungen, Einstellungen, Entstellungen: Jüdische Museen in Deutschland und Österreich.* Berlin: Philo, 2000.

Ostow, Robin. "From Displaying 'Jewish Art' to (Re)Building German-Jewish History: The Jewish Museum Berlin." In *Interrogating Race and Racism,* edited by V. Agnew, 298–320. Toronto: University of Toronto Press, 2007.

———. "The Jewish Museum in Vienna: An Anti-Heimat Museum? Or a Heimat Museum with an Accent?" In *Der 'Virtuelle Jude'—Konstruktionen des Jüdischen,* edited by K. Hödl, 53–70. Innsbruck: Studien Verlag, 2004.

———. "Mokum Is Home: Amsterdam's Jewish Historical Museum." *European Judaism* 38, no. 2 (2005): 43–68.

———. "Religion as Treasure: Exhibits of Rituals and Ritual Objects in Prague's Jewish Museum." In *Die Kanon und die Sinne: Religionsästhetik als akademischer Disziplin,* edited by S. Lanwerd. Luxembourg: Études Luxembourgeoises d'Histoire et de Sciences des Religions, 2003.

———. "Remusealizing Jewish History in Warsaw: The Privatization and Externalization of Nation Building." In Ostow, *(Re)Visualizing,* 157–80.

———, ed. *(Re)Visualizing National History: Museums and National Identities in Europe in the New Millennium.* Toronto: University of Toronto Press, 2008.

Purin, Bernhard. "Building a Jewish Museum in Germany in the Twentieth-First Century." In *(Re)Visualizing,* ed. Ostow, 139–56.

Purin, Bernhard, and Jutta Fleckenstein. *Jüdisches Museum München.* Munich: Prestel, 2007.

———. "Theodor Harburger und das jüdische Museumswesen in Bayern." In *Theodore Harburger: Die Inventarisation jüdischer Kunst- und Kulturdenkmäler in Bayern,* edited by the Central Archives for the History of the Jewish People and the Jewish Museums of Franconia in Fürth and Schnaittach, 51–62. Fürth and Jerusalem: Central Archives for the History of the Jewish People and the Jewish Museums of Franconia in Fürth and Schnaittach, 1998.

Seligmann, Rafael. *Rubinsteins Versteigerung.* Frankfurt am Main: Eichborn-Verlag, 1989.

Spiegelman, Art. *Maus: A Survivor's Tale.* New York: Pantheon, 1986.

Stadtarchiv München. *Das Jüdische Museum München 1989 bis 2006—ein Rückblick.* Munich: Stadtarchiv München, 2006.

Sucharewicz, Leo. *Israelische Geschichten.* Munich: Roman Kovar Verlag, 1992.

Welz, Gisela. *Inszenierungen kultureller Vielfalt: Frankfurt am Man und New York City.* Berlin: Akademie Verlag, 1996.

17

Real Imaginary Spaces and Places

Virtual, Actual, and Otherwise

Ruth Ellen Gruber

Much of my writing in recent years has dealt with what I have come to call "new authenticities" and the related notion of "real imaginary" spaces: spaces, be they physical and/or within the realm of thought or idea that are, so to speak, both "real" and "imaginary" at the same time. They are sites, experiences, and even mind-sets evoked by the past or by another culture or possibly one's own personal heritage, which through study, enactment, yearnings, and potentially fantasies are transformed into something that is quite different but often just as "real" as the original. They may evoke or represent something—a time, place, experience, or memory—but are themselves *not* that actual place, experience, or memory, nor are they necessarily a continuation or outgrowth of it. They are thus "imaginary" or at least "re-imagined," but nonetheless real in their own contemporary context; they exist as real physicalities, real identities, real experiences. These are places (and spaces) where what is *meant* or *signified* by them, rather than their actual apparent (or apparently actual) content, can be paramount—where, perhaps along the lines of Jean Baudrillard's "simulacra," the simulation can become the thing in itself; where the process can be seen as "creating" something new rather than "re-creating" something that once existed, thus leading to the formation of its own models, stereotypes, modes of behavior, and even traditions.[1]

These notions are an outgrowth of my thoughts on what I have called the "virtually Jewish world"—an "intense, visible, vivid [Jewish] presence in places where few Jews live today."[2] This in turn was rooted in the concept of "Jewish space" as it was articulated in the mid-1990s by the historian Diana Pinto, a pioneering thinker whose views have set many of the parameters of discussion regarding post-communist Jewish developments. At that time, Pinto used the term

"Jewish space" to describe the place occupied by Jews, Jewish culture, and Jewish memory within mainstream European society, regardless of the size or activity of the local Jewish population:

> Indeed, it is possible that the larger the "Jewish space," the smaller the number of actual Jews. In countries with sizeable Jewish communities, such as Britain and France, there is a lively and active Jewish community but perhaps less of that "Jewish space" that is distinct from the community itself. Conversely, Germany—where the Jewish community is small by pre-war standards, and is not composed of descendants of the old German Jewish community—has without doubt the most impressive "Jewish space" in Europe. That space appears to be limitless.[3]

I still remember how strongly I was struck by this notion when I first heard Pinto elaborate it in 1995 at a conference in Prague entitled "Planning for the Future of European Jewry": the "Jewish thing," she stated, "is becoming universal." I became fascinated by how, along with efforts to revive Jewish communal life and reclaim and reassert Jewish identity in post-Holocaust, post-communist countries, the "Jewish space" in Europe was often being "filled" by non-Jews: they documented synagogue buildings, Jewish cemeteries, and other abandoned Jewish heritage sites and spearheaded restoration projects; they formed klezmer bands and opened Jewish museums; they used Jewish themes in artworks, theater performances, and literary compositions; they administered, taught, and studied in academic Jewish studies programs; they created the infrastructure for Jewish-themed tourism, opened "Jewish-style" cafés and restaurants—and more. It was this phenomenon—the non-Jewish filling of the Jewish space—that formed what I have referred to as a "virtual Jewishness," or a "virtual Jewish world," peopled by "virtual Jews" who create, perform, enact, or engage with Jewish culture from an outsider perspective, often in, and often as a result of, the absence of local Jewish populations.

Around the same time that Pinto introduced her formulation, the sociologist Y. Michal Bodemann used other terminology to describe the phenomenon as he observed it in Germany, writing of a sort of "Judaizing terrain" made up of "converts to Judaism, of members of joint Jewish-German or Israeli-German associations, and of many 'professional almost-Jews' outside or even inside the apparatuses of the Jewish organizations and [Jewish communities]." Jewish culture, he wrote,

> is being manufactured, Jewish history reconstructed, by these Judaizing milieux—by German experts of Jewish culture and religion [who] enact Jewish culture from within German biographies and from within German history; this has an important bearing on the type of Jewish culture that is actually being produced: a culture that is not lived, that draws heavily from the museum, and that is still no less genuine for that.[4]

My own first major exploration of this phenomenon was a monograph, *Filling the Jewish Space in Europe,* published in 1996; it dealt with developments

mainly in Germany, Poland, and the Czech Republic. Eight years later, my further research resulted in a book, *Virtually Jewish: Reinventing Jewish Culture in
Europe,* which was a much broader exploration of these phenomena across much
of Europe.[5]

In the years since then, these and related issues have inspired considerable
debate, debate that has encompassed discussions of issues ranging from whether
an overarching "Jewish space" actually does exist in Europe to a more narrow
focus on what entails more clearly defined "Jewish spaces" and specific "Jewish places," as well as to what may or may not be regarded as "actually Jewish"
amid constantly evolving local circumstances. New terminology has emerged as
outside researchers, participant observers, philanthropic foundations, journalists,
and indeed local Jews and non-Jews themselves examine and try to make sense
out of shifting paradigms and generational and other changes.

As part of this, scholars, researchers, and others have explored the idea of
"Jewish space" from within the Jewish world, too, investigating the relevance of
space and place as environments and experiences shaped by Jews within living
Jewish culture, communities, identity, and society.[6] Pinto's work has continued to
influence these discussions: in later writings, she described Jewish space as a cultural but also material space of encounter between Jews and non-Jews, "an open
cultural and even political agora where Jews intermingle with others qua Jews,
and not just as citizens. It is a virtual space, present anywhere Jews and non-Jews
interact on Jewish themes or where a Jewish voice can make itself felt."[7]

Within the debates around these concepts, the expression "virtual" has caught
on in far more, and in more far-reaching, ways than I ever anticipated. Often,
however, I feel that my own understanding of a concept I formulated has been
subject to misinterpretation. Some commentators use the term "virtual" as a pe-
jorative or as a shorthand synonym for "fake." Some have attributed to my views
a dismissive attitude toward non-Jewish engagement with or interest in Jewish
culture or see me as having somehow disparaged as "inauthentic" the "purveyors
of … virtual Jewishness"—that is, people I have described as "virtual Jews"—
and their activities.[8] My punning reference to the fluid cyberspace concept of
"virtual" communities and worlds on the Internet has at times been missed or
overlooked. And on occasion my concept has also been stood on its head; I did
not, for example, posit that non-Jewish interest in Jewish culture in Europe *has
produced* a "virtual" Jewish world devoid of "real" Jews, but instead suggested essentially the opposite—that in post-Holocaust places *already now devoid or nearly
devoid* of living Jews, non-Jewish interest in Jewishness has had this effect. In the
Polish context, the Polish writer and Jewish activist Konstanty Gebert and sociologist Helena Datner have provided a much more apt description of my concept
of "virtually Jewish" as "a place where Jewish culture is no longer Jewish property,
but rather an open field in which anybody can use the props and as they see fit."[9]

For me, despite the passage of years and the emergence of wide-ranging discussions, Pinto's original formulation of "Jewish space" remains the most powerful; I find it almost revolutionary even, and it continues most strongly to influ-

ence my thinking. The work of the late semiologist and author Umberto Eco has also been important, particularly his essay "Travels in Hyperreality," in which he wrote of times and places where "absolute unreality is offered as real presence"[10] and where "the boundaries between game and illusion are blurred ... and falsehood is enjoyed in a situation of 'fullness.'"[11] Eco was writing, in long-ago 1975, about the United States and certain manifestations of its popular taste and culture, but I have found this assessment quite applicable to Europe and Europeans today—not just in relation to interest in, adaptation of, and creative refiguring of Jewish traditions and culture, but also in other spheres.

Homelands of the Soul

These other spheres include, for example, what I call the "imaginary Wild West" in Europe, a colorful and multifaceted subculture of music, study, lifestyle, commercialism, and entertainment that creates its own "real imaginary spaces" that in some ways echo some of those that I have described as part of the "virtually Jewish world": elements, for example, of enactment (and re-enactment); the employment of stereotypes and literary imagery; the use of another culture to inform or create identities.[12] There are major differences, of course, between the "virtually Jewish" phenomenon and the "imaginary Wild West." But both have to do with identity, and in some cases they both manifest an implicit spiritual, moral, or political message. Indeed, under communism, both the exaltation of the American West and engagement with Jewish culture, history, and heritage had implicit, if often oblique, political meaning. Both could be seen as indirect or nonconfrontational ways of expressing opposition to the Cold War communist state. "Every time I put on my cowboy hat and boots I was giving the finger to the regime," the Polish country-and-western singer Michael Lonstar told me. He echoed individuals who have described seeking out and discovering ruined synagogues and abandoned Jewish cemeteries under communism in much the same way.

In Kraków, a privately run Wild West–style saloon flourished for years, run by a passionate country music fan, Lucjan "Luciano" Żeleński, and his buckskin-clad father, who was known as "Long Bob."[13] Its swinging-door-style décor featured Wild West and country music memorabilia, with country music playing constantly in the background. I couldn't help comparing it to the Jewish-style cafés on Szeroka Street in nearby Kazimierz, Kraków's old Jewish quarter, which also feature rather standardized décor to evoke a certain imagined past: in this case an almost off-the-shelf set of decorative catering conventions such as dark wood, lacy tablecloths, candlesticks, old (or faux-old) pictures, and mismatched furniture. Here many Gentiles (tourists and locals alike) enjoy, purvey, enact, and engage in and with Jewish culture—and sometimes with Jews.

In Kazimierz, klezmer music and Yiddish songs, not country, form the soundtrack. But not unlike some "virtual Jews" and their engagement with "things Jewish," Luciano and Long Bob, in their distinctive "Wild West" garb, clearly enhanced, or created, identities based on Wild West dreams and affinities. They, like many non-Jews who find their comfort zone in Jewish culture or its trappings, were vivid illustrations of people seeking what Italian singer-songwriter Luigi Grechi described to me as a *patria dell'anima,* or "homeland of the soul."

Small Numbers and Big Presence

Jewish space and Jewish spaces, whether real or imaginary, actual or virtual, exist throughout Europe. But the issue has been particularly spotlighted in Poland, in part because of the role Poland plays in Jewish history, memory, and culture (including the Holocaust and its commemoration); in part because of the striking extent of the "virtually Jewish" world of non-Jewish interaction as compared to the small number of people who claim Jewish identity or engage directly in Jewish communal or secular life. "Small doesn't mean irrelevant," Barbara Kirshenblatt-Gimblett, chief curator of the core exhibit of the expansive new POLIN Museum of the History of Polish Jews, told me in 2011. "[The] Jewish presence in Polish consciousness is vast. There is a kind of inverse relationship between the numbers of Jews living in Poland and what we call the Jewish presence in Polish consciousness."

The new POLIN Museum, its two-decade development, and its physical, social, and political impact are a case in point and a focus of recent discussion on these issues. The museum's elaborate core exhibition, a "theater of history" that uses state-of-the-art interactive technology and dazzling installations to narrate one thousand years of Polish-Jewish history, opened with a high-profile ceremony in October 2014. But even before that, the museum had already manifested symbolic and objective power as a Jewish place and space encompassing many facets of the ongoing debates and definitions. The facility, costing more than $100 million, was an unprecedented public-private venture; the state provided $60 million for the construction of the building, the city of Warsaw donated the land it stands on, and more than five hundred private and institutional donors, many of them Jewish, raised about $50 million for the core exhibition.[14] Hundreds of thousands of people visited the museum after the building itself, without the core exhibition, opened in April 2013, hosting a range of cultural and educational activities.

Designed by the Finnish architect Rainer Mahlamäki, the museum's award-winning building is located in a Warsaw neighborhood known as Muranów, on the site of both Warsaw's prewar Jewish district and the Nazi Warsaw Ghetto. It directly faces the dramatic sculptural monument to ghetto heroes that was erected in 1948 atop the rubble left when the Nazis crushed the 1943 Warsaw Ghetto uprising. The building, shaped like a flattened cube, is steeped in symbolism. A vast irregular gap that symbolizes the rupture caused by the Holocaust breaks its shimmering glass-and-copper façade, and the narrow, vertical glass exterior panels are etched over and over with the word "POLIN" in Hebrew and

Latin characters—the word chosen as the museum's name. *Polin* is the Hebrew word for Poland; but it also refers to a legend that when the first Jews reached Polish lands they heard birds chirping the welcoming expression *Po-lin—po* (here) *lin* ([you should] dwell)—an interpretation that underlies the museum's message that Jewish history in Poland is inextricably entwined with the history of Poland itself.[15]

Inside, the swooping walls of the foyer form a canyon-like void designed to evoke the parting of the Red Sea. Looking out, the soaring glass entryway frames the Ghetto Heroes Monument, making a direct connection between past, present, and future. Carrying this through, the mezuzah placed on the main entryway was formed from a brick found on the site of the destroyed ghetto. The emptiness of the foyer is crossed by a bridge—another instance of the symbolic. "I see the museum as a meeting place, a forum, and a bridge across the chasm of the Holocaust," Kirshenblatt-Gimblett told me.

The grand opening of the core exhibition introduced even more symbolic reverberation. The opening ceremony was treated as an epochal national, even international, event. Attended by the presidents of Poland and Israel and other Polish and foreign dignitaries, it was presented as far more than the opening of a museum, rather as a tangible milestone in Polish-Jewish and Polish-Israeli relations, as well as in post-communist Poland's development as a democratic state. It all served to place the events within the context of what Clemson University scholar Michael Meng has termed "redemptive cosmopolitanism"—a "performative embrace of the Jewish past that celebrates the liberal, democratic nation-state rather than thinking critically about its past and present failures."[16] The museum, Meng wrote, several years before its opening, "offers a powerful symbol of Poland embracing its rich Jewish heritage as it returns to 'Europe.' The new Poland—democratic, tolerant, cosmopolitan—showcases its normality by recovering the Jewish past."[17]

Polish president Bronisław Komorowski made this explicit, stating that the opening of the museum made history and constituted "an eloquent sign of change that has been occurring ever since Poland won its freedom twenty-five years ago."[18] In his own speech, Auschwitz survivor Marian Turski, deputy chair of the Association of the Jewish Historical Institute, one of the institutional founders of the museum, repeated over and over the final words of the World War II Jewish partisan anthem: "*Mir zenen do!* (We are here!)," saying that it was a sign that Jews now feel empathy and solidarity in Poland.

The core exhibition's eight elaborate galleries are both thematic and chronological, telling the complex story of Jews in Poland through images, interactive technology, videos, projections, sound, and purpose-built physical installations. The most dramatic of the latter forms the physical centerpiece of the exhibition: the extraordinary, 85-percent-scale replica of the painted ceiling and bimah of the destroyed wooden synagogue in the town of Gwoździec (now in Ukraine)—so strikingly elaborate and inspirational that some observers dubbed it a "Jewish Sistine Chapel."

In hundreds of articles of all sorts, commentators and reviewers have praised the magnificence of the POLIN exhibition's presentation, the wealth of narrative it conveys, and the emotional and intellectual power of the almost overwhelming created environments that envelope, engage, and connect the visitor. "Words such as 'exhibits' or 'galleries,' which connote viewing a spectacle apart from oneself, do not capture the emotion elicited by the place," wrote Arnold M. Eisen, the chancellor of the Jewish Theological Seminary.[19] Photographer and writer Jason Francisco described POLIN as "an extravaganza" that was "part treasure chest, part fairytale theater, part high-tech expo, part sound and animation lab, part scholarly pop-up book, part multimedia kindergarten, and part solemn carnival. More than a museum per se, it is better called a museum-experience, a sequence of encounters rather than a temple of precious objects (though it does contain many remarkable objects)."[20]

The exhibition in fact displays relatively few original historic artifacts or objects. This was a deliberate part of its strategy, and a point that was discussed and debated for years while the museum was under development. Kirshenblatt-Gimblett and others involved in the creation of the core exhibition have noted that in the absence of a wide selection of physical artifacts to include, they chose to use words—that is, contemporary quotations by and about Jews.[21] Eisen appreciated this course: "Its point is not to preserve and display objects, but to tell a story that it wants its visitors to carry forward," he wrote in the commentary quoted above. But other critics regretted the lack of tangible "authenticity." Though she appreciated the power of the museum and its message, Abigail Morris, the director of the Jewish Museum in London, felt a void surrounding the museum's rich fullness, a "hollow sense of absence" that "points to a bigger, truly existential problem":

> There were over three and half million Jews who were murdered, and alongside their deaths was the annihilation of the culture and communities they had created. Maybe the museum should have highlighted, rather than tried to cover up, this absence. Maybe there should have been a hole in the centre of the building where you could see the real rubble of the ghetto remains. You would have seen the rubble, it would have been real and you'd have seen a hole. Because that was what I came away with. The museum is trying to fill a hole which is vast and void-like.[22]

Jason Francisco, too, raised questions—questions that in fact are an integral part of the intangible context in which the physical museum is set. Francisco has spent much time in western Ukraine; a collection of his photographs from there forms part of the permanent exhibition of another Jewish museum in Poland, the privately run Galicia Jewish Museum in Kraków. In a lengthy analysis of the POLIN Museum, Francisco wrote about the replica painted bimah and ceiling of the destroyed wooden synagogue in Gwoździec and recalled his own recent visit to "the forlorn site where the wooden synagogue once stood" in the town, now known as Hvizdets:

No such contextual view is presented in the museum, and indeed, virtually no contemporary photographs of places of Jewish heritage appear in the exhibition's last section,
where they would be appropriate. ... I venture that the Jewish nothingness that defines
contemporary Hvizdets—and so much of the contemporary geography of the Jewish past
in Poland—simply does not fit into the museum's interpretive parameters.

He concluded:

I don't quarrel with Polin's so-called "narrative" premise, as distinct from, say, artifact-
based exhibitions—even as, in practice, Polin's narrative is more narrative effect than narrative proper. ... But I ask: what lurks behind Polin's show, behind its surfaces and its
radiating energy? I can't help but answer: a fear of empty spaces, and the still-palpable
nearness of Jewish oblivion.[23]

Space and Place

But POLIN itself is a new reality, a palpable and authentic new place, whose
mission is to be more than a museum; as such, it aims to develop its role as a
catalyst for scholarship and hosts a variety of cultural and educational activities:
a permanent genealogy research center, films and concerts, international conferences, even occasional Jewish communal events. It forms a Jewish space, but as
a public institution, it also provides a "neutral" setting where people of all backgrounds can enter, explore, and engage with Jewish history, heritage, and culture
on a variety of levels and terms. The museum also has an impressive presence on
the Internet: its vast "Virtual Shtetl" portal provides news and information on
the Jewish heritage and history of some twenty-three hundred or more places in
historic Polish lands, and one can take a "virtual tour" of POLIN's core exhibition via the Google Cultural Institute. POLIN extends beyond its confines in a
physical sense too, with a "museum on wheels" project that takes its exhibits into
far-flung towns around the country.

The extension of spatial identity beyond the museum's walls has been at the
center of recent debates concerning the symbolic identity and spatial significance of the parklike area around the museum building. The dominating fact
of the museum, on top of the layers of wartime and prewar history, had defined
the large square, bordered by communist-era apartment buildings, as a "Jewish
space" and a "Jewish place." But must the identity of the square as a Jewish space
and place exclude an appreciation of the site as a more generalized "zone of memory"? This question has sparked heated polemics, particularly since 2013, when
plans were announced to erect a memorial between the museum building and the
Ghetto Heroes Monument it faces to honor non-Jewish Poles who rescued Jews
during the Shoah. Monuments to individual non-Jewish heroes had already been
sited near the POLIN Museum; these include a sculptural monument to Jan
Karski, the Polish resistance hero who brought word to the West of conditions

in the wartime Warsaw Ghetto, and a pathway named for Irene Sendler, a Polish woman who saved Jewish children from the ghetto. Moreover, since 2000, a monument to West Berlin's postwar mayor and former West German chancellor Willy Brandt has stood in one corner of the area to commemorate his visit to Warsaw in 1970, when he dropped to his knees in front of the Ghetto Heroes Monument in apology for Nazi Germany's crimes.

Plans for the new general memorial to rescuers polarized local and Jewish interests. The Remembrance and Future Foundation, headed by Sigmund Rolat, a New York–based philanthropist and Holocaust survivor who has contributed generously to many Jewish and memorial projects in Poland, including the POLIN Museum itself, has been spearheading this initiative.[24] Yet the plan has drawn vehement opposition from other Jews, in Poland and abroad—in part because of how such a monument would change the spatial and symbolic identity of the location.[25] Critics stress that the issue is not about whether to create a monument to the Righteous or whether to place the monument somewhere on the territory of the World War II Warsaw Ghetto. Rather, explained POLIN's Kirshenblatt-Gimblett, the issue "is about the impact of this location on the memorial space surrounding the museum and, most important, on the museum itself":

> To place the major monument to the Righteous in that park makes the museum the window for it. This not only redefines the memorial space around the museum, but also the museum itself.... Moreover, once in the building, visitors [would] not have a clear view of the Monument to the Ghetto Heroes from the large glass window at the entrance. This placement was intentional, because POLIN Museum stands in relation to the Monument to the Ghetto Heroes, but is not a Holocaust museum. Now, when in the building, visitors will have a constant reminder of the Holocaust thanks to the view of the Monument to the Righteous from the large window in the museum's main hall. This is the critical location issue.[26]

A Broader Movement

During the grand opening of the POLIN Museum's core exhibition in October 2014, the museum's director, the historian Dariusz Stola, stressed to me that POLIN was part of a wider movement that has emerged since the fall of communism "to reconnect with the past, including the Jewish past.... Without the broader movement it wouldn't have happened." In essence, POLIN is a very big tip of a very big iceberg.

This broader context encompasses that part of the "virtually Jewish world"—in small towns as well as big cities—that the Kraków-based anthropologist Annamaria Orla-Bukowska has referred to as "Gentiles doing Jewish stuff": Jewish studies courses at Polish universities, new or revamped museums, permanent exhibitions and memorials on Jewish or Holocaust themes, numerous educational programs, and scores of grassroots initiatives ranging from Jewish cemetery cleanup actions to festivals of Jewish culture.[27]

A chart compiled in the summer of 2013, for example, lists more than fifty Jewish culture festivals of one sort or another that take place each year in towns, cities, and villages all around Poland. Some are one-day events; others—like the Kraków Jewish Culture Festival (founded in 1988)—last a week or more and draw major international Jewish artists and acts. However, given that only fifteen to twenty thousand Jews live in Poland today, the vast majority of them are organized by non-Jews for a non-Jewish, often local, audience, and most take place in locales where no Jews have lived since the Holocaust.[28] It is important to note that Jewish culture and other festivals take place in many countries around Europe, organized both by Jewish communal or other Jewish institutions and by non-Jewish groups. But nowhere are there so many as in Poland.[29] "I often joke that now the mayor of every small town feels obliged to make excuses [if] he/she has no Jewish Festival in his/her town," Anna Dodziuk, a psychotherapist who is also a Jewish activist and editor, told me in 2009.

The scholar Joanna Beata Michlic has pointed out that this can be problematic; that these many far-flung Jewish-festivals-without-Jews in Poland can "sometimes, perhaps inadvertently, reinforce old stereotypes of Jews as an exotic, alien ethnic group that in the pre-1939 period intruded itself into a Polish nation that understood itself, not the Jews, to be the legitimate majority, the 'owners' of the nation-state."[30] This, of course, is not always the case. And it was one of these "virtually Jewish" culture festivals that prompted David Mazower, the great-grandson of the noted Yiddish writer Sholom Asch, to reflect on the complexity of Polish attitudes to the country's Jewish past and present. This was a Jewish culture festival held in 2014 in Asch's home town, Kutno:

> There's plenty of lingering anti-Semitism, a kind of kitsch theme park nostalgia, pride in Poles who saved Jews during the war, and guilt about those who collaborated. To complicate the picture further, there is a deep-rooted conviction that Poles themselves are victims of a turbulent history. But even that's not the whole story. Something else has emerged in the last 20 years: a growing number of Poles who feel profound loss about the Jewish nation that vanished from their midst. ... It might not be a mass movement, but as Kutno shows, it's more than just a token effort. And in a world where bitter sectarian conflicts grab most of the headlines, a Jewish festival in a town with no Jews is surely something worth celebrating.[31]

A Living Laboratory

Kraków's Kazimierz has been the focus of much writing, including quite a bit of my own, on Jewish space in the extensive absence of Jews.[32] This is partly because it is something of a highly accessible living laboratory, a district that went from being a dense pre-Holocaust Jewish neighborhood to a desolate under-inhabited communist-era slum—the archetypical "Jewish ghost town," "haunted" by the unquiet spirits of the murdered. In an evolution that began in the early 1990s, it has become a popular center of tourism, whose attraction is based in large part on its Jewish history, sites, and significance, as well as a hub of a lively youth culture of bars, cafés, clubs, and nightlife. It is also a place where there is a new surge of Jewish identity and Jewish religious and communal life whose multi-faceted parameters reflect changed and changing realities, as well as new definitions of "Jewish" that mesh, and sometimes confront, Jewish and non-Jewish input and understanding. These combinations, and the intensity with which they are played out, make Kraków something of a special case. Indeed, Michlic has rightly stressed that "outside the Kazimierz memory space, recalling Jewish history and culture does not inevitably lead to the meaningful encounters between Poles and visiting Polish Jews and their descendants that would put an end to the old negative and prejudicial stereotypes and create a 'conciliatory heritage.'"[33]

That said, Kazimierz is a historic Jewish space *par excellence*: the largest and best-preserved old Jewish quarter in Central Europe, with seven synagogues, two Jewish cemeteries, and an intact urban infrastructure of streets, alleys, courtyards, and buildings. It is thus an undeniable physical Jewish place. But it is also a Jewish space in more subjective contemporary terms. Only a few hundred Jewish people live in all of Kraków, but in Kazimierz, embedded now in this concrete historical context, are Jewish-style cafés, three Jewish bookstores, a Jewish publisher, three Jewish museums, a center for Jewish culture, the headquarters of the Jagiellonian University's Jewish Studies Program, preserved or newly painted Yiddish street signs, souvenir stands selling Jewish-themed knickknacks, and more. The cobbled Kazimierz streets and squares see a stream of tourists, pilgrims, and other visitors, Jewish and otherwise, who mingle in Diana Pinto's "agora." Hasidim taking selfies in front of the Remuh Synagogue may themselves be photographed by secular tourists, while motorized open tourist carts follow their routes around them, trailing a soundtrack of recorded audio commentary. Each summer, the annual Kraków Jewish Culture Festival, famously organized by a non-Jewish director and staff for a primarily non-Jewish audience, draws tens of thousands to its concerts, workshops, lectures, and performances. It also schedules cultural and educational events throughout the year at its "Cheder" café, a cozy place with a Yiddish name and "vibe" whose menu branches out in other Jewish directions, specializing in "Israeli, Mediterranean and Middle Eastern tastes," such as hummus, baklava, and sweet mint tea.

Like the festival, most Jewish-themed commercial, cultural, and educational initiatives are run by and/or aimed at non-Jews. However, particularly over the past decade, boundaries in Kazimierz have blurred, as definitions of what is "Jewish" have expanded beyond traditional parameters and as a varied living Jewish presence has become more visible and also more vocal. In Kazimierz, the term "Jewish" has come to take on a variety of meanings, encompassing not just what has traditionally or historically been designated by the word, but all of the very different new components that now make up religious, secular, commercial, and "non-Jewish" Jewish life. Anyone in his or her mid-twenties or even somewhat older has grown up with these complex and constantly evolving configurations of what "Jewish" means: just as young Krakóvians never knew Kazimierz as the teeming Jewish quarter it was before the Holocaust, they also never knew it as the phantom-haunted Jewish ghost town it was until the early 1990s. For them, there has always been the festival, the Jewish cultural events, the Jewish-style cafés, the tourists (including religious pilgrims), and the opportunity to engage—or not, as the case may be. The mixed character of present-day Kazimierz forms a new authenticity, a new way of understanding "Jewish," at a time and place where traditional boundaries, definitions, and self-definitions meet and merge. The Jewish Culture Festival "is seen as not only 'Jewish' but humanistic, universalistic," Annamaria Orla-Bukowska and Krzysztof Tomanek wrote in "Who Is Doing Jewishness in Poland?," an article based on a survey of people in attendance. One of the messages it sends, they wrote, is about ways of thinking "about the coexistence of various cultures."[34]

(Re)definitions

In 2004, the London-based Institute for Jewish Policy Research (JPR) published a report about changes in how European Jews in various countries think of their identities. "What appears to be happening is a general *redefinition* of what being Jewish is all about," observed the author, David Graham. "People are finding new ways to express their identity as Jews by adapting traditional practices, customs, and behaviours to fit in with their new social realities. ... Moreover, the actual meaning of being Jewish (the *Jewishness* people are consciously ascribing themselves to), also seems to be changing."[35]

Graham's focus was Europe as a whole. And although he was writing about how Jews themselves define their identity, his analysis resonates well beyond the traditional boundaries of the conventionally recognized Jewish community.

Remaining in Kraków, one can clearly see elements of his analysis. A case in point is Kraków's Jewish Community Centre (JCC), which opened in 2008. It is not a religious congregation, though an Israeli-based Modern Orthodox rabbi visits regularly to advise, hold classes, and lead religious events. The JCC offers membership to anyone who is a resident of Kraków or the region and fulfills Israel's "law of return" criteria, but it also empowers the JCC director to grant membership to anyone (Jewish or not) who makes "a significant contribution to the Jewish community, or indeed cares about the positive image of the Jewish community."[36]

The JCC puts its membership at about 600 people of all ages and all shades of Jewish identity, origin, or engagement. But it relies on dozens of young non-Jewish volunteers to run operations and help out with its broad public engagement programs and other activities. These volunteers—"virtual Jews"—have taken to calling themselves "Meshugoyim," a pun combining the Yiddish terms *meshugah* (crazy) and *goyim* (non-Jews): the Jewish world of the JCC is clearly where they have found their "homeland of the soul." One of these volunteers, Agnieszka Giś, founded a blog called *Jews of Kraków,* which invites readers to "take a peek inside modern Jewish life in Kraków!"[37] In a post on the blog entitled "Volunteers and Dybbuks," Canadian Lizy Mostowski, who volunteered at the JCC in 2013, described the non-Jewish JCC volunteers as experiencing a sort of hybrid identity: they are "young people who care about the Jewish past and the Jewish future and what Jewish life means and represents in Polish life and culture, [who are] encouraging new Jewish life to exist regularly, if only by deconstructing the stereotype that only Jews walk around Kraków wearing Stars of David and JCC t-shirts." She imagined these young non-Jewish volunteers as inhabited by the souls of Jews killed in the Holocaust.[38]

Traditional definitions and boundaries continue to exist, of course. Kraków's official Jewish religious community of about 140 members (known as the Gmina) until recently engaged only in limited outreach to outsiders; indeed, under the bylaws of the Union of Jewish Religious Communities in Poland, Jews who are not Polish citizens may not join an officially recognized Polish Jewish religious community. Recently, though, the Gmina moved its premises to a prominent location, across the street from the JCC. Chabad Lubavitch, the Hasidic group whose hallmark is outreach, has operated in Kraków since the mid-2000s and runs a kosher shop and other facilities aimed at the many Orthodox Jewish groups who visit Kraków, as well as the small number of religiously observant local people. At what may be seen as the opposite end of the religious spectrum, a small Reform congregation also operates in Kraków.

These contexts, too, reinforce the range of new authenticities in existence. "If a few years ago the production and consumption of Jewish culture in Kazimierz were often referred to as the creation of 'virtual Jewishness' mainly by non-Jews for a non-Jewish audience, in recent years the participation of the real, redefined Jewish community in discovering and defining the Jewish character of the quarter has been increasingly visible," wrote Monika Murzyn-Kupisz, a long-time observer of the development of Jewish Kraków. Moreover,

> thanks to [this participation], creating some continuity with pre-war traditions but at the same time not making them the main point of reference, Kazimierz may anew become a quarter with a unique, ethnic character, not only a quarter of the Jewish past but also of Jewish present and presence. Moreover, its spaces and Jewish heritage may ... create opportunities for renegotiating and redefining both Polish and Jewish identity, as well as serv[ing] as spaces of encounters and brokering between the two nations.[39]

Jonathan Ornstein, the New York-born executive director of the Kraków JCC (who has described himself as Jewish, vegetarian, and atheist), feels that this transition is already well under way. "People talk about Kazimierz as being the 'former' Jewish quarter of Kraków," he told me a few years back. "I say, why former? I think that it is the present Jewish quarter of Kraków."

Life after Life

An event in the summer of 2014 entwined many of these strands. It was a *brit milah*—a bris: the circumcision ceremony for the eight-day-old son of a local Jew. The baby's birth happened to coincide with the Kraków Jewish Culture Festival. This meant that while technically a private affair, the bris, just by happening where and when it did, took on the guise of a festival event, being "performative" as well as the heartfelt performance of a traditional religious rite welcoming a son into the bosom of a living people and observant Jewish family. As such it involved a number of people who were either participants in festival concerts and workshops or others, like me, who had come to Kraków to attend festival events and enjoy the atmosphere. The bris itself took place in the Izaak Synagogue, a graceful seventeenth-century building that was restored in the 1980–90s and was now being used by Chabad Lubavitch; prior to Chabad's administration, it had been used since the fall of communism by the Ronald S. Lauder Foundation, which fosters Jewish education, and also as the site of an exhibition that sought to "re-populate" the synagogue with life-size cutouts of photographs of pre–World War II Jews. After the ceremony, many of the guests moved on to a kosher luncheon at one of the local establishments that offer kosher catering. The baby was the son of a religious local man who found his way to observant Judaism in the 1990s, studying in the Jewish Studies Program at the Jagiellonian University and then at a yeshiva in Berlin; he was serving as the *mashgiach* of the restaurant where the bris luncheon was held. I had met him in the late 2000s when he was an editor at Austeria, a Jewish publishing house in Kraków run by Wojtek and Małgosia Ornat, the couple who were the first people to open a "Jewish-style" café in Kazimierz and who now run the hotel/café/restaurant Klezmer Hois. The circumcision was carried out by Cantor Benzion Miller, a Brooklyn-based Bobover Hasid whose father was a Holocaust survivor from Oświęcim, the town where Auschwitz was built, and who himself was born in a DP camp in Germany; as most years, Miller was in Kraków as a performer and workshop teacher at the Jewish Culture Festival. When in Kraków he lodges at the Hotel Eden, an establishment, run by an American Jew, that regards itself as a "Jewish hotel" and caters to religious guests; its restaurant is no longer kosher (as it was when the hotel opened), but the hotel has a mezuzah on the doorpost of each of its rooms and a private *mikveh*.

Many other guests at the bris also appeared to be out-of-towners. Several rabbis were in attendance, among them Michael Schudrich, the American-born chief rabbi of Poland; Kraków's then official rabbi, the Chabad rabbi Eliezer Gurary; Yehoshua Ellis, the young American-born Modern Orthodox rabbi serving Katowice; and Edgar Gluck, a German-born Orthodox rabbi in his seventies who for years had divided his time between Brooklyn and Poland. In New York, Gluck has a long history of political activism. In Kraków he holds an honorific but very "real imaginary" (or virtual) position—chief rabbi of Galicia, that is, the symbolically titled rabbi of a long-gone province of the Austro-Hungarian

Empire that is now divided geographically between Poland and western Ukraine. On Shabbat and holidays, in his long black coat, fur streimel, and long wispy beard, Gluck walks the Kazimierz streets, like a prewar patriarch straddling time and space amid a milling mix of tourists, hipsters, and "heritage brokers" ranging from tour guides and interpreters to souvenir sellers, gallery owners, restaurateurs, musicians, curators, craftsmen, and more.

Other people too clustered around Miller as he ritually cut the baby's foreskin. They included several whose presence provided a living timeline of some of the post-communist developments of Jewish Kazimierz: Jonathan Ornstein, the executive director of the JCC; Tadeusz Jakubowicz, the head of the official Jewish Gmina; British scholar Jonathan Webber, who researched traces of Jewish memory in Poland in the 1980s and now was teaching in the Jewish Studies Program at the Jagiellonian University; and his wife, Connie, who is the managing editor of the Littman Library of Jewish Civilization. One of the Littman Library's publications is *POLIN*, a journal of Polish-Jewish studies founded in 1986 and mainly edited by historian Antony Polonsky, who now holds the title chief historian of the new POLIN Museum in Warsaw. Allen Haberberg, the American Jew who opened the Hotel Eden in the year 2000, was there, and so was the award-winning Polish-Jewish-American filmmaker Sławomir Grünberg and his then wife, Katka Reszke, a Polish scholar, photographer, and filmmaker and author of *Return of the Jew*, an analysis of young Poles who have claimed or reclaimed a Jewish identity since the fall of communism.[40] The book grew out of Reszke's own story: as a teenager in Wrocław in the 1990s she was inexplicably drawn to Judaism, convinced on a "hunch" that she was Jewish; she converted to Judaism, lived in Israel for a while, and became involved in Jewish studies before finding out, in her mid-thirties, that she was indeed Jewish according to halacha (Jewish law).

Preserving the bris for posterity was Chuck Fishman, an American photographer whose first forays into Jewish Poland came in the 1970s, when no such event could ever have been imagined. At that time, Fishman was working on what became a photo essay on Jews in Poland; tellingly, it was entitled *Polish Jews: The Final Chapter*.[41] He returned to Poland in 2013 after a thirty-year absence to begin work on a documentation of the contemporary Jewish scene in Poland.

And that chapter, in all its complexity, is still unfolding.

Ruth Ellen Gruber is a writer, photographer, public speaker, and consultant. She coordinates the website www.jewish-heritage-europe.eu

Notes

1. Gruber, "New Authenticities," 490–91. In writing parts of this chapter, I have drawn on and incorporated material from some of my earlier writing, including my *Jewish Quarterly Review* essay; my essays in Bronner, *Framing;* my article, "Balancing the Real," and *Virtually Jewish*.

2. Gruber, "Monuments," 336. See also Gruber, *Virtually Jewish.*
3. Pinto, *New Jewish Identity.*
4. Bodemann, "Reemergence," 57–58.
5. Gruber, *Filling the Jewish Space;* Gruber, *Virtually Jewish.*
6. See, e.g., the essays in Brauch, Lipphardt, and Nocke, *Jewish Topographies.*
7. Pinto, "Jewish Challenges." One of Pinto's most recent elaborations of this topic is her 2015 essay "Jewish Spaces."
8. Lehrer, "Virtual, Virtuous, Vicarious, Vacuous?," 383–85. See my response to Lehrer in the same volume: "The Last Word," 399–400.
9. Gebert and Datner, *Jewish Life in Poland.*
10. Eco, *Travels in Hyperreality,* 7.
11. Ibid., 8.
12. See, e.g., Gruber, "New Authenticities," from which parts of this discussion are drawn. Also see my blog on this topic: http://sauerkrautcowboys.blogspot.com.
13. The saloon was there at least until 2010. I believe I last saw Long Bob, then eighty-three, in 2009; I do not know if he is still alive at this writing.
14. POLIN Museum website, http://www.polin.pl/en/about-museum/public-private-partnership.
15. Some of this description is drawn from my article "A Museum in the Void."
16. Meng, *Shattered Spaces,* 250.
17. Ibid., 218.
18. See full text of Komorowski's and Turski's speeches by downloading the files at http://www.polin.pl/en/media/press-releases.
19. "Betting on Hope," Jewish Theological Seminary blog, 30 October 2014. http://blog.jtsa.edu/chancellor-eisen/2014/10/30/betting-on-hope/.
20. Jason Francisco, http://jasonfrancisco.net/polin (accessed 21 March 2015). The description of his critique, as well as the critique by Abigail Morris, are taken from my 1 December 2014 post on the Jewish Heritage Europe website, http://www.jewish-heritage-europe.eu/focus/museums/thought-provoking-new-reviews-of-polin-museum (accessed 21 June 2016).
21. See interviews in video on POLIN website, http://www.polin.pl/en/exhibitions-core-exhibition/our-way-of-showing-1000-years-of-history (accessed 29 April 2015).
22. Morris, "Empty Museum."
23. Francisco, http://jasonfrancisco.net/polin.
24. Remembrance and Future Foundation, http://raff.org.pl/en. Tensions heightened after Rolat and the foundation board rejected the winning monument design chosen by the international jury. See "The Monument Is a Dilemma," http://politicalcritique.org/cee/poland/2016/monument-history-of-polish-jews-museum/ (accessed 21 June 2016).
25. See, e.g., the anti-monument Facebook page, https://www.facebook.com/pages/Czy-upamiętnić-Sprawiedliwych-na-terenie-byłego-getta/1457448724499165?sk=info&tab=page_info (accessed 21 June 2016).
26. B. Kirshenblatt-Gimblett in a Facebook post, 6 February 2015. For other comments and articles on the controversy, see Rukhl Schaeckter, "Polish Jews Protest Monument to Righteous Gentiles in Warsaw Ghetto," *The Forward,* January 26, 2015, http://forward.com/articles/213519/polish-jews-protest-monument-to-righteous-gentiles/; and Donald Snyder, "Poland's Dueling Holocaust Monuments to 'Righteous Gentiles' Spark Painful Debate," *The Forward,* April 27, 2014, http://forward.com/articles/197120/polands-dueling-holocaust-monuments-to-righteous-g/? Also Michlic, "Many Faces," 158–62.

27. Orla-Bukowska, "Gentiles Doing Jewish Stuff." See also Orla-Bukowska and Tomanek, "Who Is Doing Jewishness," viii.
28. "Long List of More than 40 Jewish Culture etc Festivals in Poland," *Jewish Heritage Europe* website August 27, 2017, http://www.jewish-heritage-europe.eu/2014/08/27/long-list-of-jewish-culture-etc-festivals-in-poland/%E2%80%9D. It is difficult if not impossible to estimate the number of Jews living in Poland today. Some sources estimate fewer than 10,000, others as many as 100,000. It depends on the definition of "who is a Jew" or "who can be counted as Jewish."
29. See annual lists of Jewish culture and other festivals in various locations that I have compiled at the blog *Jewish Heritage Travel,* http://jewish-heritage-travel.blogspot.com. These are far from complete, but as far as I know, there is no other website or place that compiles them.
30. Michlic, "Many Faces," 150; she gives specific examples.
31. "A Jewish Festival in a Town without Jews," transcript of a "From Our Own Correspondent" radio report first broadcast on BBC Radio 4 on 6 December 2014, http://www.bbc .com/news/magazine-30214204. "Virtual Jews" in Poland who carry out activities relating to Jewish culture have in fact been celebrated. Since 1998, annual award certificates have been presented to non-Jewish Poles who foster, care for, preserve, or promote Jewish heritage and culture; to date, about two hundred people have been so honored.
32. One of the most recent is Lehrer, *Jewish Poland Revisited.*
33. Michlic, "Many Faces," 150.
34. Orla-Bukowska and Tomanek, "Who Is Doing Jewishness," 170–71.
35. Graham, "European Jewish Identity."
36. Membership regulations on the JCC website: http://www.jcckrakow.org/attachments/ article/181/regulamin%20członkostwa%20w%20JCC.pdf.
37. *Jews of Kraków* (blog), http://jewsofkrakow.com/. Ms. Giś put together the list of Jewish culture festivals in Poland mentioned previously.
38. Lizy Mostowski, "Volunteers and Dybbuks," accessed 21 June 2016, http://jewsofkrakow .com/index.php/25-volunteers-and-dybbuks.
39. Murzyn-Kupisz. "From 'Atlantis.'"
40. Reszke, *Return of the Jew.*
41. Fishman and Vinecour, *Polish Jews.*

Bibliography

Bodemann, Y. Michal. "A Reemergence of German Jewry?" In *Reemerging Jewish Culture in Germany: Life and Literature since 1989,* edited by Sander L. Gilman and Karen Remmler, 46–61. New York: New York University Press, 1994.

Brauch, Julia, Anna Lipphardt, and Alexandra Nocke, eds. *Jewish Topographies: Visions of Space, Traditions of Place.* Aldershot: Ashgate Publishing, 2008.

Bronner, Simon J., ed. *Framing Jewish Culture: Boundaries and Representations.* Oxford: Littman Library of Jewish Civilization, 2014.

Eco, Umberto. *Travels in Hyperreality.* Trans. William Weaver. New York: Harcourt Brace, 1986, 1976.

Fishman, Carles, and Earl Vinecour. *Polish Jews: The Final Chapter.* New York: New York University Press, 1977.

Gebert, Konstanty, and Helena Datner. *Jewish Life in Poland: Achievements, Challenges and Priorities since the Collapse of Communism.* London: Institute for Jewish Policy Research, 2011.

Graham, David. "European Jewish Identity at the Dawn of the 21st Century: A Working Paper." London: JPR, 2004. Accessed 21 June 2016. http://www.bjpa.org/Publications/details.cfm?PublicationID=4334.

Gruber, Ruth Ellen. "Beyond Virtually Jewish … Balancing the Real, the Surreal and Real Imaginary Places." In *Reclaiming Memory: Urban Regeneration in the Historic Jewish Quarters of Central European Cities,* edited by M. Murzyn-Kupisz and J. Purchla, 63–79. Kraków: ICC, 2009.

———. "Beyond Virtually Jewish: Monuments to Jewish Experience in Eastern Europe." In Bronner, *Framing Jewish Culture,* 335–55.

———. "Beyond Virtually Jewish: New Authenticities and Real Imaginary Spaces in Europe." *Jewish Quarterly Review* 99, no. 4 (2009): 487–504.

———. *Filling the Jewish Space in Europe.* New York: American Jewish Committee, 1996.

———. "The Last Word." In Bronner, *Framing Jewish Culture,* 397–402.

———. "A Museum in the Void." *Hadassah Magazine,* October/November 2013. http://www.hadassahmagazine.org/2013/10/19/arts-museum-void/.

———. *Virtually Jewish: Reinventing Jewish Culture in Europe.* Berkeley: University of California Press, 2002.

Lehrer, Erica T. *Jewish Poland Revisited: Heritage Tourism in Unquiet Places.* Bloomington: Indiana University Press, 2013.

———. "Virtual, Virtuous, Vicarious, Vacuous? Towards a Vigilant Use of Labels." In Bronner, *Framing Jewish Culture,* 383–95.

Meng, Michael. *Shattered Spaces.* Cambridge, MA: Harvard University Press, 2011.

Michlic, Joanna B. "'The Many Faces of Memories': How Do Jews and the Holocaust Matter in Postcommunist Poland?" In *Lessons and Legacies XI: Expanding Perspectives on the Holocaust in a Changing World,* edited by Hilary Earl and Karl Schleunes, 144–79. Evanston, IL: Northwestern University Press, 2014.

Morris, Abigail. "The Empty Museum." *Jewish Chronicle,* 27 November 2014. http://www.thejc.com/arts/arts-features/126106/the-empty-museum.

Murzyn-Kupisz, Monika. "From 'Atlantis' to … ? The Past and Present of Jewish Kraków in the Second Decade of the New Millennium." Paper presented at the Eleventh International Conference on Urban History—Cities & Societies in a Comparative Perspective, organized by the European Association for Urban History at the Charles University in Prague, 29 August–1 September 2012.

Orla-Bukowska, Annamaria. "Gentiles Doing Jewish Stuff: The Contributions of Polish Non-Jews to Polish Jewish Life." In *Rethinking Poles and Jews: Troubled Past, Brighter Future,* edited by Robert D. Cherry and Annamaria Orla-Bukowska, 197–214. Lanham, MD: Rowman & Littlefield, 2007.

Orla-Bukowska, Annamaria, and Krzysztof Tomanek. "Who Is Doing Jewishness in Poland?" *New Eastern Europe* 3, no. 8 (2013): 162–71.

Pinto, Diana. "Epilogue: Jewish Spaces and Their Future." In *Jewish Space in Contemporary Poland,* edited by Erica T. Lehrer and Michael Meng, 280–86. Bloomington: Indiana University Press, 2015.

———. "The Jewish Challenges in the New Europe." In *Challenging Ethnic Citizenship: German and Israeli Perspectives on Immigration,* edited by Daniel Levy and Yfaat Weiss, 239–52. New York: Berghahn Books, 2002.

————. *A New Jewish Identity for Post–1989 Europe.* London: JPR, 1996. http://www.jpr
.org.uk/documents/A%20new%20Jewish%20identity%20for%20post-1989%20Europe
.pdf.

Reszke, Katka. *Return of the Jew: Narratives of the Third Post-Holocaust Generation of Jews in
Poland.* Boston: Academic Studies Press, 2013.

Index

Aachen, 170
acculturation, 90, 93, 94, 95, 98,
   99–102
Adenauer, Konrad, 64
Adler, Herman, 181
Adorno, Theodor, 113
aesthetic reforms, 146
agency, 127, 130, 134
Agudas Achim, 43
Alexandria, 129, 130
Alphen, Ernst van, 291, 293
Alsheikh, Moses, 218, 222, 224, 226
Altona, 220, 222
Amsterdam, 126, 127, 128, 134n10,
   135n12, 217, 218
Anhalt, Henriette Catharina von, 223
Annales, 24
anti-Semitism, 73, 75, 77, 90, 94, 97, 146,
   163, 182, 186, 202, 204, 246, 265,
   270, 271, 272, 275, 282, 289
appropriation, 265
architecture, 141, 144, 145, 146, 150, 152,
   154, 160–173
   Bet Tfila-Research Unit for Jewish
     Architecture in Europe, 161
   house architecture, 24, 26–27, 30
   German style, 164, 166, 171
   Gothic influences, 164
   Moorish, 164, 171
   Neues Bauen movement, 165
   postwar modernism, 169
   Romanesque, 164, 165
Arendt, Hannah, 94, 108–109
Arnim, Dietloff von, 217
Arnsberg, Paul, 111
Asch, Sholom, 307
Ascher, Felix, 166
Ashkenazic Jews, 3, 8, 9, 17n27

assimilation, 15, 40, 43, 44, 45, 46, 47, 48,
   50, 93, 146, 153, 155
Assmann, Jan, 56
astronomy, 219, 220
Auerbach, Ellen, 250
Auerbach, Philipp, 272–273
Austria-Hungary. *See* Habsburg Empire
Austria, 246, 250, 254, 256
Austro-Hungarian Empire. *See* Habsburg
   Empire
authenticity, 89, 94, 98, 101
   New Authenticities, 298, 300, 304, 309,
     310
autonomy, 142, 144
Avraham, Israel bar, 216–227
Avraham, Moses ben, 217

Baden, 64, 170
   Kippenheim, 170
Bakan, David, 76–77, 79
Baku, Azerbaijan, 47
Bałaban, Majer, 46
Bamberger, Jakob Koppel, 151
Bamberger, Ludwig, 60
Bartetzko, Dieter, 167
Baudrillard, Jean, 298
Bavaria, 64, 218, 281, 285, 286, 289, 290
   Bavarian Association of Jewish
     Communities, 286
   Bayern München soccer team, 285, 290
Beckmann, Max, 166, 168
Begegnung, 291–292
Begum, Halima, 240
beit midrash, 127
Belgium, 255
   Antwerp, 199
   Brussels, 184, 185
   Brussels Conference (1872), 185

Beller, Steven, 75–76, 80
Ben Chorin, Schalom, 282
Benjamin, Walter, 56
Berger, Joel, 237
Berger, John, 240
Berlin, 10, 12, 60, 61, 73, 82, 89, 90, 92,
    99, 127, 128, 141, 162, 164, 165, 171,
    172, 184, 185, 197–208, 218, 221,
    222, 223, 234–236, 247, 251, 282,
    286
  alt-Berlin, 207
  Berlin airlift, 248–249, 253, 256
  Berlin Wall, 247
  Gesamtarchiv der deutschen Juden, 60
  Oranienburgerstraße, 236
  Prenzlauer Berg, 61
  Scheunenviertel, 10, 12, 61, 197–208
  Verein jüdische Lesehalle und Bibliothek,
    61
Bhabha, Homi, 81
Biermann, Aenne, 250
Bing, Ilse, 246, 250
Binnenkade, Alexandra, 11
Bitburg controversy, 114
black market, 13, 265, 266, 268, 269, 270
Blankenburg, 218, 223
Bloch, Joseph, 186
Blumenfeld, Erwin, 257
Bodemann, Y. Michal, 299
Bohemia, 43
Boix, Francisco, 259n40
Bomberg, Daniel, 215
Bonaparte, Napoleon, 42
Bonn, 169
book market, 215, 222
Börne, Ludwig, 42
boundaries, 2–7, 9–10, 12–15
Bourdieu, Pierre, 7
Brahe, Tycho, 221
Brandt, Willy, 306
Braunschweig, 131, 132
Brecht, Bertolt, 257
Breitenbach, Joseph, 256, 257
Bremen, 97
Bremerhaven, 199
Breslau, 57, 131, 165
Breslauer, Marianne, 250
Brody, Austria-Hungary (Brody Affair),
    185–186
Bronner, Simon J., 232
Brück, Wolfram, 114
Brumlik, Micha, 113–14

Bubis, Ignantz, 116
Buchau, Germany, 33
Budapest, 237, 251
Bukovina, 187
Bürgerlichkeit. See middle class
burial grounds, 12, 143–144, 148–151, 154

calendars, 222
Capa, Robert, 253
Carnegie, Andrew, 180, 181, 182
Caspary, Eugen, 63
Cassel, David, 60
cemeteries. See burial grounds
censorship, 216, 221–224
Central Europe, 125, 128, 129, 246, 253,
    256
Chabad Lubavitch, 234
Chagis, Moses, 220, 222
Chemnitz, 170
Chicago, 47
Christian Church, 142
  bells, 33
Clark, (Sir) Kenneth, 249
clergy (Christian), 145
collective unconscious, 78
Cologne, 63–64
comics, 288–289
commerce, 216, 220, 221–224
communication, 13, 125–130, 133, 134n8
Connolly, William, 284
Conservative Judaism, 131, 132, 164
Constantinople, 219
contact zones, 6, 11, 23, 24, 31, 32, 34, 36
Copernicus, Nicolaus, 221
Cousineau, Jennifer, 238
Crémieux, Aldophe (1796–1880), 130
criminality, 264, 265, 268, 270, 273
culture
  entangled, 3, 4, 6, 10
  practices, 231, 235, 239
currency reform, 264, 268
Czaplicka, John, 56

Damascus, 129–130
  Damascus affair (1840), 129–130, 185
Darmstadt, 165, 170
Davidsohn, Georg, 203–204
Dawison, Bogumil, 92
de Certeau, Michel, 25
de Hirsch, Baron, 12
Dembitzer, Salomon, 201
Demski, Eva, 112

Deppner, Martin, 247
Dessau, 162, 165, 172, 216, 217, 218, 220, 222, 223, 224
Deutsch, Gerti, 256
Diaspora, 2, 3, 5, 15, 43, 45, 46, 47, 48, 49, 50, 51, 79, 171, 202
displacement, 89, 100, 102, 281, 286, 289
  displaced persons, 169, 267, 268, 269, 272, 274, 275
  Trutzhain DP camp, 169
distribution, 125, 127, 129, 134, 134n5
diversity, 208
Döblin, Alfred, 62
  Berlin Alexanderplatz, 62–63
domination, 265
doorstep. See threshold
Dresden, 144–145, 163, 171, 172, 251
Dreyfus, Alfred, 43
Dreyfus, Emil, 32–33, 34
Duisburg, 171, 172
Duntze, Klaus, 240
Dupont, E(wald) A(ndré), director, 11, 88–102
  Atlantis (1929), 89
  Das alte Gesetz (1923), aka The Ancient Law; Israel's Son, 88–99
  Die Geier-Wally (1921), 98
  Moulin Rouge (1928), 89, 106
  Peter Voß, der Millionendieb (1932), 88–89, 92, 99–102
  Piccadilly (1928/1929), 89, 106
  Two Worlds (1930), 90
  Varieté (1925), 89, 106
Dyhernfurth, 218

early modern period, 140–145, 154
Eastern Europe, 7, 128, 129, 135n17, 135n19, 135n25
Eco, Umberto, 301
economy, 268
Edwin Oppler, 150
Einstein, Albert, 202
Eisen, Arnold M., 304
Eisenstaedt, Alfred, 253, 256, 257
Eisenstein, Elizabeth L., 215, 225–226
Eisner, Kurt, 290
El Alaoui, Hicham Ben Abdallah, 240
Eliot, George, 59
  Daniel Deronda, 59
Elon, Amos, 108–109
emancipation, 140, 141, 144, 161, 163–164, 165

empowerment, 142, 145
Endingen, 23, 24, 25
Engel, Semmy, 165
Enlightenment, 3, 4, 9
Eastern Europe, 246, 255
  distinction between Central and Eastern Europe, 246
  "Eastern question," 129
Erfurt, 169
Erinnerungsorte. See memory, places of
Ermy, Hans, 200
Ernst, Eugen, 201–202
Ernst, Max, 257
eruv, eruvim, 34–35, 233, 238–41, 243n47
Essen, 165
essentialism, 78, 79, 94
Estonia, 255
Ettenheim, 146
Ettlinger, Jakob (1798–1871), 133
Eulenfeld, Robert, 203
exclusivity, 147, 150, 151

Fassbinder, Rainer Werner, 113
  Garbage, the City, and Death, 113
Federman, Raymond, 117
Fehr, Gertrude, 250
Feinberg, Anat, 235
festivals, 306, 307, 308, 309, 311
  Krakow Jewish Culture Festival, 308, 311
film, 11, 88–102
  as a mediated space, 11
films
  Jud Süss (1940), 97
  Das Cabinet des Dr. Caligari (1919/1920), 97
  Hamlet, 90, 91, 93, 97
  Hintertreppe (1921), 97
  Nosferatu (1922), 97
  Raskolnikow (1922/1923), 97
  Romeo and Juliet, 95
Fireside, Harvey, 251
Fishman, Chuck, 312
Florence, 47
Fonrobert, Charlotte, 238
forced laborers, 251
foreignness, 140, 142–144, 149, 152–154, 166, 167, 265–275
Forrester, John, 82
Foucault, Michel, 80, 241
France, 255
Francisco, Jason, 304
Franconia, 218

Fränkel, David, 220, 223, 224, 226
Frankel, Zacharias, 58
Frankfurt am Main, 12, 41, 42, 109–122,
    128, 131, 132, 141, 142, 143, 144,
    145, 147, 148, 150, 152, 162, 164,
    166, 170
  Börneplatz, 108, 119n22
  Dominikanerplatz, 111
  Judengasse, 12, 107–122, 141, 164
  Staufenmauer, 110
Frankfurt/Oder, 219, 223
Frankl, Pinkus Friedrich, 60
Franz Joseph, Emperor, 46
Franzos, Karl Emil, 60
Frege, Gottlob, 74
Freud, Sigmund, 108, 110–111, 117
Freund, Gisela, 250, 256
Friedmann, Robert, 166
Froben, Ambrosius, 215
Fürst, Julius, 132, 136n43
Fürth, 218

Gaisbauer, Adolf, 41
Galeen, Henrik, 90
Galicia, 11, 40, 43–51, 187, 199, 201, 203
Galilei, Galileo, 221
Gans, David, 220, 223
Gebert, Konstanty, 300
Geiger, Abraham (1810–74), 59, 60, 131,
    148
Geiger, Lazarus, 148, 149
Geiger, Ludwig, 59, 60
gemilut chasadim, 182
German Democratic Republic, 169
German Judaism, 130–133, 134n7, 135n23
German Psychoanalytic Society, 82
Gernsheim, Helmut, 249, 256, 257
Gerz, Jochen, 171
ghetto, 10, 11, 92, 96, 97, 115–116,
    141–142, 144, 147, 148, 149, 150,
    198–205, 251–252, 257, 302–303,
    304, 306
  literature, 150
  origin of term, 41–42
  Radom (Poland), 251
  Warsaw, 302, 306
Gibbons, James, 181
Gladstone, William Ewart, 180, 181, 182
Gontser, Meir, 202, 203
Gordon, Juda Leib, 99
Gorfinkel, Jordan, 288, 290, 292
Goro, Fritz, 253

Gotfryd, Bernard, 251–253
Gottfried Semper, 152
Grabowsky, Adolf, 200
Graetz, Heinrich, 43
Graham, David, 309
Grasmüller, Andreas, 272
Greeks, 268, 269, 271, 272, 273
Green, Nancy L., 188
Greenberg, Reesa, 288
Grimm, Richard, 282
Gromova, Alina, 9
Grosser, Alfred, 116
Grschebina, Lisolette, 250
Guggenheimer, Ernst, 169
Gutmann, Simon, 256
Gutwein, Daniel, 127
Gwoździec, 162, 303

ha-Cohen, Tuviya, 219, 221
Haas, Willy, 94
Habermas, Jürgen, 113
Habsburg Empire, 41, 46, 77, 146,
    185–186, 188, 201
Halakhah, 3
Halberstadt, 217, 222
Halevi, Uri Faybesh (1625–1715), 134n11
Halle, 217, 223, 226
Hamburg, 199, 220, 222, 250
  Bornplatz, 165, 170, 172
  Eimsbüttel, 169
  Grindel Quarter, 160, 162, 164, 165, 169
  Jewish quarter, 169, 170, 172
  Neustadt, 160
Hammer, Schenk, Harold, 161
Hanau, Salomon, 219, 226
Hanover, 164, 165, 169, 172
Harburger, Theodor, 64–65
Hardt, Hermann von der, 224
Harlan, Veit, 97
Harris, Diane, 236
Harvey, David, 265
Haskalah (Jewish Enlightenment), 4, 9, 44,
    127, 128, 216, 221
Hebenstreit, Johann Christian, 223
Hebrew. *See* languages, Hebrew
Hecker, Zvi, 171
Hegel, Georg Wilhelm Friedrich, 76
Heidegger, Martin, 117
Heidingsfelder, 151
Heilbronn, 235
Heimann, Eduard, 75
Heimat, 96–97, 98, 151, 240

Heine, Heinrich, 56, 57
  Rabbi of Bacharach, 57
Heinert, Felix, 9
Helena Datner, 300
Helmstedt, 224
Herzl, Theodor, 45, 49
Hessel, Franz, 61–62
  Spazieren in Berlin, 62
heterotopia, 80–81, 82
Heyde, Jürgen, 10
Hildesheim, 150
Hirsch, Baron de, 179–189
Hirsch, Baroness Clara de, 187
Hirsch, Nikolaus, 170
Hirsch, Samson Raphael, 150
Hirsch, Samson Raphael, 58–59
histoire, 107–108, 119n5
Hitler, Adolf, 100, 102
Hoheisel, Horst, 171
Hokhmat Israel (Wisdom of Israel), 128
Holdheim, Samuel, 58
Holian, Anna, 13
Holländer, Ludwig, 64
Holocaust (Shoah), 4, 107–108, 116–17,
  169, 197, 205–207, 247, 252, 254,
  257, 282, 286–288, 291
holy community, 144, 150
Holy Roman Empire, 141
home, 232, 233, 235
homeless foreigners, 269, 274
houshold, 232, 233, 234
Hughes, Hugh Price, 181
Hungary, 146, 246, 251
Hvizdets, 304–305

identification/identity, 31, 101, 198, 207
  elective affinities, 255
  mingled, 216
Immler, Verena, 286
industrialization, 180
informal economy. *See* black market
integration, 145, 146, 148, 152, 154, 165
Iram, Yaacov, 184
Israel, 49
  Tel Aviv, 162, 234, 235
Isserlin, Max, 74
Italiener, Bruno, 166
Italy, 42

Jacobi, Lotte, 250, 253
Jacobi, Ruth, 250
Jacobson, Jacob, 65

Jacoby, Alfred, 170
Jellinek, Adolf, 186
Jerram, Leif, 82n1
Jerusalem, 282
Jessner, Leopold, 234
Jessnitz, 13, 215–227
Jewish heritage, 299, 301, 303, 305, 308,
  310, 312
Jewish holidays, 285–286
  Purim, 95, 97
  Sukkot, 33–34
Jewish museums, 281
  Amsterdam, 286, 288, 290, 293
  Berlin, 172–173, 282, 284, 286, 290, 291
  Frankfurt, 170
  Galicia Jewish Museum, 304
  Munich, 13, 280–294
    Café Makom, 280, 292
  Paris, 286, 288, 290, 291, 293
  Prague, 286, 288, 290, 293
  Vienna, 250, 288
  Warsaw (POLIN), 14, 282, 284, 286,
    291, 302–306, 312
Jost, Markus Isaak, 43
Judendorf, 24, 36
Judenhaus, 26
Jung, C. G., 78–79

Kacyzne, Alter, 250
Kahl, Margit, 170, 172
Kalish (Kalisz), 238
Kalmus, Gerd, 251
Kant, Immanuel, 79
Kaplan, Chaim. *See* Kaplanskis, Chaimas
Kaplan, Feitska, 255
Kaplanskis, Chaimas (Chaim Kaplan), 255,
  256
Karlsruhe, 162, 163, 169
Karpeles, Gustav, 60
Karski, Jan, 305
Kassel, 163, 170, 171
Kastan, Erich, 250
Kauders, Anthony D., 11
Kaunas (Kovno), 255
  Ninth Fort, 255
Kazimierz, 238, 301, 307, 308, 309, 310,
  311, 312
Keßler, Katrin, 236
Kimchi, David, 220, 221
Kirshenblatt-Gimblett, Barbara, 302, 303,
  304, 306
Klein, Dennis, 79–80

Klesser, George Friedrich, 224, 225
Klesser, Johann Ehrenfried, 224
Knoblauch, Eduard, 59, 164
Knobloch, Charlotte, 283
Koethen, 218, 219, 226
Kohl, Helmut, 114, 170
Kollek, Teddy, 282
Komorowski, Bronisław, 303
Korn, Solomon, 111
Koselleck, Reinhart, 108, 117–18
Kovno. *See* Kaunas
Kraków, 92, 128, 162, 201, 218, 238, 301,
    304, 307, 308, 309, 310, 311
Kraus, Karl, 75
Krefeld, 237
Kreutzmüller, Christoph, 250
Krone, Kerstin von der, 12
Kronfeld, Arthur, 74, 75
Krüger, Lore, 250

languages
    Hebrew, 127–129, 132
        as national language, 128
        Hebraism, 215–218, 224–226
        Hebrew grammar, 219, 220, 221,
            226
        Hebrew printing, 13
    Jewish languages, 127–28, 135n15
    Ladino, 3, 134n10
    multilingualism, 135n19
    vernacular languagues, 128
    Yiddish, 3, 35, 127–129, 135n18, 199,
        202, 203, 204, 206
L'viv, 43, 44, 47, 128
Ladino. *See* languages
Lamm, Hans, 282
Land, Edwin, 249
Landauer, Fritz, 165
Landauer, Kurt, 285
Lasker, Eduard, 60
Lässig, Simone, 188
Laube-Kramer, Franz, 28
Lazare, Bernard, 43
Lebenswelt, 231
Lefebvre, Henri, 10, 162, 265
Lehmann, Berend, 222
Leica company, 250
Leipzig, 128, 218, 222, 223, 224
Leitz, Ernst, 250
Lemberg. *See* L'viv
Lengnau, 11, 23, 24, 25
Lewandowski, Louis, 60
Lewysohn, Ludwig, 58, 151

Lezzi, Eva, 162
Licht, Hugo, 60
Liedtke, Rainer, 188
Lifschitz, Sharone, 283–284, 285, 290, 291,
    292, 293
Linz, 251
Lipphardt, Anna, 8, 179
Lithuania, 255–256. *See also* Kaunas
    Kelme, 256
    Jurbarkas, 256
    Ponar, 255
Lodz, 203, 257
Loeb, Yehuda Arie, 219
Loewe, Heinrich, 202
Loewy, Hanno, 290
London, 184
Lonstar, Michael, 301
Lorant, Stefan, 253
Lorsch, Wolfgang, 170
Lublin, 218
Lubtisch, Ernst, 90
Luce, Henry, 253

Mahlamäki, Rainer, 302
Maimonides, Moses, 216–217, 220, 222,
    223, 225, 226–227
Mainz, 3
makom, 8, 125
Mann, Barbara E., 8, 125, 162, 179
Mannheim, 256
Mannheimer, Moses, 58
Manning, Henry Edward, 180, 181
Margulies, Samuel Margulies, 48
Marty, Martin, 232
Maskil/Maskilim, 125–127, 130
Maurer, Hans-Rudolf, 29
Mauthausen (concentration camp), 251
May, Joe, 90
May, Klaus, 169
Mazower, David, 307
medinah/medinot, 35, 37n30
Meier, André, 197
Meisenheim, 237
Memel, 255
memory
    archaic, 78
    boom, 7
    cultural, 56
    cultures of, 7–8, 208
    culture of remembrance, 56, 169
    environments of, 56
    national, 55
    places of, 149, 155, 161, 168

place of remembrance, 56, 198
politics of, 12
sites of, 56
spaces of, 10, 56
spaces of remembrance, 10
Mendelssohn, Moses, 60, 99, 217, 226
Meng, Michael, 303
Menorah, 233
Mezuzah, 232–35
Michlic, Joanna Beata, 307, 308
middle class, 4, 8, 10 , 140, 145, 153, 154,
 161, 164
 Bürgerlichkeit, 8, 10, 188
minorities, 2, 4, 5, 7, 11, 12, 14, 15
mobility, 125, 126, 128, 135n19
modernity, 2–5, 11, 12, 13, 14, 16n6, 141,
 144, 145, 146, 148, 150, 153, 207
modernization, 55–56, 90, 95, 146, 154,
 180, 181–183, 189
Moholy, Lucia, 250
Montefiore, Moses (1784–1885), 130
Morocco, 100
Morris, Abigail, 304
Mortara affair, 185
multiculturalism, 208
Munich, 13, 14 Jewish Museum, 13 Munich
 Möhlstrasse, 13
Munich, 145, 157n40, 162, 171, 172, 173,
 250, 253, 256, 257, 263–275, 281,
 282–283, 285, 293
 City Museum, 256
 Jewish Center Munich, 280, 283, 285
 Jewish community of Munich, 282, 283,
 292
 Jewish history of, 282–283, 285, 289,
 292
 Jewish Museum. *See* Jewish museums,
 Munich
 Möhlstrasse, 13, 263–275
 Munich Olympics (1972), 285
museum models
 didactic encyclopaedic model of
 representation, 292
 Heimat museum, 290
 Schatzkammer, 286
 migration museums, 290
 narrative exhibition model, 291
museums, musealization, 7, 13, 149. *See also*
 Jewish museums
 City Museum of Munich, 256
 exhibition effect, 291, 293
 United States Holocaust Memorial
 Museum, 114

nation building, 2
National Socialism, 12–14, 166, 168, 250
natural sciences, 219, 220
 Wissenschaft, 151
Necker, Sylvia, 12
Netter, Charles, 186
networks (family), 143, 150
Neuländer, Else Ernestine (Yva), 246, 250,
 253
neuropolitics, 284
New York, 47, 48, 184
Nöckel, Willy, 169
Nora, Pierre, 55
Nordau, Max, 49
Nossig, Alfred, 45

Ochs, Vanessa, 233, 237
Olimsky, Fritz, 92, 99, 103
Oppenheim, Hermann, 74
Oppler, Edwin, 164
Orenstein, 252–253
Orientalism, 153, 166
Orla-Bukowska, Annamaria, 306, 309
Ornstein, Jonathan, 310
Orthodox Judaism, 130, 131, 132, 133,
 146–153, 156n20, 161, 164
Ostjude, 200, 246, 254
Ostow, Robin, 13
Ostwald, Hans, 199
Ottoman Empire, 186

Padua, 42
Palestine, 2, 187
Paquda, Bachiya Ibn, 219, 220
Paris, 184, 186, 237
patriotism, 145
Penslar, Derek J., 183
Perez-Farayn (labor union), 203
periodicals
 Allgemeine Zeitung des Judenthums
 (1837–1922), 131, 132, 136n43
 Arbeiter-Illustrierte-Zeitung, 204
 Berliner Beobachter, 205
 Bikkurei Halttim (1820–31), 128
 C.V. Zeitung (1893–1938), 131
 Der Israelit des 19. Jahrhunderts (1840–
 48), 136n38
 Der Orient (1840–50/51), 132
 Der Treue Zionswächter (1845–54), 133,
 136n51
 Deutsche Zeitung, 201
 Die Neuzeit, 185
 Die rote Fahne, 204

Die Welt, 49
Dinstagishe und Fraytagishe Kurant
    (1686–87), 127
Dirnfurter prifilegirte Tzaitung (1771),
    127–128
Gazeta de Amsterdam (1675–1702),
    126–127, 134n11
Ha-Mazkir, 44, 45
Ha-Melits (1860–1904), 135n18
HaLevanon (1863–86), 129
HaMaggid (1856–1903), 129
HaMeassef (1784–1811), 127, 135n14
HaShahar (1868–84), 129
Israelit. Organ des Vereins Schomer Israel,
    44
Israelitisches Familienblatt, 204
Israelitisches Predigt- und Schulmagazin
    (1834–36), 131
Izraelita. Organ Szomer Izrael, 44
Jedność. Organ Żydów Polskich, 44, 46, 48
Keren Hemed (1833–56), 128
Kol Mevaser (1862–72), 135n18
Lancet, 187
Life magazine, 253
New York Times, 186, 187, 189
New York Times, 254
North American Review, 180, 183
Ojczyzna, 43, 48
Pri Ets Haim (1721–61), 127
Przyszłość, 45, 46
Sulamith (1806–48), 131
Wissenschaft des Judentums, 136n37
Wissenschaftliche Zeitschrift für Jüdische
    Theologie (1835–47), 131
Wschód, 47, 49
Zeitschrift für die religiösen Interessen des
    Judenthums (1844–46), 132
Zeitschrift für die Wissenschaft des
    Judenthums, 136n37
permanence, 150, 155
philanthropy, 12, 179–189
    elite, 188
    global, 184
    Jewish, 179, 181–183, 185, 188, 189
philanthropic organizations
    Alliance Israélite Universelle (AIU), 183,
        184, 186, 187
    Anglo-Jewish Association (AJA), 183
    Baron Hirsch Stiftung (BHS), 187
    Baronin Clara von Hirsch-Kaiser
        Jubiläums-Stiftung, 187
    Hilfsverein der Deutschen Juden, 183

    Israelitische Allianz zu Wien (IAzW),
        183, 184, 185, 186, 187
    Jewish Colonization Association (JCA),
        184
    Universal Beneficent Society, 180
    West London Methodist Mission, 181
Philippson, Ludwig, 131, 132, 136n43
philosophy, 217, 219–221, 226, 227
photography, 13, 246–257
    association with egoism, 249
    association with pornography, 249
    lack of respectability, 249
    Polaroid company, 249
    Soviet photojournalism, 247
Pinto, Diana, 298–299, 300, 308
Pitzer, Franz Xaver, 271
Planck, Max, 74
Plauen, 165
Poincaré, Henri, 74
Poland, 43, 44, 45, 47, 199, 200, 251, 255,
    300, 302, 303, 304, 305, 307, 310, 312
police, 263–265, 269, 271, 272–274
Polonsky, Antony, 312
positivism, 75
postal system, 126, 135n22
practice, 24, 25, 31, 34, 36
Prague, 141, 166, 218
preservationism, 109
press, 125–134
    as mediated space, 11
    circulation, 125
    newspapers, 125–134
    printing, 215–227
Pringsheim, Alfred, 285
Protestantisches Konsistorium, 142
Prussia, 7, 129, 136n36, 201, 202, 207
psychoanalysis, 11, 72–82
    Psychological Wednesday Society, 79
Purin, Bernhard, 282, 288, 289, 292

rabbinical conference(s), 131, 132
raids (by police), 205, 263, 266, 269,
    272–275
railway system, 126
Ranke, Leopold von, 55
Rathenau, Walther, 88
Raumer, Johan Georg von, 224
Rawicz, Victor Meyer, 146
Raz-Krakotzkin, Amnon, 216
real imaginary, 298, 301, 311
récit, 107–108, 119n5
redemptive cosmopolitanism, 303

Reform Judaism, 58, 126, 131–133,
    136n37, 136n43, 146–147, 150, 164,
    165, 166
Philanthropin (School), 147
refugees (quota) (*Kontingentflüchtlinge*), 171
Reiner, Helmut, 252
Reiss-Engelhorn Museum (Mannheim), 256
religious law, 231–32
Reno, Paul, 92
Republik der Neuen Mitte, 207
Reszke, Katka, 312
Rhine-Main Region, 141
Rhineland, 63–64
Ries, Henry, 247–249
    German Faces, 247–249, 250
Riess, Frieda, 250
Riga, Latvia, 254
ritual objects, 233, 234
Roemer, Nils, 11
Rolat, Sigmund, 306
Romania, 201
Romanticism, 166
Rome, 47, 110–111, 117, 217
Rosengarten, Albert, 163
Rosenheim, Jacob, 59
Rosenthal, Berthold, 64
Rosenthal, Leo, 246, 253–254
Roth, Joseph, 61, 202
Rothschild, Amschel Meyer (Rothschild
    family), 147–148, 149–150, 187
Rotterdam, 184, 199
Rückheim, Ulrich, 170
Ruhleben, 199
Russia, 46, 128, 129, 135n18, 199, 200,
    201, 255

Sabbath, 162, 233, 234, 238
    Sabbath practices, 34, 35
Sabbioneta, 217
Sachs, Yolla Niclas, 250
Sachsse, Rolf, 250
Sadowski, Dirk, 13
Salomon, Erich, 246, 253, 254, 256
Samogitia, 255
San Francisco, 254
Sander, August, 247, 249
Saß, Anne-Christin, 12–13
Satanow, Isaak, 221
Sattelzeit, 7
Schiller, Salomon, 46, 52n29
Schipper, Ignacy, 49
Schlesinger, Philip, 241

Schlör, Joachim, 13
Schnock, Frieder, 285
Schorske, Carl, 75
Schröder, Gerhard, 207
Schroer, Markus, 5–6
Schulmann, Ludwig, 150
seclusion, 141, 147, 151, 154, 155
Seeliger, Ewgar, 99
segregationism, 150
Seiffert, Ernst, 200
Semitic origin, 153
Semper, Gottfried, 163–164, 172, 174n26
Sendler, Irene, 306
Sepharad, 3
Sephardic Jews, 3, 9, 17n27
settlement, 231, 232, 237, 238, 240
Sforim, Mendele Mojcher, 49
sh'elot ut'shuvot (responsa), 127
Shalev-Gerz, Esther, 171
Shapiro, Michael, 234
Shemtov, Vered, 8, 179
Shneer, David, 247
Shoah. *See* Holocaust
shtadlanut, shtadlan, 130, 136n32
shtetl, 89, 93, 95, 96, 97, 208
Shylock, 234
Siao, Eva (Sandberg), 250
Sigmund Freud, 72–82
Silesia, 218
Simmel, Georg, 56
simulation, 90, 94, 100
Singer, I. J., 250
Siodmak, Robert, 90
Slavet, Eliza, 78–79
Snopkowski, Simon, 286–288
Sobibor, 253
Sonnenfeld, Leni, 250
Sorkin, David, 80, 188
soundscapes, 32, 33
space, 82–83n1
    as an analytical category, 1, 5, 7, 9, 10,
        24, 25
    gendered, 165
    historicization of, 148
    hybrid, 6, 216, 221, 226
    identity with, 95, 96, 162
    imagined, 4, 207
    layers of, 24, 32, 35, 36
    liminal, 6, 7
    Jewish, 1, 4–10, 12, 13, 141–143, 160,
        172, 200, 246, 249, 253, 299, 300, 308
    materiality of, 5, 6, 24, 27, 32

memory, 161. *See also* memory, spaces of
of longing, 166–168
open, 145, 153
public, 144, 149, 152, 153, 155, 165, 240
Sattelräume (saddle spaces), 7
third, 81
vs. place, 2, 5, 8, 160, 173, 179
spatial turn, 1
Speigelman, Art, 288
Speyer, 3
Spitzer, André, 285
Spitzweg, Carl, 166
Stamm, Fridolin, 29, 30
standardization, 126
Stein, Leopold, 147–149
Steiner, Lisl, 256
Steinschneider, Moritz, 60
Stern, Greta, 250
Stern, William, 74
Stih, Renata, 285
Stockholm, 235
Stola, Dariusz, 306
Stölting, Erhard, 291
Straus, O. S., 184
Stringer, Ann, 247
Stühler, Friedrich August, 59
Stuttgart, 97, 169
Sulzbach, 218
Surbtal, 28, 29, 31, 32
Switzerland, 23
 Basel, 215
 Zurich, 35
synagogues, 12, 140–155, 160–173, 236–238, 299, 301, 303, 304, 308, 311
 bells, 33
 seats, 144, 147

Talmudic Jews, 9
Tananbaum, Susan L., 189
Tartas, David de Castro (1630–98), 127, 134n11
Täubler, Eugen, 60
Telsiai (Telz), 255
temporality, 126
Theologenversammlung, 132
Thon, Osias, 45, 49, 50
threshold, 13, 232, 233–236, 240
Tibbon, Jehuda Ibn, 220
Tiktin, Salomon (1791–1843), 131
timescapes, 33
Tomanek, Krzysztof, 309
tourism, 150, 162, 299, 301, 307, 308, 309

trade, 266, 268, 270–272, 274
traditionalism, 146, 149, 150, 151, 153, 154, 156n20
transmigrancy, 232
transnationalism, 183–185
trauma, 286
Trier, Salomon Salman, 147
Turin, 184
Turner, Victor, 240
Turski, Marian, 303
typography, 126
tzedaka, 182

Ukraine, 303, 304, 312
urbanism, 5, 9, 12, 55–56, 94, 96, 98, 142, 161, 162, 166, 168, 170, 198, 207
 urban modernism, 108, 112

Veneziani, Emmanuel Felix, 186
Venice, 41, 47, 61, 215, 217, 224
 Mestre, 42
Verga, Salomon Ibn, 219
Vergangenheitsbewältigung, 256
Vienna, 44, 73, 75, 77, 81, 82, 88, 92, 93, 95–98, 101, 183, 184, 250, 251, 252, 256, 258n34
 Berggasse, 19, 81, 82
"virtual Jewishness," 14, 298–302, 306, 307, 309, 310
Vishniac, Roman, 257
visibility, 144–145, 149, 152–153, 154, 161, 164, 169, 170
Voigt, Sebastian, 9
Vowinckel, Annette, 250

Waller, Lore Lizbeth, 250
Wallmann, Walter, 114
Wandel, Andrea, 170
Wandsbek, 220, 222
Warsaw, 14, 47, 128, 162, 250, 282, 302, 306
Wegener, Paul, 96
 Der Golem, wie er in die Welt kam (1920), 96
Weimar Republic, 11, 12, 74, 90, 100, 164, 165, 168, 202, 204, 207
Weinbrenner, Friedrich, 163, 174n26
Weinreich, Otto, 234
Weltsch, Robert, 94
Welzing-Bräutigam, Bianca, 254
Wertham, Fredric, 289
Wertheimer, Joseph von, 183

Wetzlar, 251
White, Arnold, 187–188
Whiteread, Rachel, 288
Wiener, Max, 75
Wilhermsdorf, 218
Wimmer, Thomas, 273
Wirth, Louis, 41
Wistrich, Robert, 76, 80
Wohnen, 240
Wolff-Schneider, Ursula, 250
Wongel, Karl Heinz, 169
Worms, 3, 57, 58, 150–151
    Austrittsgesetz, 58
Wulff, Moses Benjamin, 217, 223

xenophobia, 208, 265, 275

Yagel, Abraham, 224–225
Yerushalmi, Yosef, 77–78
Yiddish. *See* languages, Yiddish
Yugoslavs, 268, 271, 272
Yva (*See* Else Ernestine Neuländer)

Zamosc, Israel, 220, 221
Zederbaum, Alexander (1816–1893),
    135n18
Zeiss factories, 251
Zionism, 3, 5, 40, 41, 45, 47, 49, 128, 130,
    202, 235, 286
Zunz, Leopold, 58, 60
Zweig, Arnold, 94, 95
Zweig, Stefan, 96, 98